Study Guide/Lab Manual

to Accompany

Refrigeration and Air Conditioning Technology

6th Edition

Concepts, Procedures, and Troubleshooting Techniques

William C. Whitman
William M. Johnson
John A. Tomczyk
Eugene Silberstein

DELMAR
CENGAGE Learning™

Australia • Brazil • Japan • Korea • Mexico • Singapore • Spain • United Kingdom • United States

Study Guide/Lab Manual to Accompany Refrigeration & Air Conditioning Technology, 6th Edition
William C. Whitman,
William M. Johnson,
John A. Tomczyk,
and Eugene Silberstein

Vice President, Career and Professional Editorial: Dave Garza

Director of Learning Solutions: Sandy Clark

Senior Acquisitions Editor: James Devoe

Managing Editor: Larry Main

Senior Product Manager: John Fisher

Editorial Assistant: Thomas Best

Vice President, Career and Professional Marketing: Jennifer McAvey

Marketing Director: Deborah S. Yarnell

Marketing Manager: Jimmy Stephens

Marketing Coordinator: Mark Pierro

Production Director: Wendy Troeger

Production Manager: Mark Bernard

Content Project Manager: David Plagenza

Art Director: Bethany Casey

Technology Project Manager: Christopher Catalina

Production Technology Analyst: Thomas Stover

© 2009 Delmar, Cengage Learning

For product information and technology assistance, contact us at **Professional & Career Group Customer Support, 1-800-648-7450**

For permission to use material from this text or product, submit all requests online at **cengage.com/permissions.** Further permissions questions can be e-mailed to **permissionrequest@cengage.com.**

Library of Congress Control Number: 2008000465

ISBN-13: 978-1-4283-1937-0

ISBN-10: 1-4283-1937-9

Delmar
5 Maxwell Drive
Clifton Park, NY 12065-2919
USA

Cengage Learning products are represented in Canada by Nelson Education, Ltd.

For your lifelong learning solutions, visit **delmar.cengage.com**

Visit our corporate website at **cengage.com.**

Notice to the Reader

Printed in the United States of America
1 2 3 4 5 08

Contents

This *Study Guide/Lab Manual* was prepared by the authors as a companion to *Refrigeration and Air Conditioning Technology, 6th Edition.* Each of the 50 units correlates with a unit in the text. Each unit is divided into the following sections to help students organize their study efforts and proceed through their learning activities:

- Unit Overview
- Key Terms
- Review Test
- Lab Exercises (Not all units have lab exercises.)

UNIT OVERVIEW

The **Unit Overview** provides a brief summary. The unit overviews are intended only as a brief review. They are not intended to provide enough information for the student to satisfactorily meet the objectives and complete the learning activities for that unit.

KEY TERMS

The **Key Terms** are extracted from the corresponding unit in the text. Definitions are not provided. It is felt that this learning activity will be more beneficial if the student defines from memory or looks up the terms in the text where they are used in the proper context.

REVIEW TEST

The **Review Test** is provided in an objective format. Students may be assigned this activity as an open book exercise or as an end-of-unit test.

LAB EXERCISE

Lab Exercises are provided for most units. Each lab exercise provides the following:

- Objectives
- Introduction
- Text References
- Tools and Materials
- Safety Precautions
- Procedures
- Maintenance of Workstation and Tools
- Summary Statement
- Questions

Objectives are clearly stated so that the student will understand before starting what will be accomplished and learned.

Introductions provide a brief description of the labs and the lab activities.

Text References are provided so that the student may review appropriate paragraphs in the text for a better understanding of the lab and specific subject material involved.

Tools and Materials are listed for the convenience of the instructor and student. When equipment, tools, and materials are readily available, the lab exercises can be pursued in an efficient, organized manner.

Safety Precautions are presented for each lab. Emphasis on safety is provided throughout the text and this *Study Guide/Lab Manual*. The authors feel strongly that when "hands-on" activities are pursued, all safety precautions should be observed for the entirety of each activity.

Procedures are stated in a clear systematic manner so that the student may proceed through the exercise efficiently.

Maintenance of Workstation and Tools is important to develop good work habits.

Summary Statements provide the students the opportunity to describe a particular aspect of the lab exercise.

Questions are provided as an additional learning activity regarding related subject material.

Many units in this manual have three different types of exercises for the student. The instructor may choose to use any or all of the exercises for a particular unit.

- Lab Exercises
- Troubleshooting Exercises
- Situational Service Ticket

Lab exercises were chosen to be practical for the majority of institutions offering these types of programs and instructive for the students. These exercises were developed to utilize equipment that will most likely be available in institutional labs. An attempt has been made to provide service-type exercises found in normal service calls.

There are two different types of troubleshooting exercises. One type is labeled "Troubleshooting Exercise." It is intended that the instructor place a particular type of problem in the unit. The type of problem and how to place this problem is covered for the instructor in the *Instructor's Guide*. This exercise is designed to guide the student through a logical procedure of finding the problem. The other type of troubleshooting exercise is labeled "Situational Service Ticket" and is also an exercise where the instructor places a problem in the equipment. This exercise has no guided instruction. Part of the exercise consists of the student listing the symptoms once the problem has been placed in the equipment. Many times, the diagnosis appears to the student while he or she is stating the problem. Some hints are given at the bottom of the service ticket to keep the student from straying from the intended direction.

The authors believe this student study guide can be a valuable tool for a student to use while interviewing for a job. The student should be encouraged to take good care of this book and show the work to the prospective employer. A book that is filled out with care indicates to the employer the scope of the student's knowledge and the student's ability to handle paperwork. As the industry moves closer to certifying technicians, this book may be used as a guideline for competency-based education.

INSTRUCTOR'S GUIDE

Answers to questions in the text and the *Study Guide/Lab Manual* can be found in the *Instructor's Guide*. Information for the instructor regarding troubleshooting exercises in this manual can also be found in the *Instructor's Guide,* along with suggestions pertaining to organization of the material, audio/ video, and general teaching.

ONLINE COMPANION GAME SHELL

A Game Shell with text questions is available for students online. Test questions will be either true/false, multiple choice, or fill-in-the-blank. To locate the Game Shell go to www.delmarlearning.com, and click on Online Companions (under Instructors and Teachers). You can then find the Game Shell by searching by discipline, title, author, or ISBN.

Operating Systems, Air-Conditioning and Refrigeration Service Equipment, Tools, Materials, and Supplies List

This list contains most operating systems, service equipment, tools, materials, and supplies needed to complete the lab exercises in this manual. You will need only those items pertaining to the lab exercises assigned for the topics being studied. Your instructor may also make substitutions appropriate for the specific type of systems and equipment available.

OPERATING SYSTEMS

Air-cooled commercial refrigeration systems
    High temperature
    Medium temperature
    Low temperature
Systems with capillary tube and thermostatic
        expansion valve metering devices
    Systems with gas-cooled compressors
      with receivers and service valves
      with crankcase pressure regulators
Water-cooled, wastewater refrigeration systems
        with water-regulating valve
    Package ice maker
    Electric furnaces—multistage and single-stage
    Gas furnaces
      with thermocouple safety pilots
      with spark ignition
      with temperature-operated fan-limit control
Oil furnace with cad-cell and primary control
Hot-water heating system
Central air-conditioning systems
    with fixed-bore metering device
    with thermostatic expansion valve
Package and split systems
All-weather systems with electric air conditioning
    with electric heat
    with gas heat
    with oil heat
Heat pump (air-to-air and water or ground source)
      systems with fixed-bore metering device
    with manufacturer's performance chart
    with wiring diagram
    Domestic refrigerator
    Domestic freezer
    Room air conditioner

AIR-CONDITIONING AND REFRIGERATION SERVICE EQUIPMENT

Halide leak detector
Electronic leak detector
Gage manifold
Air acetylene unit
Vacuum pump (two-stage)
Electronic vacuum gage
Refrigerant recovery system
U-tube mercury manometer
Scales—standard and electronic
Graduated charging cylinder
Thermometers
    Dial-type with immersion stem
    Glass stem
    Electronic
    Room thermostat cover with thermometer
Ammeter
Voltmeter
Volt-Ohm-Milliameter (VOM)
Milli-voltmeter
Wattmeter
Test cord for checking compressor start circuit
Meter with power cord splitter for measuring
        power in a power cord
Pulley puller
Vacuum cleaner
Flue-gas analysis kit
Water manometer
Vacuum or compound gage
Pressure gage, 150 psig
Draft gage
Prop-type velometer
24 V, 115 V, 230 V power sources
Oil service pressure gages

TOOLS

Goggles
Gloves
Calculator
Portable electric drill with drill bits

Tube cutter with reamer
Flaring tools set
Swaging tools
Tubing brushes, $1/4$", $3/8$", $1/2$"
Vise to hold tubing
Service valve wrench
Sling psychrometer
Oil furnace nozzle wrench
Two 8" adjustable wrenches
Two 10" pipe wrenches
Two 14" pipe wrenches
Allen wrench set
Open or box end wrenches, $3/8$", $1/2$", $9/16$", $5/8$"
Small socket wrench set
Needle-nose pliers
Slip joint pliers
Side cutting pliers
Straight-blade screwdrivers
Phillips screwdrivers
Nut drivers, $1/4$" and $5/16$"
Soft face hammer
Tape measure
Inspection mirror with telescoping handle
Flashlight

MATERIALS AND SUPPLIES

Appropriate manufacturer's charging charts for
    systems being used
Appropriate wiring diagrams for systems being used
Comfort chart
Psychrometric charts
Wire size chart
Pressure-temperature chart
Liquid-line drier
2-way liquid-line drier
Pressure taps
$1/4$" tubing tees
$1/4$" Schrader valve fittings
Flare nuts, $1/4$", $3/8$", $1/2$"
Reducing flare unions, $1/4$" $\times$ $3/8$", $3/8$" $\times$ $1/2$"

Flare plugs, $1/4$", $3/8$", and $1/2$"
Flare unions, $1/4$" $\times$ $1/4$", $3/8$" $\times$ $3/8$", and $1/2$" $\times$ $1/2$"
Plugs for gage manifold lines
Refrigerant recovery machine
DOT-approved refrigerant recovery cylinders
Cylinders of appropriate refrigerants for
    systems used
Soft copper ACR tubing, $1/4$", $3/8$", $1/2$" OD
Cylinder of nitrogen with regulator
Calculator
Resistance heater
Fan-limit control
Remote bulb thermostat (line voltage)
Gas furnace thermocouple
Room thermostat
Adjustable low-pressure control
Relay, contactor, motor starter
Thermocouple adapter (for millivoltmeter)
Adapters to adapt gages to oil burner pump
Several resistors of known value
20,000-ohm, 5-watt resistor
Start and run capacitors
Shallow pan
100 W troublelights
150 W flood lights
Refrigerant oil
Light lubricating oil
Low-temperature solder with flux, 50/50, 95/5, or
    low-temperature silver
High-temperature brazing filler metal with flux
Sandcloth (approved for hermetic compressor)
Emery cloth (300 grit)
Low-voltage wire
Electrical tape
Insulation material for insulating thermometers
Cardboard for blocking a condenser
Colored pencils
Compressed air
Thread seal compatible with natural gas
Rags
Chlorine solution
Matches

Heat and Pressure

Unit Overview

Temperature can be thought of as a description of the level of heat or the intensity of molecular activity. Everything in the universe is made up of molecules. The Fahrenheit scale of temperature measurement is currently used in the United States by most technicians in the air conditioning, heating, and refrigeration industry. The Celsius scale is used to some extent in the United States but is the scale used most frequently in the rest of the world. Pure water boils at 212°F (100°C) at sea level when the atmospheric pressure is 14.696 psi. Pure water freezes at 32°F (0°C). Molecules in matter are constantly moving; the hotter the substance, the faster they move.

The British thermal unit (Btu) describes the amount of heat it takes to raise the temperature of 1 lb of water 1°F. When there is a temperature difference between two substances, heat transfer will take place. Heat flows from a warmer substance to a cooler substance.

Conduction is the transfer of heat from molecule to molecule. Convection is the movement of heat by a liquid or gas from one place to another. Radiation is the transmission of heat through air, a gas, or a vacuum without heating it but heating the solid objects with which it comes in contact.

Latent heat is the heat added to or rejected by a substance, with no temperature change. Sensible heat is when heat is added to or rejected by a substance and there is a temperature change.

Pressure is the force applied to a specific unit of area. This is normally expressed in pounds per square inch (psi). The earth's atmosphere has weight and applies a force of 14.696 psi at sea level at 70°F. Barometers measure atmospheric pressure, and gages measure pressure in enclosed systems. Gage pressures do not include the pressure of the atmosphere and are known as pounds per square inch gage (psig). If the atmospheric pressure is added to this, it is known as pounds per square inch absolute (psia).

Key Terms

Absolute Temperature
Atmospheric Pressure
Barometer
Bourdon Tube
British Thermal Unit
  (Btu)
Celsius

Conduction
Convection
Fahrenheit
Gram
Heat Transfer
Joule
Kelvin Scale

Kilopascal
Latent Heat
Molecular Motion
Molecule
Pascal
Pressure
Radiation

Rankine Scale
Sensible Heat
Specific Heat
Temperature
Thermometer
Vapor

REVIEW TEST

Name	Date	Grade

Circle the letter that indicates the correct answer.

1. The standard atmospheric conditions for water to boil at 212°F are:
   A. atmospheric pressure: 15.969 psi, temperature: 70°F.
   B. atmospheric pressure: 16.696 psi, temperature: 68°F.
   C. atmospheric pressure: 14.696 psi, temperature: 70°F.
   D. atmospheric pressure: 15.696 psi, temperature: 68°F.

2. At standard conditions on the Celsius scale, water will boil at:
   A. 100°C.
   B. 212°C.
   C. 32°C.
   D. 0°C.

3. If water weighs 8.34 pounds per gallon, it will take _____ Btu to raise the temperature of 4 gallons of water from 70°F to 72°F.
   A. 66.7   C. 14.7
   B. 10   D. 106

4. Heat flows from a warmer substance to a cooler substance.
   A. true
   B. false

5. When heat transfers by conduction, it:
   A. moves through space and air and heats solid objects.
   B. is transported from one place to another by air or liquid.
   C. moves from molecule to molecule.
   D. all of the above.

6. When heat transfers by convection, it:
   A. moves through space and air and heats solid objects.
   B. is transported from one place to another by air or liquid.
   C. moves from molecule to molecule.
   D. all of the above.

7. When heat transfers by radiation, it:
   A. moves through space and air and heats solid objects.
   B. is transported from one place to another by air or liquid.
   C. moves from molecule to molecule.
   D. all of the above.

8. Sensible heat energy levels:
   A. can be measured with a thermometer.
   B. are known as hidden heat levels.
   C. are always in the comfort range.
   D. diminish by the square of the distance.

9. Latent heat energy levels:
   A. can be measured with a thermometer.
   B. are determined with a barometer.
   C. are always very warm.
   D. cannot be measured with a thermometer.

10. The atmosphere will support a column of mercury _____ inches at sea level.
    A. 32.00
    B. 28.69
    C. 14.696
    D. 29.92

11. A Bourdon tube is often found in a:
    A. mercury barometer.
    B. aneroid barometer.
    C. pressure gage.
    D. mercury thermometer.

12. Psig indicates:
    A. pounds per square inch of pressure plus atmospheric pressure.
    B. pounds per square inch of gage pressure minus atmospheric pressure.
    C. pounds per square inch of pressure.
    D. pounds per square inch of gravity.

13. 98°C converted to degrees Fahrenheit equals:
    A. 214.6.
    B. 196.0.
    C. 119.8.
    D. 208.4.

14. 70°F converted to degrees Celsius equals:
    A. 21.1.
    B. 22.6.
    C. 98.6.
    D. 20.3.

15. 200°F converted to degrees Celsius equals:
    A. 92.6.
    B. 94.7.
    C. 93.3.
    D. 91.0.

16. 28°C converted to degrees Fahrenheit equals:
    A. 27.3.
    B. 26.7.
    C. 81.6.
    D. 82.4.

17. A compound gage indicates pressures:
    A. above and below atmospheric pressure.
    B. above and below 500 psi.
    C. that are absolute pressures.
    D. that are measured with a bellows.

18. When a substance is heated, the molecules:
    A. move slower.
    B. move faster.
    C. move at the same speed.
    D. do not move.

19. A metric term for pressure is:
    A. joule.
    B. kelvin.
    C. rankine.
    D. kilopascal.

20. Psia at sea level under standard conditions is _____ psi.
    A. 0
    B. 14.696
    C. 29.92
    D. 144

Matter and Energy

Unit Overview

Matter takes up space, has weight, and can be in the form of a solid, a liquid, or a gas. Solids exert pressure downward. A liquid exerts pressure downward and outward. Gas molecules seem to travel at random. Mass has to do with the weight of matter. The density of a substance describes its mass- or weight-to-volume relationship. Specific gravity compares the densities of various substances. Specific volume is the amount of space that a pound of vapor or gas will occupy.

Electricity operates controls and motors that control and drive pumps, compressors, and fans. Heat is considered energy because of the movement of molecules within substances. Much of the energy we use is converted to usable heat from fossil fuels. Examples of fossil fuels are natural gas, oil, and coal. Electricity is sold by the kilowatt-hour, natural gas by the cubic foot, oil by the gallon, and coal by the ton.

The term *work* can be explained as a force moving an object in the direction of the force. The formula for work is:

$$\text{Work} = \text{Force} \times \text{Distance}.$$

Power is the rate of doing work. One horsepower is the equivalent of lifting 33,000 lb a height of 1 ft in 1 min, or some combination totaling the same.

The unit of measurement for electrical power is the watt. One kilowatt is equal to 1000 watts.

746 watts (W) = 1 horsepower (hp)
3413 British thermal units (Btu) = 1 watt (W)
1 kilowatt (kW) = 3413 Btu

Key Terms

Density

Energy

Fossil Fuels

Gas

Horsepower

Liquid

Magnetism

Mass

Matter

Power

Solid

Specific Gravity

Specific Volume

Watt

Work

REVIEW TEST

Name	Date	Grade

Circle the letter that indicates the correct answer.

1. **Matter is:** Pg 14, 2.1
   A. a form of energy.
   B. used to determine the amount of work.
   C. a substance that takes up space and has weight.
   D. a solid.

2. **All matter:**
   A. has mass.
   B. is a solid.
   C. has molecules that travel at random.
   D. freezes at 32°F.

3. **The density of a substance:** Pg 15, 2.3
   A. is used to measure its energy.
   B. describes its mass-to-volume relationship.
   C. is used to determine the transferability of heat.
   D. determines its ability to attract unlike magnetic charges.

4. **Specific volume is a term used to:** Pg 15, 2.5
   A. compare the densities of various liquids.
   B. compare the densities of various solids.
   C. indicate the space a weight of gas will occupy.
   D. describe the energy characteristics of a substance.

5. If the specific gravity of copper is 8.91, it weighs:
   - A. 558.7 lb/cu ft.
   - B. 555.98 lb/cu ft.
   - C. 23.88 lb/cu ft.
   - D. 556 lb/cu ft.

6. An example of a fossil fuel is:
   - A. natural gas.
   - B. hydrogen.
   - C. wood.
   - D. liquid oxygen.

7. How many ft-lb of work is done when a 500 lb condensing unit is lifted to the top of a 60 ft building?
   - A. 560
   - B. 440
   - C. 8.33
   - D. 30,000

8. How many watts of electrical power are equal to 1 horsepower?
   - A. 33,000
   - B. 15,000
   - C. 746
   - D. 640

9. How many Btu are there in 8 kW of electrical power?
   - A. 26,286
   - B. 5968
   - C. 27,304
   - D. 68,260

10. The unit of energy the power company uses to charge customers is the:
    - A. British thermal unit.
    - B. joule.
    - C. horsepower.
    - D. kilowatt-hour.

11. Gas molecules travel in:
    - A. a downward direction.
    - B. an upward direction.
    - C. a sideways direction.
    - D. all of the above.

12. The only new energy we get is from:
    - A. natural gas from the ground.
    - B. coal from the ground.
    - C. electricity.
    - D. the sun.

13. Work is equal to:
    - A. force × power.
    - B. force × energy.
    - C. force × distance.
    - D. distance × energy.

Refrigeration and Refrigerants

Unit Overview

Bacterial growth that causes food spoilage slows at lower temperatures. Below 0°F, food-spoiling bacteria stops growing.

Refrigerated box temperatures above 45°F and below room temperature are considered high-temperature refrigeration; between 35°F and 45°F, medium-temperature; and between 0°F and −10°F are considered low-temperature refrigeration.

One ton of refrigeration is the amount of heat it would take to melt 1 ton of ice in 24 hours. Because it takes 144 Btu of heat energy to melt one pound of ice at 32°F, it will take 288,000 Btu to melt one ton, or 2000 pounds, of ice in a 24-hour period. This equates to 12,000 Btu/hour (288,000 ÷ 24), or 200 Btu/min (12,000 ÷ 24).

The boiling point of liquids can be changed and controlled by controlling the pressure on the liquid. When the pressure is increased, the temperature and boiling point are raised. When the pressure is decreased, the temperature and boiling point are lowered.

A compressor is a pump that compresses refrigerant vapor and pumps the refrigerant through the refrigeration system. Refrigerant flow is created by the difference in pressure between the outlet and inlet of the compressor. The compressed refrigerant moves to the condenser through the discharge line, where heat is rejected and the vapor is condensed into a liquid. From the condenser, the liquid refrigerant flows to the metering device, where the amount of refrigerant moving into the evaporator is controlled. The metering device causes part of the refrigerant to vaporize as it moves into the evaporator, and the pressure and temperature are greatly reduced. This reduction in pressure causes a severe lowering in temperature inside the evaporator. The evaporator absorbs heat from the surrounding air, which cools the air. The refrigerant totally vaporizes or boils within the evaporator. This vapor proceeds through the suction line to the compressor, where it is compressed and the cycle starts over.

Refrigerants are designated with an "R" and a number. A refrigerant must be safe. It must be detectable, have a low boiling point, and have good pumping characteristics.

Refrigerants should not be allowed to escape into the atmosphere. It is believed that chlorofluorocarbon (CFC) and hydrochlorofluorocarbon (HCFC) refrigerants deplete the stratospheric ozone layer, allowing harmful ultraviolet wavelengths from the sun to reach the earth. Refrigerants should be recovered and/or stored while a refrigeration system is being serviced.

Key Terms

Azeotropic
Boiling Point of the
   Refrigerant
Centrifugal
   Compressor
Chlorofluorocarbons
   (CFCs)
Compressor
Condenser
Dehumidifying
Enthalpy
Evaporator

Flash Gas
Hot Gas Line
Humidity
Hydrochlorofluoro-
   carbons (HCFCs)
Hydrofluorocarbons
   (HFCs)
Isobars
Isotherms
Leak Detection
   Devices
Liquid Line

Metering Device
Near-azeotropic
Net Refrigeration
   Effect
Ozone Layer
Pressure-Enthalpy
   Diagram
Temperature/Pressure
   Relationship
Pure Compound
Reciprocating
   Compressor

Refrigerant
Refrigerant Reclaim
Refrigerant
   Recovery
Refrigeration
Rotary Compressor
Screw Compressor
Scroll Compressor
Suction Line
Superheat
Temperature Glide
Ton of Refrigeration

REVIEW TEST

Name	Date	Grade

Circle the letter that indicates the correct answer.

1. Ice melts in ice boxes because:
A. heat flows through the insulated walls.
B. heat flows through the doorway when it is opened.
C. heat is moved into the box when warm food is placed in it.
D. all of the above.

P 23 3.2

2. High-temperature refrigeration is that produced by:
A. heat from a furnace.
B. an air-conditioning system.
C. the vegetable cooling system of a refrigerator.
D. the freezer section of a refrigerator.

P 22 3.1

3. A ton of refrigeration is equal to:
A. 1 ton of ice.   C. 144 Btu/h
B. 288,000 Btu/24 h   D. a large refrigerator.

P 24 3.5

4. The evaporator in a refrigeration system:
A. meters the refrigerant.
B. condenses the refrigerant.
C. is the component where refrigerant boils and absorbs heat.
D. compresses the vaporized refrigerant.

P 30 3.7

5. The compressor in a refrigeration system:
A. compresses the refrigerant.
B. pumps the refrigerant throughout the system.
C. both A and B.
D. none of the above.

P 32 3.8

6. The condenser:
A. rejects the heat from the refrigerant.
B. meters the refrigerant throughout the system.
C. causes superheat in the refrigerant.
D. all of the above.

P 34 3.9

7. Only liquid refrigerant should enter the metering device.
A. true   B. false

8. Reciprocating compressors have:
A. a piston and a cylinder.
B. two scrolls that mesh together.
C. a rotor to compress the refrigerant.
D. a large fan-type compressor component.

P 32 3.8

9. R-22 is a refrigerant most commonly used in _____ systems.
A. residential air conditioning
B. light commercial refrigeration
C. industrial refrigeration
D. household refrigeration

Pg 27 #6 para

10. R-12 is a refrigerant most commonly used in _____ systems.
A. medium-temperature
B. residential air-conditioning
C. primarily low-temperature
D. all of the above

Pg 40 3.12

11. The cylinder color code for R-410A is:
A. rose.
B. white.
C. yellow.
D. green.

Pg 43

12. The cylinder color code for R-134a is:
A. red.
B. yellow.
C. light blue.
D. green.

Pg 43

13. The cylinder color code for R-12 is:
A. orchid.   C. purple.
B. white.   D. green.

Pg 43

14. The cylinder color code for R-22 is:
A. orchid.   C. purple.
B. white.   D. green.

P 43

15. Good ventilation is important when working around refrigeration equipment because modern refrigerants:
A. are toxic and can poison you.
B. have a very unpleasant odor.
C. may get on your skin and cause a rash.
D. are heavier than air and may displace the oxygen around you.

Pg 44

16. The boiling point of a refrigerant should be low enough at atmospheric pressure so that low temperatures can be reached and the refrigeration system will:
A. not overheat.
B. not explode.
C. not go into a vacuum.
D. not ice up.

Pg 41 3.15

17. It is believed that certain refrigerants such as CFCs and HCFCs, when allowed to escape into the atmosphere, will:
    A. destroy the hydrogen layer and allow more of the sun's rays to overheat the earth.
    B. deplete the stratospheric ozone layer and allow harmful ultraviolet rays from the sun to reach the earth.
    C. reduce the nitrogen content of the air surrounding the earth.
    D. reduce the carbon dioxide necessary for plant life.

18. It is important for technicians to recover refrigerants so that they:
    A. will not be released into the atmosphere.
    B. may be used again.
    C. may be reclaimed for future use.
    D. all of the above.

19. Enthalpy describes:
    A. the moisture content in air.
    B. the amount of refrigerant needed in a system.
    C. the amount of heat a substance contains from some starting point.
    D. the amount of harmful ultraviolet rays in the atmosphere.

20. A near-azeotropic refrigerant blend replacing R-22 in residential and commercial air-conditioning applications is:
    A. R-404A.   C. R-410A.
    B. R-407C.   D. R-134a.

21. A refrigerant that has properties very similar to R-12 and is used primarily in medium- and high-temperature refrigeration applications, refrigerators and freezers, and automotive air conditioning is:
    A. R-22.   C. R-404A.
    B. R-134a.   D. R-407C.

22. A refrigerant replacing R-502 in low- and medium-temperature refrigeration applications, which is a near-azeotropic refrigerant blend with a small temperature glide, is:
    A. R-404A.   C. R-134a.
    B. R-407C.   D. R-500.

23. The cylinder color code for R-404A is:
    A. bright green. C. aqua.
    B. rose.   D. orange.

24. Near-azeotropic refrigerant blends all exhibit some:
    A. temperature glide and fractionation.
    B. oil problems.

C. high boiling points.
D. low condensing pressures.

25. The quantity of heat expressed in Btu/lb that the refrigerant absorbs from the refrigerated space to produce useful cooling is:
    A. a calorie.
    B. enthalpy.
    C. the net refrigeration effect.
    D. a degree.

26. The amount of heat, in Btu/lb, that is given off by the system in the discharge line and the condenser is referred to as the:
    A. net refrigeration effect.
    B. heat of work.
    C. heat of compression.
    D. total heat of rejection.

27. When the portion of a pressure-enthalpy diagram is beneath the saturation curve, then:
    A. the refrigerant is at a constant temperature.
    B. the refrigerant is at a constant pressure.
    C. the refrigerant follows a temperature/pressure relationship.
    D. the refrigerant is 100% vapor.

28. According to the pressure-enthalpy chart, as refrigerant flows through the compressor:
    A. the pressure, temperature, and heat content of the refrigerant all increase.
    B. the pressure and temperature of the refrigerant increase, while the heat content of the refrigerant decreases.
    C. the pressure and heat content of the refrigerant increase, while the temperature of the refrigerant decreases.
    D. the pressure, temperature, and heat content of the refrigerant all decrease.

29. Lines of constant pressure on a pressure/enthalpy diagram are referred to as:
    A. isotherms.
    B. isometrics.
    C. isobars.
    D. glide lines.

30. Lines of constant temperature on a pressure/enthalpy diagram are referred to as:
    A. isotherms.   C. isobars.
    B. isometrics.   D. fractionation lines.

General Safety Practices

Unit Overview

The following are safety statements that all technicians should follow at all times:

Pressurized Cylinders and Systems. Move a pressurized cylinder only while a protective cap is on it, if it is designed for one.

If it is necessary to apply heat to a refrigerant cylinder to keep the pressure of the refrigerant up, only do this by setting it in a container of warm water with a temperature no higher than 90°F.

Keep your skin and eyes away from all refrigerant. Liquid R-22 boils at −41°F when released to the atmosphere.

Wear gloves and goggles when transferring refrigerant.

Never try to stop a refrigerant leak with your finger or hand.

Never try to refill a disposable cylinder.

Pressures inside nitrogen and oxygen cylinders are very high. Never move these cylinders without the protective cap in place.

Never use nitrogen directly from the cylinder. Ensure that a regulator is in place, with which the cylinder pressure can be reduced.

All oxygen fittings must be kept oil free. When in contact with high-pressure oxygen, oil can cause an explosion.

Acetylene cylinder pressures are not as high as oxygen but still must be equipped with a regulator to control the release of the acetylene gas because it is highly explosive.

When moving cylinders of oxygen or acetylene, use an approved cart with a chain, and ensure that the protective caps are in place.

Electrical Safety. Electrical power should always be shut off at the distribution or entrance panel when installing equipment and whenever possible when servicing equipment. Lock and tag the unit disconnect panel when the power is off. Keep the only key on your person.

When the power must be on when troubleshooting an electrical system or circuit, extreme care must be taken.

Make sure that you know the voltage to be measured and that the range selector on your test equipment is set properly.

Do not stand in a wet or damp area when making these tests.

Do not let your body become a conductor between two live wires or between a live wire and ground.

Always use properly grounded or protected power tools.

Use nonconducting ladders such as those made from fiberglass or wood.

Do not wear jewelry such as rings and watches when working around electrical circuits.

Only use a screwdriver or other tool when the electricity is turned off.

Ladder Safety. Use a nonconducting ladder when working around electricity or where overhead electrical lines are located. Side rails of a portable ladder should extend 3 ft above the upper landing surface. The angle of the ladder should be such that the horizontal distance at the bottom from the ladder to the upright surface is approximately ¼ the distance of the working length of the ladder.

Soldering or Brazing. When using a torch to solder or braze, be sure that all combustible materials are safely away from the area. If these materials must be there, ensure that they are protected with a shield.

A fire extinguisher must always be nearby when you are using a torch.

Learn to use the fire extinguisher before you have to use it.

Never solder tubing connections that are sealed and pressurized. Service valves or Schrader ports should be open before you start to solder. Wear goggles and gloves to protect yourself against pressurized refrigerant.

Be careful when working around hot refrigerant lines, heat exchangers, and motors as they can become hot enough to burn your skin.

Cautions When Working In and Around Cold and Freezing Equipment. Liquid refrigerant can freeze your skin instantaneously.

Long exposure can also be harmful. Working outside in below-freezing weather or in low-temperature freezers can cause you to not think properly and can cause frostbite. Wear cold-weather gear.

Rotating Equipment. Never wear ties, jewelry, or loose clothing around rotating equipment.

Never try to stop coasting rotating equipment such as fans or motors.

Never grab a moving belt or pulley.

Be sure that all set screws are tight, that all coupling fasteners are tight, and that wrenches and other tools are away from this equipment before it is started.

Moving Equipment. Use brain power rather than muscle power when moving heavy objects.

Use special tools such as a pry bar, lever truck, hand truck, portable dolly, or the lift gate on a truck.

Roll equipment on short lengths of pipe when possible.

Use your legs, not your back, when lifting heavy objects. Use a back brace belt.

Refrigerant Vapors. Refrigerant vapors are heavier than air and can displace oxygen in a closed space.

If refrigerant has been accidentally released in a closed space, see that it is properly ventilated.

Proper ventilation should be set up before starting a job.

Refrigerant vapors that have been heated while soldering or heated by some other means may be toxic.

When soldering, keep your head below rising fumes and make sure to have plenty of ventilation.

Refrigerant and oil from a compressor motor burnout can be very toxic and harmful. They can be hazardous to your skin, eyes, and lungs as they contain acid. Keep your distance if any amount of this escapes.

Chemicals. Chemicals are used for cleaning equipment and for water treatment. When using them, follow the manufacturer's directions.

If any of these chemicals get into your eyes, wash them out carefully and go to a doctor. Follow first-aid treatment on the package.

Key Terms

Double-Insulated
  Power Tools
Electrical Shock
Frostbite
Fusible Plug
Grounded Power Tools

Hand Truck
Lever Truck
Nonconducting Ladder
Package Refrigeration
  Equipment
Portable Dolly

Temperature/Pressure
  Relationship Chart
Pressure Vessels and
  Piping
Relief Valve
Rotating Equipment

Three-to-Two-Prong
  Adapters
Transferring
  Refrigerant

REVIEW TEST

Name	Date	Grade

Circle the letter that indicates the correct answer.

1. **An example of a pressure vessel is:**
   A. a small ship containing refrigerant cylinders.
   B. the compressor in a refrigerant system.
   C. a cylinder containing refrigerant for a refrigeration system.
   D. a portable dolly.

2. **If you were a technician about to service an air-conditioning system using R-22 refrigerant, and the temperature of the refrigerant in the cylinder in your truck were 105°F, what would be the pressure of the refrigerant (psig) in the cylinder? (Use the chart at the end of this guide.)**
   A. 126.6   B. 67.0   C. 210.8   D. 63.8

3. **A fusible plug is made from:**
   A. material with a low melting temperature.
   B. plastic.
   C. copper fused with brass.
   D. a metal alloy that will burn when overheated.

4. **It is possible that your hand will suffer frostbite if it comes in contact with liquid R-22 when accidentally released into the atmosphere.**
   A. true   B. false

5. **When transferring refrigerant from a cylinder to a refrigeration system, you should always wear:**
   A. gloves and goggles.
   B. heavy protective clothing.
   C. rubber gloves.
   D. industrial-type shoes and insulated gloves.

6. **Oil must never be used around a (an) _____ cylinder.**
   A. nitrogen   C. refrigerant
   B. oxygen   D. acetylene

7. **Before using your voltmeter, always:**
   A. zero the ohms indicator.
   B. touch the two leads to insure continuity.
   C. set the range selector to a setting higher than the voltage you will be reading.
   D. turn off the power.

8. **Properly grounded power tools should be used:**
   A. so that if the motor or other electrical part of the tool becomes shorted to the frame, the electric current will follow the grounding circuit.
   B. so that it will not create sparks at the plug.
   C. so that the technician will not be burned from the tool overheating.
   D. so that the technician will not have to use a nonconducting ladder.

9. **A double-insulated electric drill:**
   A. has two ground wires to ensure proper grounding.
   B. generally has a plastic case with the electrical connections and motor insulated within the tool.
   C. is used with a three-to-two-prong adapter.
   D. has rechargeable batteries.

10. **When soldering a replacement fitting in a refrigeration circuit, it is important that:**
    A. related valves or Schrader ports are open.
    B. you take into consideration the type of refrigerant used in the system.
    C. you know the voltage used for the compressor.
    D. all of the above.

11. **Nitrogen must be used with a regulator:**
    A. to reduce the very high pressure in the cylinder to a safer pressure.
    B. to help keep the nitrogen from exploding.
    C. because it would create a fire hazard otherwise.
    D. because there is not always a protective cap available.

12. **What two types of injury are caused by electrical energy?**
    A. shock and burns
    B. nausea and choking
    C. possible loss of limb and hearing
    D. none of the above

Tools and Equipment

Unit Overview

The air-conditioning, heating, and refrigeration technician uses many hand tools common to most trades. There are also many tools and a considerable amount of equipment that are designed specifically for this field. These tools and this equipment help the installation and service technicians do their jobs as efficiently as possible.

Key Terms

Air Acetylene
Air-Conditioning and
    Refrigeration
    Reversible Ratchet
    Box Wrench
Air Velocity Measuring
    Instruments
Belt Tension Gages
Carbon Dioxide and
    Oxygen Indicators
Charging Cylinder
Clamp-on Ammeter
Compressor Oil
    Charging Pump

Draft Gage
Electronic Charging
    Scale
Electronic Combustion
    Analyzer
Electronic Leak
    Detector
Electronic Vacuum
    Gage
Fin Straighteners
Flare Nut Wrench
Flaring Tool
Gage Manifold
Halide Leak Detector

Inclined Water
    Manometer
Inner-Outer Reamers
Inspection Mirrors
Leak Detectors
Megohmmeter
Nut Driver
Nylon Strap Fastener
Oxyacetylene Unit
Plastic Tubing Shear
Refrigerant Recovery
    Station
Sling Psychrometer
Stapling Tackers

Swaging Tool
Thermometer
Tube Benders
Tube Brushes
Tube Cutter
Tubing Piercing Valves
Tubing Pinch-Off Tool
U-Tube Mercury
    Manometer
U-Tube Water
    Manometer
Vacuum Pump
Volt-Ohm-Milliameter
    (VOM)

REVIEW TEST

Name	Date	Grade

Circle the letter that indicates the correct answer.

1. A wiring and crimping tool is used to:
A. cut and strip wire.
B. crimp tubing.
C. staple low-voltage wire.
D. fasten insulation to duct.

2. Air-conditioning and refrigeration reversible ratchet box wrenches are used to:
A. tighten hex head screws on refrigeration cabinets.
B. tighten flare nuts over tubing.
C. open and close refrigeration valves.
D. tighten nuts that cannot be reached with other wrenches.

3. Flare nut wrenches are used to:
A. tighten or loosen nuts over tubing flares.
B. open or close refrigeration valves.
C. tighten nuts that cannot be reached with other wrenches.
D. remove hex head screws from refrigeration cabinets.

4. Tube cutters are used to:
A. cut the insulation from tubing.
B. cut iron pipe used for natural gas.
C. cut copper tubing used in refrigeration piping systems.
D. cut sheet metal duct.

5. **A flaring tool has a:**
   A. bar to hold tubing.
   B. yoke.
   C. feed screw with cone and handle.
   D. all of the above.

6. **The gage manifold is a very important piece of equipment for the air-conditioning and refrigeration technician and it may include:**
   A. compound gage, high-pressure gage, manifold, valves, and hoses.
   B. either two or four valves.
   C. a four-valve design, including separate valves for the vacuum, low-pressure, high-pressure, and refrigerant cylinder connections.
   D. all of the above.

7. **The charging cylinder:**
   A. is used to store refrigerant.
   B. is used to reclaim refrigerant.
   C. is used to measure the refrigerant charge into a refrigeration system.
   D. all of the above.

8. **The vacuum pump is used to:**
   A. vacuum dust from the evaporator coil.
   B. evacuate nitrogen and carbon dioxide from the area around the refrigeration system.
   C. remove air and noncondensible gases.
   D. all of the above.

9. **A fluorescent leak detection system:**
   A. detects refrigerant leaks in a refrigeration system.
   B. detects water vapor leaks in a hot water boiler.
   C. detects nitrogen leaks at the pressure regulator.
   D. all of the above.

10. **A sling psychrometer:**
    A. uses the wet-bulb/dry-bulb principle to determine relative humidity.
    B. measures the air velocity in a duct system.
    C. uses a microprocessor to provide information for proper air balancing in a duct system.
    D. indicates the presence of carbon dioxide.

11. **A draft gage is used to:**
    A. check the pressure of flue gases.
    B. measure oxygen concentrations in flue gases.
    C. determine the carbon monoxide in flue gases.
    D. analyze the unburned natural gas in the flue of a gas furnace.

12. **An inclined water manometer:**
    A. determines the level of vacuum when a refrigeration system is being evacuated.
    B. determines natural gas and propane gas pressures.
    C. determines the vacuum when a refrigeration system is being evacuated.
    D. is used to analyze airflow in air-conditioning and heating systems.

13. **Two types of tube benders are:**
    A. centrifugal and pipe.
    B. pipe and plastic.
    C. spring and plastic.
    D. spring and lever.

14. **The electronic vacuum gage checks system pressures:**
    A. down to 2 inches of Hg.
    B. down to 25 inches of Hg.
    C. up to 5 psig.
    D. down into the micron range.

15. **Thermometers used in the field may be:**
    A. dial type.
    B. glass stem type.
    C. electronic type.
    D. all of the above.

16. **The pressures in flue gas installations for oil or gas may be checked using:**
    A. a manifold gage.
    B. a high-pressure gage.
    C. a draft gage.
    D. an inclined manometer.

Fasteners

Unit Overview

An air-conditioning and refrigeration technician should be aware of different types of fasteners and fastening systems so that the fastener most appropriate for the job will be used to securely install all equipment and materials. Some of these fasteners are used in many different construction trades, and others are more specific to this field.

Key Terms

Anchor Shields
Common Nail
Cotter Pin
Finishing Nail
Grille Clip
Machine Screws
Nylon Strap

Outward Clinch Tacker
Perforated Strap
Pipe Hook
Pipe Strap
Powder-Actuated
   Fasteners
Rivets

Roofing Nail
Set Screws
Staple
Tapping or Sheet Metal
   Screws
Threaded Rod and
   Angle Steel

Toggle Bolts
Wall Anchor
Wood Screws

REVIEW TEST

Name	Date	Grade

Circle the letter that indicates the correct answer.

1. **The term _____ is used to indicate the size of a common nail.**
   A. "inch"
   B. "diameter"
   C. "length"
   D. "penny"

2. **An outward clinch tacker is used to:**
   A. bend wire on a pipe hook.
   B. install a pin rivet.
   C. drive roofing nails.
   D. spread the ends of a staple outward.

3. **Wood screws generally have a:**
   A. flat head.
   B. round head.
   C. oval head.
   D. any of the above.

4. **The hole drilled in sheet metal for a tapping screw should be the approximate size of:**
   A. the thread diameter.
   B. the thread root diameter.
   C. the shank diameter.
   D. none of the above.

5. **Hollow rivets are assembled in a pin often called a:**
   A. mandrel.
   B. nozzle.
   C. anchor.
   D. clinch.

6. **In the tapping screw dimension "6-20 × $\frac{1}{2}$ type AB, slotted hex," the term "20" indicates the:**
   A. length.
   B. number of threads per inch.
   C. thread diameter.
   D. head diameter.

7. **In the machine screw dimension "$\frac{5}{16}$-18 UNC—2," $\frac{5}{16}$ indicates the:**
   A. length.
   B. number of threads per inch.
   C. outside thread diameter.
   D. thread diameter.

8. **Set screws may:**
   A. have square heads.
   B. have hex heads.
   C. be slotted headless.
   D. all of the above.

9. **Anchor shields, wall anchors, and toggle bolts may all be used in:**
   A. wood.
   B. masonry walls.
   C. hollow walls.
   D. all of the above.

10. **Which of the following heads may identify a machine screw?**
    A. flat head
    B. round head
    C. fillister head
    D. any of the above

11. **Pin rivets are installed using a:**
    A. drill screw.
    B. pin rivet gun.
    C. pin rivet drill.
    D. hammer.

12. **Which of the following nails are made of hardened steel?**
    A. finishing nails
    B. masonry nails
    C. roofing nails
    D. none of the above

13. **Stapling tackers are operated by:**
    A. air.
    B. water.
    C. electricity.
    D. hand.

14. **A nylon strap and strapping gun is used to:**
    A. fasten pipe together.
    B. fasten flexible duct to the fittings.
    C. fasten drain lines together.
    D. none of the above.

Tubing and Piping

Unit Overview

The use of correct tubing, piping, and fittings along with proper installation is necessary for a cooling, heating, refrigeration, or air-conditioning system to operate properly. Copper tubing is generally used for plumbing, heating, refrigeration, and air-conditioning piping. Copper tubing is available in soft- or hard-drawn copper. Air-conditioning refrigeration (ACR) tubing, normally used for refrigeration and air conditioning, can be purchased in rolls or straight lengths or as line sets, which are charged with refrigerant and sealed on both ends. The tubing on the low-pressure side of an air-conditioning or refrigeration system is often insulated to keep the refrigerant from absorbing excess heat and to prevent condensation from forming on the line.

Tubing is normally cut with a tube cutter or a hacksaw. It is recommended that tubing used to carry refrigerant be cut with a tubing cutter to prevent filings from entering the piping material. Soft tubing can be bent. Tubing bending springs or lever-type benders may be used. Copper tubing can be soldered or brazed. Temperatures below 800°F are used for soldering and temperatures over 800°F for brazing. Tubing and fittings must be cleaned where solder will flow. Filler material will move into the space between fittings and tubing by capillary attraction. Flare joints with fittings and swaged joints may be used to join tubing.

Steel pipe and plastic pipe may also be used for certain plumbing and heating applications. Steel pipe can be joined by threading and using fittings or it can be welded. The pipe is threaded on the end with a tapered thread. The thread diameter refers to the approximate inside diameter of the pipe. Plastic pipe can be cut with a special plastic tube shear, a tube cutter, or a hacksaw. Plastic pipe is joined with fittings using adhesive.

Key Terms

ABS
(Acrylonitrile Butadiene
   Styrene)
ACR (Air-Conditioning
   Refrigeration)
   Tubing
Air Acetylene

Brazing
Burr
Butane
Capillary Attraction
CPVC (Chlorinated
   Polyvinyl Chloride)
Double Flare Joint

Flux
Line Sets
Mapp Gas
National Pipe Taper
   (NPT)
PE (Polyethylene)
Propane

PVC (Polyvinyl
   Chloride) Reamer
Soldering
Swaged Joint
Die Threader
Tubing Insulation

REVIEW TEST

Name	Date	Grade

Circle the letter that indicates the correct answer.

1. Which of the following types of tubing is generally used for refrigerant piping?
A. type K copper   C. type L copper
B. ACR   D. steel

2. Which type of tubing or piping would be used for natural gas on a gas furnace?
A. Type L copper   C. steel or wrought iron
B. Type K copper   D. ACR

3. The size of Type L copper tubing is the approximate:
A. inside diameter.   C. wall thickness.
B. outside diameter.   D. none of the above.

4. The size of ACR tubing is the approximate:
A. inside diameter.
B. outside diameter.
C. wall thickness.
D. none of the above.

5. Tubing used for air-conditioning and refrigeration installations is usually insulated on the:
   A. low-pressure side.
   B. discharge line.
   C. high-pressure side from the condenser to the metering device.
   D. all of the above.

6. When tubing is cut with a tubing cutter, a _____ is often produced from excess material pushed into the end of the pipe from the pressure of the cutter.
   A. burr
   B. chip
   C. filing
   D. flux

7. The maximum temperature at which soldering (soft soldering) is done is:
   A. 400°F.
   B. 212°F.
   C. 600°F.
   D. 800°F.

8. Brazing is done at _____ soldering.
   A. the same temperature as
   B. lower temperatures than
   C. higher temperatures than
   D. temperatures that can be higher or lower than

9. A common filler material used for soldering is composed of:
   A. 50/50 tin-lead.
   B. 95/5 tin-antimony.
   C. 15%–60% silver.
   D. both A and B.

10. A common filler material used for brazing is composed of:
    A. 50/50 tin-lead.
    B. 95/5 tin-antimony.
    C. 15%–60% silver.
    D. cast steel.

11. Which of the following heat sources is often used for soldering and brazing?
    A. natural gas
    B. liquified oxygen
    C. air-acetylene
    D. nitrogen

12. A flux is used when soldering to:
    A. minimize oxidation while the joint is being heated.
    B. allow the fitting to fit easily onto the tubing.
    C. keep the filler metal from dripping on the floor.
    D. help the tubing and fitting to heat faster.

13. When soldering or brazing, the filler metal is:
    A. heated directly by the flame from the heat source.
    B. heated by the surrounding air.
    C. used to oxidize the metal.
    D. melted by heat from the tubing and fitting.

14. A flare joint:
    A. is made by expanding tubing to fit over other tubing.
    B. uses a flare on the end of a piece of tubing against an angle on a fitting, secured with a flare nut.
    C. uses a flare on the end of a piece of tubing soldered to a flare on a mating piece of tubing.
    D. is constructed using a tubing bending spring.

15. A swaged joint:
    A. uses a flare on one piece of tubing against the end of another piece of tubing.
    B. is made by expanding one piece of tubing to fit over another piece of tubing of the same diameter.
    C. is constructed using a tubing bending spring.
    D. is a fitting brazed to a piece of tubing.

16. Steel pipe:
    A. is joined with threaded fittings.
    B. may be joined by welding.
    C. is often joined with a flare fitting.
    D. both A and B.

17. Pipe threads, when used on pipe that will contain liquid or vapor under pressure:
    A. are generally tapered.
    B. should have seven perfect threads.
    C. generally have a V shape.
    D. all of the above.

18. In the thread specification $^1/_2$–14NPT, the 14 stands for:
    A. thread diameter.
    B. pipe size.
    C. number of threads per inch.
    D. length of the taper.

19. The end of a piece of steel pipe is reamed after it is cut:
    A. to remove the oil from the cut.
    B. to remove the burr from the inside of the pipe.
    C. to begin the threading process.
    D. to insure that the thread will be the correct length.

20. Plastic pipe is often used for:
    A. natural gas.
    B. oil burners.
    C. installations where it will be exposed to extreme heat and pressure.
    D. plumbing, venting, and condensate applications.

Flaring Tubing

Name	Date	Grade

Objectives: Upon completion of this exercise, you will be able to cut copper tubing, make a flare connection, and perform a simple leak check to ensure that the tubing connections are leak free and that they may be used for refrigerant.

Introduction: You will cut tubing of different sizes, make a flare on the end, assemble it with fittings into a single piece, pressurize it, and check it for leaks.

Text References: Paragraphs 7.1, 7.2, 7.5, 7.12, and 7.13.

Tools and Materials: A tube cutter with reamer, flaring tool set, refrigerant oil, leak detector (halide or electronic), goggles, adjustable wrenches, gage manifold, cylinder of R-22, cylinder of nitrogen with regulator, soap bubbles, and the following materials:

    2 each flare nuts, $1/4$", $3/8$", and $1/2$"
    1 each reducing flare unions, $1/4$" × $3/8$" and $3/8$" × $1/2$"
    1 $1/2$" flare plug
    1 $1/4$" × $1/4$" flare union
    Soft copper tubing $1/4$", $3/8$", and $1/2$" outside diameter (OD)
See Figure 7-1 for a photo showing these materials.

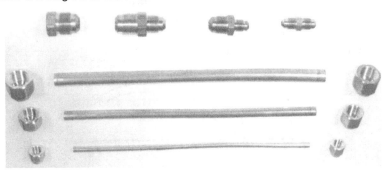

Figure 7-1 *Photo by Bill Johnson*

Safety Precautions: Care must be used while cutting and flaring tubing. Your instructor should teach you to use the tube cutter, reamer, flaring tools, leak detector, and gage manifold before you begin this exercise. Always wear goggles while working in a lab environment.

PROCEDURES

1. Cut a 10" length of each of the tubing sizes, $1/4$", $3/8$", and $1/2$".

2. Ream (carefully) the end of each piece of tubing to prepare it for flaring, Figure 7-2.

Figure 7-2 *Photo by Bill Johnson*

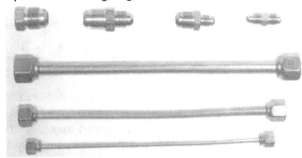

Figure 7-3 *Photo by Bill Johnson*

3. Slide the respective flare nuts over each piece of tubing before the flares are made, Figure 7-3. Be sure to apply some refrigerant oil to the flare cone before the flare is made. Make the flares.

4. Assemble the connections as in Figure 7-4 and tighten all connections using two adjustable wrenches.

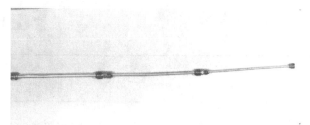

Figure 7-4 *Photo by Bill Johnson*

5. Add a small amount of R-22 refrigerant (10 psig) and apply pressure to the assembly with the cylinder of nitrogen and the gage manifold, Figure 7-5.

Figure 7-5 *Photo by Bill Johnson*

6. Leak test the assembly very slowly around each fitting, Figure 7-6.

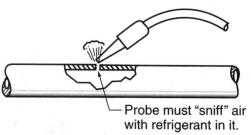

Probe must "sniff" air with refrigerant in it.

Figure 7-6

7. Examine any connection that leaks and repair it by either replacing the brass fitting or flaring the tubing again. A scratch on the brass fitting may be a possible leak. A ridge around the flare is a sign that the tubing was not reamed correctly and will probably leak, Figure 7-7.

Ridge

Figure 7-7 *Photo by Bill Johnson*

8. After all fittings have been proven leak free, take the assembly apart except for the end plug and the 1/4" × 1/4" flare union at the ends. Cut all flare nuts off except the two on the end. They will be used for pressure testing in the next exercise. Save the pieces of tubing for the next exercise, which is soldering.

Maintenance of Workstation and Tools: Wipe any oil off the tools and return all tools and supplies to their places. Store the leftover tubing lengths for the next exercise.

Summary Statement: Describe, step by step, the flaring process.

QUESTIONS

1. Why should you apply refrigerant oil to the flare cone before making a flare?

2. What happens if the tubing cutter is tightened down too fast while cutting?

3. What is the purpose of reaming the tube before flaring?

4. Why is thread-sealing compound not used on flare connections?

5. What actually seals the connection in a flare connection?

6. Why is a flare connection sometimes preferred to a solder connection?

7. What is the angle on a flare connection?

8. Give two reasons why a leak may occur in a flare connection.

9. Name two methods of leak testing a flare connection.

10. When a flare connection leaks due to a bad flare, what is the recommended repair?

Soldering and Brazing

Name	Date	Grade

Objectives: Upon completion of this exercise, you will be able to make connections with copper tubing using both low-temperature solder and high-temperature brazing material.

Introduction: You will first join together the tubing used in the previous exercise using low-temperature solder, then cut the connections apart and join them using high-temperature brazing.

Text References: Paragraphs 7.1, 7.2, 7.5, 7.7, 7.8, 7.9, 7.10, and 7.11.

Tools and Materials: An air acetylene unit with a medium tip, a striker, clear safety goggles, adjustable wrenches, light gloves, low-temperature solder and flux (50/50, 95/5, or low-temperature silver), high-temperature brazing filler material and the proper flux, sand cloth (approved for hermetic compressors) or wire brushes ($1/4$", $3/8$", and $1/2$"), a vise to hold the tubing, two each reducing couplings (sweat type, $1/4$" × $3/8$", $3/8$" × $1/2$"), a leak detector, a manifold gage, a nitrogen cylinder with regulator, and a cylinder of refrigerant.

Safety Precautions: You will be working with acetylene gas and heat. READ THE TEXT, paragraphs 7.7, 7.8, 7.9, 7.10, and 7.11, before starting. Your instructor must also provide you with instruction in soldering and brazing before you begin this exercise. Wear light gloves while soldering or brazing. Always wear goggles when working in the lab.

PROCEDURES

1. Clean the ends of the tubing with the sand cloth.

2. Clean the inside of the fittings using an appropriate fitting brush.

3. Fit the assembly together using the correct flux, Figure 7-8.

Figure 7-8 *Photo by Bill Johnson*

4. Place the assembly in a vertical position using the vise. This will give you practice in soldering both up and down in the vertical position, Figure 7-9.

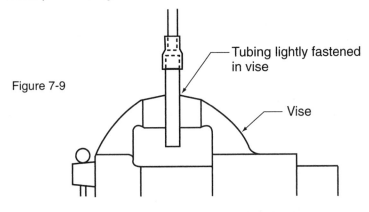

Figure 7-9

Tubing lightly fastened in vise

Vise

5. Follow procedures in paragraphs 7.7, 7.8, and 7.9 in the text. Start with the top joint at the top connection, then solder the joint underneath, Figure 7-10. After completing this, solder the top joint of the second connection, then solder underneath. Be sure not to overheat the connections. IF THE TUBING BECOMES RED OR DISCOLORS BADLY, THE CONNECTION IS TOO HOT.

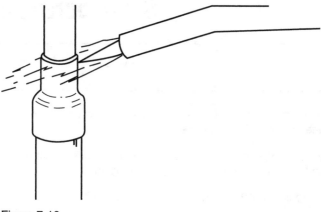

Figure 7-10

6. Let the assembly cool before you disturb it, or the connections may come loose.

7. After the assembly has cooled, perform a leak test as in Lab 7-1. You may also want to submerge the pressurized assembly in water for a test. If bubbles appear when the assembly is submerged, a leak is present.

8. Cut the reducing connectors out of the assembly and reclean the tubing ends.

9. Clean the second set of reducing couplings.

10. Assemble the pipes and connectors again using flux, and hold them upright in the vise.

11. Solder the top connections, then the bottom connections, as in the previous exercise. Only use high-temperature brazing material. **CAUTION:** THE PIPE AND FITTINGS WILL HAVE TO BE CHERRY RED TO MELT THE FILLER METAL, SO BE CAREFUL.

12. After the fittings have cooled, leak test them as before. Wear gloves while transferring refrigerant.

13. You can use a hacksaw to cut into the fittings and examine the solder penetration.

14. Upon completion of the leak test, cut off the flare nuts and scrap the tubing.

Maintenance of Workstation and Tools: CLEAN ANY FLUX OFF THE TOOLS OR FITTINGS AND THE WORK BENCH. Turn off the torch, turn off the tank, and bleed the acetylene from the torch hose. Place all extra fittings and tools in their proper places.

Summary Statement: Describe the advantages and disadvantages of low-temperature and high-temperature solder.

QUESTIONS

1. What is the approximate melting temperature of 50/50 solder?

2. What is the approximate melting temperature of 45% silver solder?

3. What is the approximate melting temperature of 15% silver solder?

4. During the soldering process, what causes the solder or filler material to melt?

5. What is the metal content of 50/50 solder?

6. What is the metal content of 95/5 solder?

7. Which solder is stronger, 95/5 or 15% silver solder? Why?

8. Which solder would be the best choice for the discharge line, 95/5 or 15% silver solder? Why?

9. What pulls the solder into the connection when it is heated? Explain this process.

10. Why is a special sand cloth used on hermetic compressor systems?

Bending Copper Tubing

Name	Date	Grade

Objectives: Upon completion of this exercise, you will be able to properly bend copper tubing according to specified guidelines.

Introduction: As part of this lab exercise, you will measure and cut sections of copper tubing to construct a model that you will use as a guide for your tubing bending exercises. You will gain experience making multiple 45- and 90-degree bends in copper tubing according to the guide you have constructed. By using bending techniques instead of performing soldering or brazing processes, the potential for refrigerant leaks is greatly reduced.

Text References: Paragraphs 7.1, 7.2, 7.5, and 7.6.

Tools and Materials: 3/8" copper tubing, tubing cutter, reamer, lever-type tubing bender, tape measure, pencil, two 45-degree 3/8" copper elbows, two 90-degree 3/8" copper elbows.

Safety Precautions: When handling and working with copper tubing, make certain that all sharp edges and burrs are removed from the tubing. Keep work area neat and clean at all times. Always follow all of your teacher's instructions.

PROCEDURES

1. Measure and cut three 5" pieces of 3/8" copper tubing, making certain that the sections are straight and free from burrs and sharp edges.

2. Using the three cut sections of tubing and two 90-degree elbows, assemble the piping arrangement, Figure 7-11. There is no need to solder or braze these sections together. This will serve as a guide for bending your tubing project.

3. Using the guide created in step 2, obtain the measurements necessary to fabricate a similar design from a single piece of tubing.

4. Transpose your measurements to the tubing and bend the tubing accordingly.

5. Make the appropriate bends.

6. To check your work, measure the overall length and width of the project and compare them to the overall length and width of the guide.

7. Repeat steps 2 through 6 using the arrangement shown in Figure 7-12.

8. After completing the fabrication, Figure 7-11 and Figure 7-12, repeat steps 2 through 6 using two 45-degree elbows instead of the 90-degree elbows. You will use Figure 7-13 and Figure 7-14 as guides for these last two exercises.

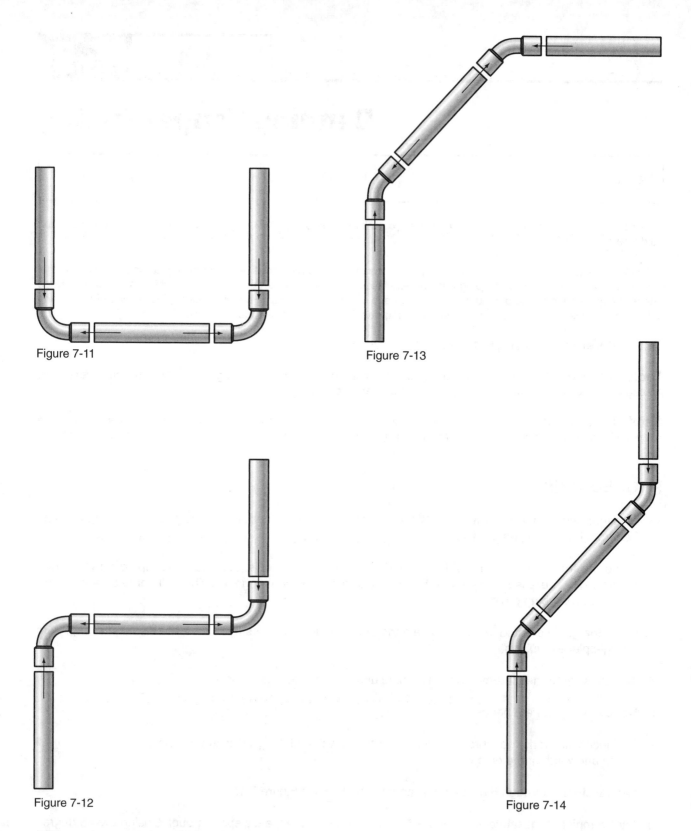

Figure 7-11

Figure 7-13

Figure 7-12

Figure 7-14

Maintenance of Workstation and Tools: Make certain that the work area is left neat and clean and that all tools have been returned to their proper places.

Summary Statement: Learning to bend soft-drawn tubing accurately takes a lot of practice, but the results obtained are well worth the initial effort. By using a single bend in a copper tubing run, you eliminate two solder joints and two potential refrigerant leaks. It is desirable to use bending techniques whenever possible, especially when the refrigerant lines will be run through walls and ceilings that will eventually be sealed. Gaining access to hidden joints in the event of a future leak often proves to be a time-consuming and costly process.

QUESTIONS

1. Explain why a technician might choose bending instead of soldering or brazing when installing refrigerant lines.

2. In addition to the lever bender, what other methods/tools can be used to bend copper or other soft-drawn tubing?

3. When bending copper tubing, why is it important to avoid kinking the line?

4. When installing refrigerant lines, why is it good field practice to install the lines so that they are as short and straight as possible?

5. Whenever possible, it is good field practice to use a wide radius when bending refrigerant lines. Why is this the case?

System Evacuation

Unit Overview

Only two materials should be circulating in a refrigeration system: the refrigerant and a lubricating oil. Noncondensible gases and moisture may get into the system during manufacturing or repair. They should not be in the system and should be removed by evacuating the system.

Evacuation is accomplished by pumping the vapors out of the system with a vacuum pump, to below atmospheric pressure. Liquids in the system must be vaporized or boiled and then removed with other vapors. Water makes such a large volume of vapor if a system has been flooded that it should be drained from the system, if possible.

Two common vacuum gages are the U-tube manometer and the electronic micron gage. Large unrestricted lines should be used. The refrigerant should be recovered using a refrigerant recovery system, not exhausted to the atmosphere. If the refrigerant is suspected to be contaminated, do not reuse it. Send it to be reclaimed.

Key Terms

Cold Trap
Compound Gage
Contaminants
Deep Vacuum
Evacuation
Leak Detection

Micron
Multiple Evacuation
Nitrogen
Noncondensible Gas
Refrigerant Reclaim or
  Recycling

Refrigerant Recovery
Schrader Valve
U-Tube Mercury
  Manometer

Vacuum
Vacuum Pump
Valve Stem Depressor

REVIEW TEST

Name	Date	Grade

Circle the letter that indicates the correct answer.

1. **Refrigeration systems are designed for what two substances to circulate in them?**
   A. refrigerants and oxygen
   B. refrigerants and nitrogen
   C. refrigerants and hydrogen
   D. refrigerants and oil

2. **To pull a vacuum means to:**
   A. raise the pressure in a refrigeration system.
   B. lower the pressure in a refrigeration system.
   C. allow refrigerants to escape into the atmosphere.
   D. condense the gases in a system.

3. **A compound gage:**
   A. indicates pressures above 14.696 psia.
   B. indicates pressures of in. Hg below atmospheric pressure.
   C. is one of two gages on a typical gage manifold.
   D. all of the above.

4. **Noncondensible gases:**
   A. are desirable in a refrigeration system.
   B. are undesirable in a refrigeration system.
   C. condense readily under normal atmospheric conditions.
   D. help in pulling a vacuum.

5. **A mercury U-tube manometer:**
   A. is used to measure a vacuum in a refrigeration system.
   B. is used to measure atmospheric pressure.
   C. is used to measure the pressure in a refrigeration system under normal operating conditions at the compressor discharge line.
   D. is used to measure atmospheric pressure at sea level.

6. **A micron gage:**
   A. is used to measure atmospheric pressure at high elevations.
   B. is used to measure atmospheric pressure at sea level.
   C. is often used to measure a vacuum at less than 1 mm Hg.
   D. is used to measure pressure on the high-pressure side of an operating refrigeration system.

7. **There are _____ microns in 1 mm Hg.**
   A. 100
   B. 1000
   C. 10,000
   D. 100,000

8. **A deep vacuum involves reducing the pressure in a refrigeration system to approximately:**
   A. 50 to 200 microns.
   B. 400 to 800 microns.
   C. 28 in. Hg.
   D. 20 psig.

9. **A multiple evacuation means that:**
   A. you connect three vacuum pumps simultaneously to a refrigeration system and pull a vacuum.
   B. you pull a vacuum on three different systems at the same time.
   C. you pull a vacuum on a refrigeration system more than once, allowing some refrigerant back into the system between each procedure.
   D. there must be more than one system operating before a vacuum can be pulled.

10. **If a vacuum has been pulled on a system and there is a leak, the vacuum gage will:**
    A. begin to rise.
    B. stay where it was when the vacuum pump was valved off.
    C. begin to fall.
    D. vibrate back and forth.

11. **The best leak detection test is when the refrigeration system is:**
    A. in a vacuum.
    B. open to the atmosphere.
    C. operating with all fans activated.
    D. not operating but in a standing pressure condition.

12. **Moisture in a liquid state must be _____ before it can be removed with a vacuum pump.**
    A. vaporized
    B. condensed
    C. cooled
    D. solidified

13. **1 mm Hg is equal to:**
    A. 1000 microns.
    B. 500 microns.
    C. 1 psig.
    D. all of the above.

14. **At 15,500 microns, water boils at:**
    A. 64.47°F.
    B. 212°F.
    C. 100°C.
    D. none of the above.

15. **Which of the following may be used to protect the vacuum pump when excess moisture is in a system to be evacuated?**
    A. cold trap
    B. changing the vacuum pump oil often
    C. using the gas ballast feature of the vacuum pump
    D. all of the above

Standing Pressure Test

Name	Date	Grade

Objectives: Upon completion of this exercise, you will be able to perform a standing pressure test on a vessel using dry nitrogen.

Introduction: Before a gage manifold may be used with confidence in a high-vacuum situation, it must be leak free. You will use dry nitrogen to test your gage manifold to assure that it is leak free. The standing pressure test is the same for any vessel, including a refrigeration system.

Test References: Paragraphs 8.2 and 8.4.

Tools and Materials: A gage manifold, plugs for the gage manifold lines, a cylinder of dry nitrogen with regulator, and goggles.

Safety Precautions: Goggles should be worn at all times while working with pressurized gases. NEVER USE DRY NITROGEN WITHOUT A PRESSURE REGULATOR. THE PRESSURE IN THE TANK CAN BE IN EXCESS OF 2000 PSI. Your instructor must give you instruction in the use of the gage manifold and a dry nitrogen setup before you start this exercise.

PROCEDURES

1. Check the gages for calibration. With the gage opened to the atmosphere, it should read 0 psig. If it does not, use the calibration screw under the gage glass and calibrate it to 0 psig.

2. Check the gaskets in both ends of the gage hoses. These are often tightened too much and the gaskets ruined. If the gasket is defaced, replace it. Figure 8-1 shows good and defective gaskets.

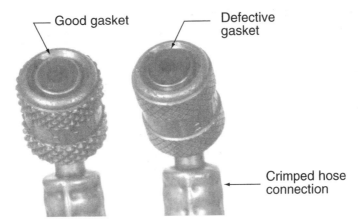

Good gasket   Defective gasket   Crimped hose connection

Figure 8-1 *Photo by Bill Johnson*

3. Fasten the center gage line to the dry nitrogen regulator and open both gage manifold handles. This opens the high- and low-side gages to the center port, Figure 8-2.

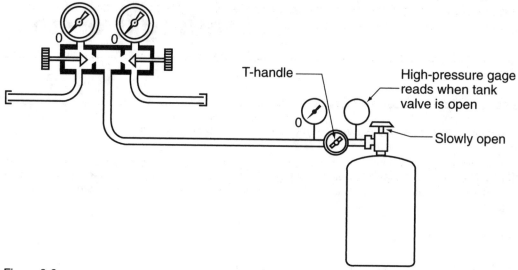

Figure 8-2

4. Place plugs in the high- and low-side gage lines. **CAUTION:** TURN THE NITROGEN REGULATOR T-HANDLE ALL OF THE WAY OUT, Figure 8-2. Now slowly open the tank valve and allow tank pressure into the regulator. DO NOT STAND IN FRONT OF THE REGULATOR T-HANDLE.

5. Slowly turn the T-handle inward and allow pressure to fill the gage manifold. Allow the pressure in the manifold to build up to 150 psig, then shut off the tank valve and turn the T-handle outward. The regulator is now acting as a plug for the center gage line. See Figure 8-3.

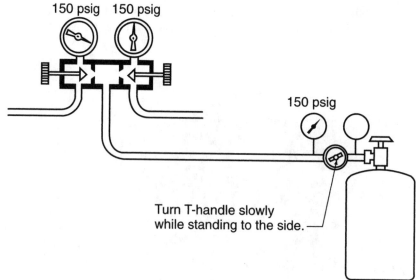

Figure 8-3

6. Allow the pressurized gage manifold to set, undisturbed, for 15 minutes. THE PRESSURE SHOULD NOT DROP.

7. If the pressure drops, pressurize the manifold again and valve off the manifold by closing the gage manifold valves (turn toward center). A drop in pressure in either gage will help to locate the leak.

   NOTE: If you cannot make the manifold leak free by gasket replacement, you may want to submerge the gage lines in water to determine where the pressure is escaping. Gage lines often leak around the crimped ends or become porous along the hoses and need to be replaced.

8. If you cannot stop the leak in a gage manifold, replace the gage lines with $1/4$" copper lines used in evacuation of a system.

Maintenance of Workstation and Tools: Place all tools in their proper places. Your instructor may suggest that you leave the gage manifold under pressure for a period of time by placing plugs in the lines. If you do this, you will know whether the gage manifold is leak free the next time you use it.

Summary Statements: Describe the reason that dry nitrogen is used in a pressure test instead of a refrigerant. Describe why a leak test while the system is in a positive pressure is more effective than using a vacuum for leak checking.

QUESTIONS

1. What can cause a gage line to leak?

2. What is the gage manifold made of?

3. What fastens the gage line to the end fittings on the end of a gage line?

4. What seals the gage line fitting to the gage manifold?

5. What can be done for repair if a gage line leaks at the fitting?

6. What would happen if a leaking gage manifold were used while trying to evacuate a system to a deep vacuum?

7. Why is refrigerant not used for a standing pressure test?

8. How much pressure does the atmosphere exert on a gage line under a deep vacuum?

9. How is a gage line tightened on a fitting?

10. What may be done for gage lines if the flexible ones cannot be made leak free?

Performing a Deep Vacuum

Name	Date	Grade

Objectives: Upon completion of this exercise, you will be able to perform a deep vacuum test using a high-quality vacuum pump and an electronic vacuum gage.

Introduction: You will perform a high-vacuum (200 microns) evacuation on a typical refrigeration system that has service valves. A high-vacuum evacuation is the same whether the system is small or large. The only difference is the time it takes to evacuate the system. Larger vacuum pumps are typically used on large systems to save time. It is always best to give a system a standing pressure test using nitrogen before a deep vacuum is pulled to assure that the system is leak free. An overnight pressure test of 150 psig with no pressure drop will assure that the vacuum test will go as planned. Performing a vacuum test on just the gage manifold and gage lines will also help assure that the test will be performed satisfactorily.

Text References: Paragraphs 8.6, 8.7, 8.8, 8.9, 8.10, 8.11, and 8.12.

Tools and Materials: A leak-free gage manifold, a cylinder of refrigerant, a nitrogen cylinder with regulator, a high-quality vacuum pump (two-stage is preferred), an electronic vacuum gage with some arrangement for valving it off from the system, a refrigerant recovery system, goggles, and gloves.

Safety Precautions: Goggles and gloves are safety requirements when transferring refrigerant. A valve arrangement such as in Figure 8-4 may be used to protect the electronic vacuum gage element from pressure. The electronic vacuum gage sensing element must be protected from any refrigerant oil that may enter it. This may be accomplished by mounting it with the fitting end down, Figure 8-5. Your instructor must give you instruction in using a gage manifold, refrigerant recovery system, vacuum pump, and an electronic vacuum gage before you start this exercise.

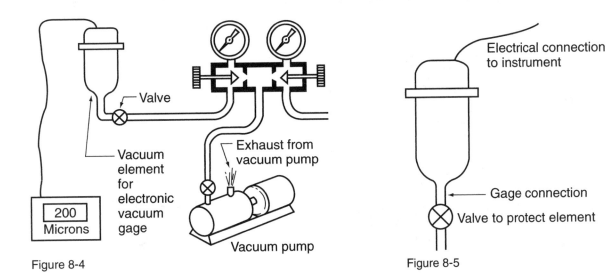

Figure 8-4

Figure 8-5

PROCEDURES

1. Fasten the gage manifold lines to the refrigeration system to be evacuated. Fasten the lines to the service valves at the high and the low side of the system if possible. DO NOT overtighten the gage line connections or you will damage the gaskets.

2. Position the service valve stems in the mid-position and tighten the packing glands on the valve stems, Figure 8-6. Be sure to replace the protective caps before starting the evacuation. Many service technicians have been deceived because of leaks at the service valves. If the system has Schrader valves, you may want to remove the valve stems during evacuation and replace them after completing the test. This will speed the evacuation procedure.

3. If the system has refrigerant in it, it must be recovered before the test begins. Your instructor will explain to you how to recover the refrigerant.

4. CHECK THE OIL LEVEL IN THE VACUUM PUMP AND ADD VACUUM PUMP OIL IF NEEDED.

5. After the system pressure is reduced to the correct level using the recovery unit, fasten the center gage line to the vacuum pump, Figure 8-6.

6. Put on your goggles and start the vacuum pump.

7. When the vacuum pump has operated long enough that the low-side compound gage has pulled down to 25 in. Hg (inches of mercury), Figure 8-7, open the valve to the electronic vacuum gage sensing element. Observe the compound gage as the system pressure reduces. It depends on the type of vacuum gage you have as to when it will start to indicate. Some of them do not start to indicate until around 1000 microns (1 millimeter of mercury vacuum) and will not indicate until very close to the end of the vacuum. Some gages start indicating at about 5000 microns (5 millimeters).

8. Operate the vacuum pump until the indicator reads 200 microns. **NOTE:** The deeper the vacuum, the slower it pulls down. It takes much more time to obtain the last few microns than the first part of the vacuum.

9. When the vacuum reaches 200 microns, you have obtained a low vacuum, one that will satisfy most any manufacturer's requirements.

10. If the system will not reach 200 microns, valve the vacuum pump off and let the system stand. If the vacuum rises, either there is a leak or liquid is boiling out of the system. You must find the problem. A starting point is to put pressure back into the system to above atmosphere and remove the gage manifold. Evacuate the gage manifold as in a system evacuation and perform a leak test on it. It must be leak free or you will never obtain a vacuum.

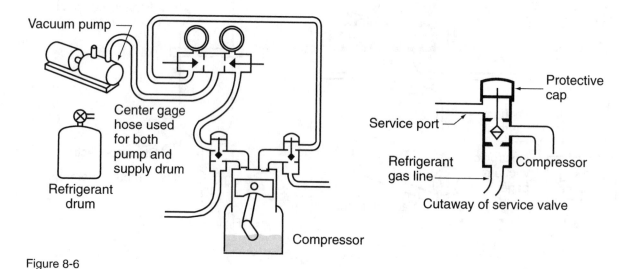

Figure 8-6

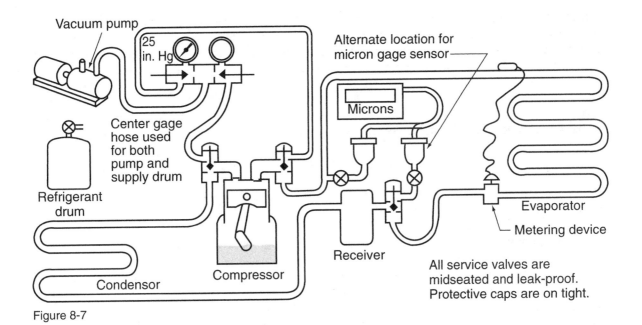

Figure 8-7

11. Upon successful completion of an evacuation, shut off the valve to the vacuum-sensing element and charge enough refrigerant into the system to get the pressure above atmospheric. IT IS NEVER GOOD PRACTICE TO LEAVE A SYSTEM IN A DEEP VACUUM FOR LONG PERIODS. A LEAK COULD CONTAMINATE THE SYSTEM AND DRIERS.

Maintenance of Workstation and Tools: Check the oil level in the vacuum pump and fill as needed with vacuum pump oil. Return the refrigeration system to the condition in which you found it. Wipe any oil off the tools and return them to their places.

Summary Statement: In your own words, explain how your evacuation procedure went. Do not leave out any mistakes you made.

QUESTIONS

1. Why should the electronic vacuum gage sensing element always be kept upright?

2. At what point did the electronic vacuum gage start indicating?

3. Which is the most accurate for checking a deep vacuum, an electronic vacuum gage or a compound gage?

4. What is the purpose of system evacuation?

5. What happens to a system that leaks while in a vacuum?

6. Is a vacuum the best method of removing large amounts of moisture from a system?

7. What will the results be if a fitting is leaking and the system is evacuated to a deep vacuum?

8. What is the approximate boiling pressure of refrigerant oil in a system?

9. Why is vacuum pump oil used in a vacuum pump instead of refrigerant oil?

10. How may water be released from below the oil level in a compressor crankcase?

Triple Evacuation

Name	Date	Grade

Objectives: Upon completion of this exercise, you will be able to perform a triple evacuation on a typical refrigeration system using a U-tube mercury manometer or a gage manifold compound gage for the vacuum indicator.

Introduction: You will use a U-tube mercury manometer if one is available for a vacuum check. If a U-tube manometer is not available, you may use a gage manifold and adequate time (depending on system size) to accomplish the same evacuation. You will use a refrigeration system that has service valves. A triple evacuation can be used as no refrigerant is allowed into the system to a pressure higher than 15 in. of Hg vacuum, so there is no violation of EPA rules.

Text References: Paragraphs 8.6, 8.7, 8.8, 8.9, 8.12, and 8.13.

Tools and Materials: A gage manifold, gloves, goggles, high-quality vacuum pump (two-stage is preferred), U-tube mercury manometer, refrigerant recovery system, and a cylinder of refrigerant of the same type as in the system to be evacuated.

Safety Precautions: Use caution while transferring refrigerant. Wear gloves and goggles. FOLLOW THE DIRECTIONS WITH THE MERCURY MANOMETER FOR TRANSPORTATION. DO NOT LAY THE INSTRUMENT DOWN WITHOUT THE SHIPPING PLUNGER IN PLACE OR THE MERCURY WILL BE LOST AND THE INSTRUMENT WILL BE RUINED. Check the oil in the vacuum pump. Do not begin this exercise without proper instruction in the use of the gage manifold, U-tube manometer, refrigerant recovery system, and a vacuum pump.

PROCEDURE A
Using a U-Tube Manometer

1. Put on your goggles and gloves. Fasten the gage lines to the service valves on the low and high side of the refrigeration system.

2. Recover all refrigerant from the system down to atmospheric pressure or the required EPA vacuum. Your instructor will tell you how to recover the refrigerant.

3. Check the oil in the vacuum pump.

4. Fasten the U-tube manometer to the system in such a manner that it may be valved off, Figure 8-8.

5. Fasten the center gage line of the gage manifold to the vacuum pump. Open the gage manifold valves and start the vacuum pump.

6. When the needle on the compound gage starts into a vacuum, you may open the valve to the mercury manometer. REMEMBER: It will not start to indicate until the vacuum is down to about 25 inches of mercury on the compound gage.

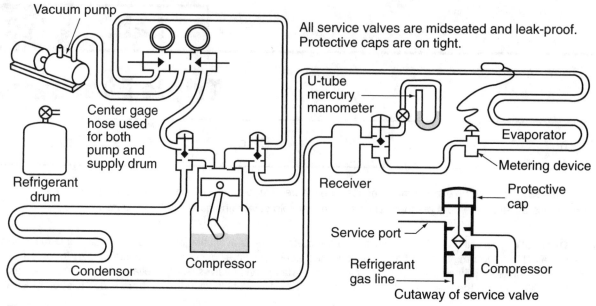

All service valves are midseated and leak-proof. Protective caps are on tight.

Figure 8-8

7. Allow the vacuum pump to run until the mercury columns read as in Figure 8-9. At this reading you can be assured that the vacuum is below 1 mm (millimeter) of mercury.

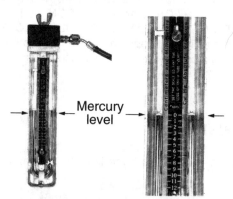

Mercury level

Figure 8-9 *Photo by Bill Johnson*

8. **IMPORTANT:** Valve off the mercury manometer.

9. Allow a small amount of refrigerant to enter the system, up to about 20 in. Hg on the compound gage. This is called "breaking the vacuum." Be sure to use the refrigerant that is characteristic to the system.

10. Now restart the vacuum procedure with the pump and open the valve to the mercury manometer. Allow the vacuum to reach below 1 mm Hg again and repeat steps 8 and 9.

11. When the mercury manometer has reached below 1 mm Hg again, allow the vacuum pump to run for some time, depending on the size of the system. There is no guideline for this time.

12. When some time has passed, VALVE OFF the mercury manometer and allow refrigerant to enter the system to cylinder pressure.

PROCEDURE B

Using a Gage Manifold Only

1. Fasten gage lines to the service valves on the high and low side of the system.

2. Recover refrigerant from the system down to atmospheric pressure. Ask your instructor for the correct procedure to do this.

3. Check the oil in the vacuum pump. Add oil if it needs it.

4. Fasten the center gage line to the vacuum pump. Open the gage manifold valves and start the pump.

   NOTE: Instead of a vacuum indicator, which may not be available to every technician, we will use the compound gage reading and the audible sound the vacuum pump makes while pumping to assure that we have a deep vacuum.

5. When the compound gage has stopped moving (it will read close to 30 in. Hg vacuum), turn the gage manifold handles to the closed position. Pay attention to the tone of the vacuum pump before and after the valves are closed. If there is a tone change when closing the valves, the pump is still pumping. This is a tone comparison with the valves open and closed. NOTE: The tone of the vacuum pump may be more evident with the exhaust cap removed, Figure 8-10. The indicator that a deep vacuum has been obtained is when there is no tone change when the valves are closed. The exhaust cap is often an oil eliminator to stop oil from leaving the pump during the early stages of the vacuum. DO NOT REMOVE IT UNTIL THE VACUUM IS BELOW 25 in. Hg. If you remove it too soon, oil will splash out of the pump with the exhaust.

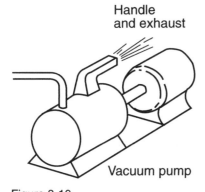
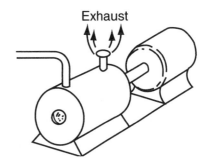

Handle
and exhaust

Exhaust

Vacuum pump

Figure 8-10

6. When you are satisfied that a deep vacuum has been obtained, valve off the vacuum pump and allow a small amount of refrigerant or nitrogen to enter the system, up to about 15 in. Hg on the compound gage. This is called "breaking the vacuum." Be sure to use the refrigerant that is characteristic to the system.

7. Restart the vacuum pump, removing the refrigerant or nitrogen from the system and follow the procedure in steps 5 and 6.

8. Perform three low-vacuum evacuations on the system and then allow the refrigerant charge to fill the system to the cylinder pressure. The system is now evacuated and ready for the proper charge.

Maintenance of Workstation and Tools: Check the oil level in the vacuum pump and add oil if needed. Clean any oil from tools and work area. Leave the refrigeration system in the manner you found it or as your instructor tells you.

Summary Statement: Describe what happens when a small amount of refrigerant is allowed to enter the system to "break the vacuum."

QUESTIONS

1. What is a deep vacuum?

2. Why must no foreign vapors be left in a refrigeration system?

3. What vapors does a deep vacuum pull out of the system?

4. What type of oil is used in a vacuum pump?

5. What comes out of the exhaust of a vacuum pump during the early part of an evacuation?

6. Why must vacuum pump oil only be used for a deep vacuum pump?

7. Why is a large vacuum pump required to remove moisture from a system that has been flooded with water?

Refrigerant and Oil Chemistry and Management—Recovery/ Recycling/Reclaiming and Retrofitting

Unit Overview

Over the years, refrigerants have leaked or have been vented from systems and have reached the stratosphere. Here they are thought to have combined with ozone, creating new chemical compounds and depleting some of the ozone layer that protects humans, animals, and plants on the earth from the sun's ultraviolet rays.

CFC refrigerants are considered the most damaging. Popular CFC-type refrigerants are R-11, R-12, and R-502. HCFC refrigerants have less potential for ozone depletion. R-22 is a common HCFC. HFC refrigerants are believed to do no harm to the stratospheric ozone layer but do contribute to global warming. R-124, R-125, R-134a, 143a, 152a, 404a, and R-410a are HFCs.

To "recover" refrigerant is to remove it from a system and store it in an external container. It may be charged back into the original system or another system owned by the same person or company.

To "recycle" refrigerant is to remove it from the system, clean it by passing it through filter driers, and return it to the system.

To "reclaim" refrigerant is to remove it from a system, process it to new product specifications, and perform a chemical analysis to ensure it meets specifications. It can then be sold as reclaimed refrigerant.

Refrigerant must be transferred only into DOT-approved cylinders. Equipment is available to recover and recycle refrigerant. Oil is removed from the system with the refrigerant and should be separated from it. Refrigerants should not be mixed.

Key Terms

Active Recovery
Air-Conditioning and Refrigeration Institute (ARI)
Alkylbenzene
Azeotropes
Chlorofluorocarbon (CFC)
Department of Transportation (DOT)
Environmental Protection Agency (EPA)

Fractionation
Global Warming
Global Warming Potential (GWP)
Greenhouse Effect
Hydrocarbon (HC)
Hydrochloro-fluorocarbon (HCFC)
Hydrofluorocarbon (HFC)
Low Loss Fitting
Montreal Protocol
Ozone

Ozone Depletion Potential (ODP)
Passive Recovery
Piercing Valve
Polyalkylene Glycol (PAG)
Polyol Ester (POE)
Refrigerant Reclaim
Refrigerant Recovery
Refrigerant Recycle
Schrader Valve
Technician
Temperature Glide

Total Equivalent Warming Impact (TEWI)
Ultraviolet
Zeotropes

REVIEW TEST

Name	Date	Grade

Circle the letter that indicates the correct answer.

1. Ozone is a type of:
   - A. chlorine.
   - (B) oxygen.
   - C. fluorine.
   - D. carbon.

2. Depletion of the earth's ozone shield will _____ the sun's ultraviolet rays on the earth's surface.
   - (A.) increase
   - B. decrease

3. Which of the following substances reacts with the earth's stratospheric ozone layer?
   - (A.) chlorine
   - B. fluorine
   - C. carbon
   - D. hydrogen

4. The refrigerant R-11 is a(n):
   - (A) CFC.
   - B. HCFC.
   - C. HFC.

5. The refrigerant R-22 is a(n):
   - A. CFC.
   - (B.) HCFC.
   - C. HFC.

6. The number assigned a refrigerant that identifies the rate at which that refrigerant is thought to deplete the ozone layer is known as the:
   - A. CFC.
   - B. DOT.
   - C. EPA.
   - (D) ODP.

7. When refrigerant is removed from a system, it must be:
   - (A.) recovered.
   - B. recycled.
   - C. reclaimed.
   - D. returned to the manufacturer.

8. Before a refrigerant that has been removed from a system can be sold to another customer, it must be processed to the _____ specifications.
   - A. EPA 700
   - B. ARI 740
   - (C.) ARI 700
   - D. ODP 720

9. Cylinders used in the recovery of refrigerant must be approved by the:
   - A. Environmental Protection Agency.
   - B. Air-Conditioning and Refrigeration Institute.
   - (C) Department of Transportation.

10. A micron is a measurement of a level of a:
    - (A) vacuum.
    - B. circle.
    - C. positive pressure.
    - D. capacitance.

11. The number one contributor to global warming is:
    - (A) carbon dioxide, CO_2.
    - B. ODP.
    - C. chlorine.
    - D. R-12.

12. Most refrigerants:
    - A. are lighter than air.
    - B. weigh the same as air.
    - (C) are heavier than air.

13. The refrigerants that are thought to be the most damaging to the earth's stratospheric ozone layer are the:
    - (A) CFC series.
    - B. HCFC series.
    - C. HFC series.
    - D. none of the above.

14. The azeotropic mixture refrigerants:
    - A. have only one boiling point.
    - B. do not fractionate when being charged in the vapor state.
    - C. include R-500 and R-502.
    - (D) all of the above.

15. The near-azeotropic refrigerant blends:
    - A. have more than one boiling point.
    - B. fractionate when being charged in the vapor state.
    - C. experience temperature glide.
    - (D.) all of the above.

16. The correct refrigerant oils must be chosen for all of the above refrigerants.
    - (A.) true
    - B. false

17. The refrigerants R-404a and R-410a are:
    - (A) HFCs.
    - B. HCFCs.
    - C. CFCs.
    - D. HCs.

18. R-507 is an azeotropic refrigerant blend.
    - (A) true
    - B. false

19. R-407c is a near-azeotropic refrigerant blend.
   A. true   B. false

20. R-401a, R-402a and R-402b, and R-409a are near-azeotropic refrigerant blends.
   A. true   B. false

21. An earth warming process caused by the atmosphere's absorption of heat energy radiated from the earth's surface is:
   A. ozone depletion.
   B. global warming or the greenhouse effect.
   C. glacial melt.
   D. all of the above.

22. Molecules that contain nothing but hydrogen and carbon are:
   A. hydrocarbon.
   B. ozone.
   C. chlorofluorocarbon.
   D. HCFCs.

23. The Total Equivalent Warming Impact (TEWI) takes into consideration both the direct and indirect _____ effects of refrigerants.
   A. global warming
   B. ozone depletion
   C. poisoning
   D. smog

24. An index that measures the direct effects of global warming created by chemicals emitted directly into the atmosphere is:
   A. the Total Equivalent Warming Impact (TEWI).
   B. the Global Warming Potential (GWP).
   C. the Montreal Protocol.
   D. the Greenhouse Index.

25. An agreement signed by the United States and other countries to control the release of ozone-depleting substances with phase-out dates and production bans is:
   A. the Greenhouse Conference.
   B. the Geneva Convention.
   C. the Montreal Protocol.
   D. the Toronto Conference.

26. Blended refrigerants whose temperature glide and fractionation exist to a larger extent than a near-azeotropic blend of refrigerants are:
   A. azeotropes.
   B. pure mixtures.
   C. pure compounds.
   D. zeotropes.

27. When one or more of the refrigerants in a blend will condense or evaporate at different rates than the other refrigerants in the same blend, _____ occurs.
   A. temperature glide
   B. fractionation
   C. sublimation
   D. dehydration

28. When a blend of refrigerants has many different temperatures when it evaporates and condenses at a given pressure, _____ occurs.
   A. fractionation
   B. system lag
   C. refrigerant slump
   D. temperature glide

29. A synthetic lubricant used in R-134a refrigeration and air-conditioning systems is:
   A. polyol ester.
   B. mineral oil.
   C. crude oil.
   D. vegetable oil.

30. A synthetic lubricant that works best with HCFC-based refrigerant blends in refrigeration applications is:
   A. polyol ester.
   B. mineral oil.
   C. alkylbenzene.
   D. petroleum oil.

31. A synthetic glycol-based lubricant often used in automotive applications with R-134a is:
   A. ester.
   B. vegetable oil.
   C. alkylbenzene.
   D. polyalkylene glycol.

32. _____ is also known as system-dependent recovery, where the internal pressure of the refrigeration system and/or the system's compressor aids in the refrigerant recovery process.
   A. Active recovery
   B. Passive recovery
   C. Powered recovery
   D. Assisted recovery

33. _____ is when refrigerant is removed from a system using a certified self-contained recovery machine.
   A. Active recovery
   B. Passive recovery
   C. Powered recovery
   D. Assisted recovery

34. Any person who performs maintenance, services, or repair that could reasonably be expected to release CFC or HCFC refrigerants into the atmosphere is:
   A. a service person.
   B. a janitor.
   C. a building manager.
   D. a technician.

35. A _____ is a fitting on a refrigeration system that either manually or automatically closes off when a service technician takes it off the pressurized system or recovery device.
   A. low-loss fitting
   B. piercing valve
   C. Shrader valve
   D. pressurized fitting

36. A _____ is a valve on a refrigeration system used for easy access having a valve core, a needle to depress, and a cap to prevent leaks.
   A. Schrader valve
   B. piercing valve
   C. low-loss fitting
   D. three-way valve

37. A _____ is a valve used for accessing a small appliance that has a sharp pin that is driven through the system's tubing.
   A. Schrader valve
   B. three-way valve
   C. low-loss fitting
   D. piercing valve

Refrigerant Recovery

Name	Date	Grade

Objectives: Upon completion of this exercise, you will be able to use one of two methods for recovering refrigerant from a system.

Introduction: You will use two methods of removing refrigerant from a system. The first is to pump all available liquid into an approved refrigerant cylinder using the refrigeration system containing the refrigerant. The second is to use refrigerant recovery equipment commercially manufactured for refrigerant recovery.

Text References: Paragraphs 9.1 and 9.17.

Tools and Materials: Gage manifold, gloves, goggles, service valve wrench, two 8" adjustable wrenches, DOT-approved refrigerant cylinder, scales, refrigerant recovery system, and a refrigeration system from which refrigerant is to be recovered.

Safety Precautions: DO NOT OVERFILL ANY REFRIGERANT CYLINDER; USE ONLY APPROVED CYLINDERS. Do not exceed the net weight of the cylinder. Wear gloves and goggles while transferring refrigerant.

PROCEDURES

Pumping the Liquid from the System Method

1. Fasten the low-side gage line to the suction connection on the system used.

2. Fasten the high-side gage line to the king valve or a liquid-line connection. **NOTE:** All systems do not have a liquid-line connection. If this system does not have one, you will need to use the second method of recovering refrigerant. Wear goggles and gloves.

3. Loosely connect the center line of the gage manifold to the refrigerant cylinder. Use only an approved recovery cylinder.

4. Bleed a small amount of refrigerant from each line through the center line and allow it to seep out at the cylinder connection. This will purge any contaminants from the gage manifold.

5. Set the cylinder on the scales and determine the maximum weight printed on the cylinder. DO NOT ALLOW THE CYLINDER WEIGHT TO EXCEED THIS MAXIMUM. Many cylinders have automatic devices which prevent the technician from exceeding dangerous levels of refrigerant in a recovery cylinder.

6. Start the system and open the valve arrangement between the liquid line and the cylinder. Liquid refrigerant will start to move toward the cylinder.

7. Watch the high-pressure gage and do not allow the system pressure to rise above the working pressure of the refrigerant cylinder. In most cases this is 250 psig.

8. Watch the suction pressure and do not allow it to fall below 0 psig. When it reaches 0 psig, you have removed all the refrigerant from the system that you can using this method.

9. Remove the refrigerant cylinder. You may need to recover the remaining refrigerant using recovery equipment. Recovery can be speeded up by setting the recovery cylinder in a bucket of ice, thus reducing the pressure in the cylinder.

Commercially Manufactured Recovery Unit Equipment Method

1. Read the manufacturer's directions closely. Some units recover refrigerants only and some recover and filter the refrigerant.

2. Connect the recovery unit to the system following the manufacturer's directions for the particular type of system you have.

3. Operate the recovery system as per the manufacturer's directions.

4. Upon completion of the operation, the refrigerant will be recovered in an approved container.

Maintenance of Workstation and Tools: Return the system to the condition your instructor requests you to. Return all tools and materials to their proper places.

Summary Statement: Describe the difference between recovery only and recycling.

QUESTIONS

1. After a hermetic motor burn, what contaminant might be found in the recovered refrigerant?

2. What would the result be if a recovery cylinder were filled with liquid and allowed to warm up?

3. Why must liquid refrigerant be present for the pressure-temperature relationship to apply?

4. If an air-conditioning unit using R-22 were located totally outside and the temperature were 85°F, what would the pressure in the system be with the unit off? (Hint: It would follow the P and T relationship.)

5. When an evaporator using R-12 is boiling the refrigerant at 21°F, what will the suction pressure be? _____ psig

6. When an evaporator using R-22 is boiling the refrigerant at 40°F, what will the suction pressure be? _____ psig

7. When the suction pressure of an R-22 system is 60 psig, at what temperature is the refrigerant boiling? _____ °F

8. When the suction pressure of an R-502 system is 20 psig, at what temperature is the refrigerant boiling? _____ °F.

9. What is the condensing temperature for a head pressure of 175 psig for an R-12 system? _____ °F

10. What is the condensing temperature for a head pressure of 296 psig for an R-22 system? _____ °F

System Charging

Unit Overview

Charging a system refers to adding refrigerant to a refrigeration system. Refrigerant may be added while it is in the vapor state or under certain conditions while it is in its liquid state.

When charging a system with refrigerant vapor, the gage manifold and refrigerant cylinder are connected so that the vapor will move from the cylinder through the gage manifold and into the low-pressure side of the system. If the refrigerant cylinder is warm, the pressure inside it will be greater than that in the low-pressure side of the system. However, as refrigerant is removed from the cylinder, the pressure in the cylinder will drop.

If a refrigeration system is in a vacuum or out of refrigerant, it may be charged with liquid refrigerant on the high side of the system. The refrigerant cylinder is turned upside down, and liquid is allowed to flow through the king valve between the condenser and evaporator. Often blended refrigerants use a dip tube to the bottom of the disposable tank and should not be turned upside down. It is always important to ensure that no liquid refrigerant reaches the compressor.

The charging or recharging of a refrigeration system using a near-azeotropic or zeotropic refrigerant blend must be done with liquid refrigerant. This will avoid fractionation of the refrigerant out of the charging cylinder. As long as virgin liquid refrigerant is used in recharging a system that has leaked, capacity losses from leak fractionation will be small and insignificant. However, if vapor is removed from the charging cylinder, fractionation will occur.

When adding liquid refrigerant to the suction line of a system that is running, the liquid, once out of the charging cylinder, has to be restricted and vaporized before entering the system to avoid any damage to the compressor. This can be accomplished by restricting the manifold gauge set or with a commercial restricting device in line with the compressor on the charging hose. Also, restricting the three-way suction service valve at the compressor by just cracking it off of its back seat will help vaporize any liquid refrigerant trying to enter the compressor.

When a system has no refrigerant, the charge should be measured. This can be done with a scale or a graduated cylinder.

Key Terms

Back Seated
Charging Near-Azeotropic Refrigerant Blends
Electronic Charging Scale (Programmable)
Electronic Scale
Front Seated
Graduated Charging Cylinder

King Valve
Liquid Refrigerant Charging
Midseated (Cracked)
System Charging
Vapor Refrigerant Charging

REVIEW TEST

Name	Date	Grade

Circle the letter that indicates the correct answer.

1. Charging a refrigeration system refers to:
   A. removing refrigerant from a system.
   B. adding refrigerant to a system.
   C. cleaning the refrigerant.
   D. vaporizing the refrigerant.

2. Vapor refrigerant is added to a system through the:
   A. receiver king valve.
   B. liquid-line service valve.
   C. compressor discharge line.
   D. low-pressure side of the system.

3. A warm refrigerant storage cylinder will have _____ the low-pressure side of a refrigeration system.
   A. a higher pressure than
   B. a lower pressure than
   C. the same pressure as

4. As refrigerant is removed from a storage cylinder, the pressure inside will:
   A. rise.   C. stay the same.
   B. fall.

5. Liquid refrigerant charging is only done when the refrigeration system is:
   A. in a vacuum.
   B. out of refrigerant.
   C. very low on refrigerant.
   D. all of the above.

6. When charging liquid refrigerant, the cylinder should be placed:
   A. upside down.
   B. right side up.
   C. on its side.
   D. in any of the positions indicated.

7. If 22 oz of refrigerant are to be added to a system and the cylinder weighs 24 lb 12 oz before any is added, what will the cylinder weigh after the system has been charged?
   A. 22 lb 12 oz   C. 22 lb 8 oz
   B. 22 lb 6 oz   D. 23 lb 6 oz

8. If 30 oz of refrigerant are to be added to a system and the cylinder weighs 26 lb 4 oz before any is added, what will the cylinder weigh after the system has been charged?
   A. 22 lb 8 oz   C. 24 lb 8 oz
   B. 24 lb 6 oz   D. 22 lb 4 oz

9. A programmable electronic charging scale:
   A. will calculate the amount of charge for you.
   B. has a microprocessor that will provide a readout of the condition of the refrigerant.
   C. has a solid-state microprocessor that stops the charging process when the programmed weight of refrigerant has been dispensed.
   D. all of the above.

10. A graduated charging cylinder is a device that:
    A. measures refrigerant by volume for precise refrigerant charging.
    B. has a provision for dialing in the temperature of the refrigerant.
    C. often has a heater to keep the refrigerant temperature from dropping when vapor is released into the system.
    D. all of the above.

11. Liquid refrigerant charging into a compressor must be done very carefully because:
    A. refrigerant is expensive.
    B. damage to the compressor may occur.
    C. it takes too long.
    D. all of the above.

12. Vapor charging refrigerant into a system is _____ than charging liquid refrigerant into the system.
    A. faster
    B. slower

13. When charging vapor refrigerant into a system and the cylinder pressure drops to the point that refrigerant will not move into the system, you should:
    A. warm the cylinder in warm water, not to exceed 90°F.
    B. wait until the next day when the cylinder pressure rises.
    C. get another cylinder of refrigerant.
    D. none of the above.

14. To add refrigerant to the system through the suction service valve, the valve stem must be:
    A. back seated.
    B. front seated.
    C. midseated.
    D. none of the above.

15. When charging a system with a low pressure control and the control keeps shutting the compressor off, you should:
    A. come back tomorrow.
    B. let the system set for awhile.
    C. use a different refrigerant cylinder.
    D. bypass the low pressure control while charging.

16. When charging refrigeration and air conditioning systems containing refrigerant blends that have the potential to fractionate, the service technician must always charge with _____ refrigerant.
    A. liquid   C. liquid and vapor
    B. vapor   D. all of the above

17. Which device may be used to throttle and vaporize liquid refrigerant before it enters the compressor?
    A. three-way suction service valve
    B. manifold gauge set
    C. commercial restrictor in charging line
    D. all of the above

Charging a System

Name	Date	Grade

Objectives: Upon completion of this exercise, you will be able to add refrigerant to a charging cylinder and charge refrigerant into a typical refrigeration system using a charging cylinder.

Introduction: You will apply gages to a typical capillary tube refrigeration system, recover the existing charge, and evacuate the system to a deep vacuum. You will then add enough charge to a charging cylinder, connect it to the system, and measure in the correct charge. You will need to use a package system or one with a known charge that has service valves or Schrader valve ports.

Text References: Paragraphs 9.1, 9.17, 10.1, 10.2, 10.3, 10.5, 10.6, and 10.7.

Tools and Materials: A vacuum pump, gage manifold, refrigerant recovery equipment, goggles, gloves, refrigerant that is characteristic to the system, graduated charging cylinder, straight-blade screwdriver, $\frac{1}{4}$" and $\frac{5}{16}$" nut drivers, and a capillary tube refrigeration system as described above.

Safety Precautions: Wear gloves and goggles while transferring refrigerant to and from a system. DO NOT overcharge the charging cylinder. Follow all instructions carefully. Your instructor must give you instruction in using the gage manifold and refrigerant recovery equipment, adding refrigerant to the charging cylinder and to the system, using a vacuum pump, and evacuating a system before you proceed with this exercise.

PROCEDURES

1. Connect the refrigerant recovery equipment to the system and recover all refrigerant from the system to be charged. Your instructor will instruct you as to where and how to recover the refrigerant.

2. Connect the gage lines to the service valves or Schrader valve ports on the high and low sides of the system.

3. Check the oil level in the vacuum pump, add oil if needed, connect it to the center line of the manifold, and start the pump. Open the gage manifold valves.

4. Determine the correct charge for the system, and write it here for reference: _____ oz

5. While the evacuation is being performed, connect the charging cylinder to the cylinder of refrigerant, Figure 10-1. When transferring any refrigerant from a refrigerant cylinder to a charging cylinder, it is illegal under Section 608 of the Clean Air Act to vent vapor from the top of the charging cylinder when it is being filled. The service technician must gently heat the refrigerant cylinder to 90°F to 100°F to increase its pressure before refrigerant can be successfully transferred to the charging cylinder. The best way to heat a cylinder is to put it in a container of warm water. Also, when charging with a refrigerant blend that has a potential to fractionate, the service technician must throttle liquid refrigerant from the refrigerant cylinder into the charging cylinder to avoid fractionation.

6. When the charging cylinder has enough refrigerant to charge the system and the vacuum has reached the correct level (deep vacuum is determined by sound or instrument), you are ready to charge the refrigerant into the system.

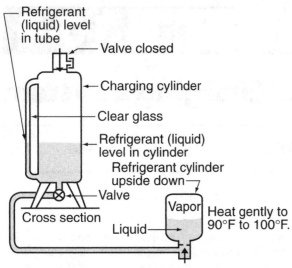

Refrigerant (liquid) level in tube

Valve closed

Charging cylinder

Clear glass

Refrigerant (liquid) level in cylinder

Refrigerant cylinder upside down

Valve

Cross section

Liquid

Vapor Heat gently to 90°F to 100°F.

Figure 10-1

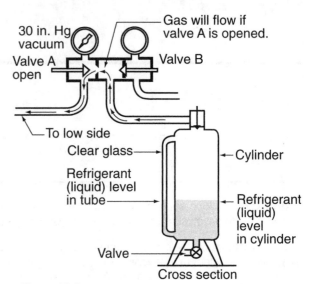

30 in. Hg vacuum

Gas will flow if valve A is opened.

Valve A open

Valve B

To low side

Clear glass

Cylinder

Refrigerant (liquid) level in tube

Refrigerant (liquid) level in cylinder

Valve

Cross section

Figure 10-2

7. Shut off the vacuum pump, and connect the charging cylinder, Figure 10-2.

8. Purge a small amount of refrigerant from the gage manifold line by loosening the center gage line at the manifold after the connection is made, because some air will be drawn into the gage line when disconnecting from the vacuum pump.

9. Turn the outside dial on the charging cylinder to correspond to the pressure in the cylinder. For example, if the pressure gage at the top of the cylinder reads 75 psig (this would be R-134a or R-12), turn the dial to the 75 psig scale and read the exact number of ounces on the sight glass of the charging cylinder, Figure 10-3. Record it here: _____ ounces

10. Subtract the amount of the unit charge from the charging cylinder level and record here: _____ ounces (This is the stopping point of the charge.)

11. Open the vapor valve at the top of the cylinder and the low-side valve on the gage manifold. The charge will begin to fill the system. Remember, when charging with a refrigerant blend that has a potential to fractionate, the service technician must also throttle liquid refrigerant from the charging cylinder into the refrigeration system to avoid fractionation.

12. When the system is above 0 psig, disconnect the high-side gage line. The reason for this is that the charging cylinder pressure will soon equalize to the same pressure as the system, and the refrigerant will stop flowing. You may have to start the system to pull in the correct charge. You do not want any condensed refrigerant in the high-side gage line. Some charging cylinders have a heater in the base to keep the pressure up, and the system may not have to be started.

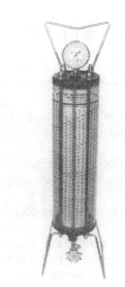

Figure 10-3 *Photo by Bill Johnson*

13. Keep adding refrigerant vapor until the correct low level appears on the sight glass on the charging cylinder, Figure 10-4, and then shut off the suction valve on the gage manifold.

Refrigeration & Air Conditioning Technology

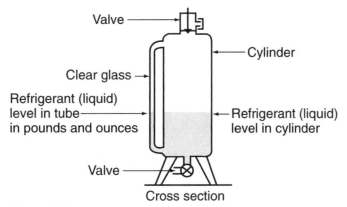

Valve

Cylinder

Clear glass →

Refrigerant (liquid)
level in tube
in pounds and ounces

Refrigerant (liquid)
level in cylinder

Valve

Cross section

Figure 10-4

Maintenance of Workstation and Tools: Check the vacuum pump oil and add more if needed. Wipe up any oil in the work area. Return all tools to their proper places. Replace panels on the unit.

Summary Statement: Describe the calculations you had to make to determine and make the correct charge for the system.

QUESTIONS

1. What are the graduations on the charging cylinder?

2. Why is it necessary to turn the dial to the correct pressure reading and use this figure for the charge?

3. What scale or graduation was the charge expressed in on the system you charged, pounds or pounds and ounces?

4. How many ounces are in a pound?

5. How many ounces are in one-fourth (0.25) of a pound?

6. Why was the high-side line disconnected before starting the unit?

7. How may refrigerant be charged into the system if the cylinder pressure drops?

8. How do some charging cylinders keep the pressure from dropping as the charge is added?

9. What type of metering device normally requires an exact charge?

10. Do all systems have sight glasses in the liquid line to aid in charging?

Charging a System with Refrigerant Using Scales

Name	Date	Grade

Objectives: Upon completion of this exercise, you will be able to charge a typical refrigeration system using scales to weigh the refrigerant.

Introduction: You will recover refrigerant, if any, from a refrigeration system, evacuate the system to prepare it for charging using refrigerant weight, then charge the system to the correct operating charge.

Text References: Paragraphs 9.1, 9.17, 10.1, 10.2, 10.3, and 10.4.

Tools and Materials: A gage manifold, gloves, goggles, refrigerant recovery equipment, a high-quality vacuum pump, a set of accurate scales (electronic scales are preferred because of their ability to count backward, but the exercise will be written around standard platform scales), and a cylinder of the proper refrigerant for the system to be charged. You will need a refrigeration system that has service valves or ports and that has a known required charge.

Safety Precautions: Wear gloves and goggles any time you are transferring refrigerant. Your instructor must give you instruction in recovering and transferring refrigerant before starting this exercise.

PROCEDURES

1. Recover any refrigerant that may be in the system down to atmospheric pressure. Your instructor will tell you how to recover the refrigerant correctly.

2. While wearing goggles, connect the gages to the high- and the low-side service valves or Schrader ports.

3. Add oil to the vacuum pump if it is not at the proper level. When the level is correct, connect it to the center gage line and turn it on.

4. Determine the correct charge for the system from the nameplate or the manufacturer's specifications and record it here: _____ ounces

5. Set the refrigerant cylinder on the scales with the vapor line up. Make a provision to secure the gage line when it is disconnected from the vacuum pump and connected to the cylinder on the scales so it will not move and vary the reading. A long gage line is preferred for this, Figure 10-5.

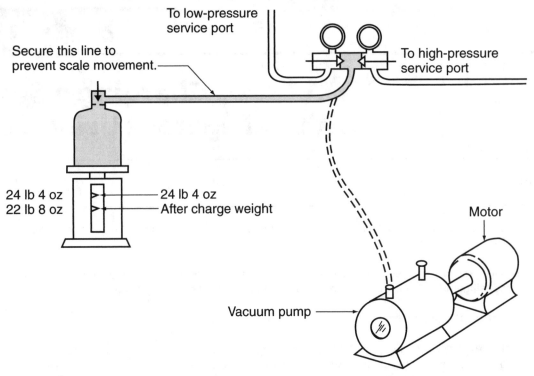

To low-pressure
service port

Secure this line to
prevent scale movement.

To high-pressure
service port

24 lb 4 oz
22 lb 8 oz

24 lb 4 oz
After charge weight

Motor

Vacuum pump

Figure 10-5

6. When the vacuum pump has evacuated the system to the correct point (the sound test mentioned in Lab 8-3 may be used if no vacuum gage is available), close the gage manifold valves, turn off the vacuum pump, disconnect the center gage line from the vacuum pump, and connect it to the refrigerant cylinder. Do not forget to secure it.

7. Allow a very small amount of refrigerant vapor to bleed out the center gage line at the manifold because when you disconnect the vacuum pump, air will be drawn in.

8. Record the cylinder weight in ounces here: _____ oz

   NOTE: The scales may read in pounds and tenths of a pound. Convert this to ounces to make calculations easier. For instance, 0.4 of a lb equals 6.4 oz (16 oz $\times$ 0.4 = 6.4).

9. Calculate what the reading on the scale should be when the system has been completely charged and enter here: _____ oz

10. Open the low-side gage manifold valve and allow refrigerant vapor to start entering the system. Remember, when charging with a refrigerant blend that has a potential to fractionate, the service technician must also throttle liquid refrigerant from the charging cylinder into the refrigeration system to avoid fractionation.

11. When the system pressure is above atmospheric, disconnect the high-side line. The reason for this is because refrigerant will condense in this high-side line when the system is started, and the charge will not be correct.

12. If all of the charge will not enter the system, the system may be started to lower the system pressure so that the rest of the charge will be pushed in. Do not let the low-side pressure exceed normal low-side pressure.

13. When the complete charge is in the system according to the scales, disconnect the low-side line and run the system if it is ready to run.

Maintenance of Workstation and Tools. Wipe any oil off the tools and work area. Check the oil in the vacuum pump and fill with approved vacuum pump oil if needed. Replace any panels on the unit. Place all tools in their respective places.

Summary Statements: Describe the procedure for determining the amount of refrigerant to be charged into the system. Describe the calculations you made regarding the use of the scales. Include any conversions from pounds to ounces.

QUESTIONS

1. Why is the high-side line disconnected before the system is started?

2. Why is the line to the refrigerant cylinder purged after evacuation and before charging?

3. Why is the gage line secured when the refrigerant cylinder is on the scales while transferring refrigerant?

4. Why would the cylinder pressure drop when charging vapor into a system?

5. What can be done to keep refrigerant moving into the system if the cylinder pressure drops and the cylinder must remain on the scales?

6. Why is only vapor recommended for charging a system?

7. What would happen if liquid refrigerant were to enter the compressor cylinder while it was running?

8. What type of oil is recommended for a high-vacuum pump?

9. How many ounces are in a pound?

10. How many ounces are in three-tenths (0.3) of a pound?

Charging a Capillary Tube Air-Conditioning System with Refrigerant

Name	Date	Grade

Objectives: Upon completion of this exercise, you will able to properly charge a split-type air-conditioning system that is equipped with a capillary tube or other fixed-bore metering device.

Introduction: As part of this lab exercise, you will install gages on a split-type air-conditioning system that is equipped with a fixed-bore metering device. Because the equipment is field installed, the exact refrigerant charge for the system is not found on the nameplate of the system. In order to properly charge this system, we will rely on the evaporator super-heat to ensure that the refrigerant charge is correct.

Text References: 9.1, 9.17, 10.1, 10.2, and 10.3.

Tools and Materials: Gage manifold, refrigerant tank, high-quality thermocouple thermometer, temperature/pressure chart, goggles, gloves, and a refrigerant scale.

Safety Precautions: When handling and working with refrigerants, be sure to protect yourself from frostbite, as escaping refrigerant is extremely cold. Wear safety goggles as well as gloves and other pieces of personal protection equipment. Make certain that the charging process is performed slowly and carefully to ensure that excessive system pressures do not result. Always follow all of your teacher's instructions.

PROCEDURES

1. Make certain that the system you are working on has been properly evacuated to the required vacuum level determined by your instructor. At this point, the gages should be on the system, with the low-side hose connected to the suction-line service port on the condensing unit and the high-side hose connected to the liquid-line service port on the condensing unit. Both valves on the gage manifold should be in the closed position.

2. If possible, connect a low-side compound gage to the suction line at the outlet of the evaporator. By taking the pressure reading as close as possible to the outlet of the evaporator, your superheat calculations will be more accurate.

3. Inspect the system and make certain that the suction line is properly insulated. If the insulation is missing or loosely installed, add and resecure insulation as needed.

4. Determine the correct refrigerant for the system. This information is provided on the nameplate of the unit.

5. Obtain the proper tank of refrigerant.

6. Place the refrigerant tank on the charging scale and turn the scale on.

7. Connect the center hose from your gage manifold to the refrigerant tank.

8. Open the valve on the refrigerant tank.

9. Very briefly, loosen the center hose connection on the gage manifold to purge any air that might be in the center hose.

10. Retighten the center hose connection on the gage manifold.

11. On refrigerant scales equipped with a zero-adjust feature, reset the scale so that the display reads "0 LBS, 0 OZ." If your scale does not have this feature, write down the weight of the cylinder in pounds and ounces.

12. Make certain that the air-conditioning system is off.

13. Open the high-side valve on the gage manifold. Refrigerant will flow from the tank through the gage manifold and into the high side of the air-conditioning system. You will note that, although you are adding refrigerant to the low side of the system, the pressure reading on the low side of the system will rise. This is because refrigerant is flowing through the capillary tube into the low side of the system.

14. Continue to allow refrigerant to flow into the system until the high- and low-side system pressures are equal and no more refrigerant is flowing from the refrigerant tank.

15. Close the valves on the gage manifold.

16. Secure the sensing bulb of your thermocouple thermometer to the top of the suction line at the outlet of the evaporator. Make certain that the sensing bulb is well insulated by wrapping the sensing element and the suction line with foam insulation tape. Poor thermal contact between the sensing element and the suction line will result in a system undercharge, as the calculated evaporator superheat will be higher than the actual evaporator superheat.

17. Inspect the system and make certain that there are no restrictions or obstructions in the air paths through the evaporator and condenser coils. This means checking for dirty air filters, dirty coils, closed supply registers, blocked returns, and other possible restrictions to airflow that might be present.

18. Energize the air-conditioning system and allow it to operate for a minimum of 5 minutes.

19. Obtain and record the following information:
    - Ambient temperature surrounding the condensing unit _____ °F
    - Indoor ambient temperature (return air to the evaporator coil) _____ °F
    - Supply air temperature coming off the evaporator coil _____ °F
    - Evaporator saturation pressure _____ psig
    - Evaporator saturation temperature _____ °F
    - Evaporator outlet temperature _____ °F
    - Condenser saturation pressure _____ psig
    - Condenser saturation temperature _____ °F
    - Evaporator superheat _____ °F
    - Compressor amperage _____ A
    - Temperature differential between return and supply air temperatures _____ °F
    - Temperature differential between outside ambient and condenser saturation temperatures _____ °F

20. Calculate the evaporator superheat:

	Evaporator outlet temp:	_____ °F
−	Evaporator saturation temp:	_____ °F
	Evaporator superheat:	_____ °F

21. If the evaporator superheat is greater than the 8°F–12°F range, add 4 ounces of refrigerant to the system, allow it to operate for at least 5 minutes, and return to step 20. If the evaporator superheat is less than the 8°F–12°F range, you will need to recover some refrigerant from the system, allow it operate for at least 5 minutes, and return to step 20. If the evaporator superheat is within the 8°F–12°F range, the system is properly charged. Proceed to step 22.

22. Now that the system is charged, obtain and record the following information:

   - Ambient temperature surrounding the condensing unit _____ °F
   - Indoor ambient temperature (return air to the evaporator coil) _____ °F
   - Supply air temperature coming off the evaporator coil _____ °F
   - Evaporator saturation pressure _____ psig
   - Evaporator saturation temperature _____ °F
   - Evaporator outlet temperature _____ °F
   - Condenser saturation pressure _____ psig
   - Condenser saturation temperature _____ °F
   - Evaporator superheat _____ °F
   - Compressor amperage _____ A
   - Temperature differential between return and supply air temperatures _____ °F
   - Temperature differential between outside ambient and condenser saturation temperatures _____ °F
   - Total amount of refrigerant added to the system _____ lb _____ oz

23. Compare the recorded values from step 19 to those recorded in step 22. For each of the listed parameters, determine whether it increased, decreased, or remained the same.

24. Provide explanations for the changes noted in step 23.

Maintenance of Workstation and Tools: Make certain that the gage manifold is properly removed from the system and that the hoses are connected to the blank ports on the manifold to prevent hose contamination. Make certain that the service port caps are replaced on the system service valves. Return all tools to their proper places. Make certain that refrigerant tanks are closed and properly stored.

Summary Statement: Explain how evaporator superheat changes with respect to the refrigerant charge in an operating air-conditioning system. Explain the effects of an overcharge/undercharge on system operation and compressor amperage.

QUESTIONS

1. Why is it necessary to purge the center hose of the gage manifold before adding refrigerant to the system?

2. In plain terms, explain what evaporator superheat is.

3. In plain terms, explain how evaporator superheat is calculated.

4. Why is it important to secure the loose ends of the service hoses to the blank ports on the gage manifold?

5. Why does the sensor on the thermocouple thermometer need to be insulated?

6. Why do you have to wait at least 5 minutes before adding additional refrigerant to the system?

7. Explain why excessive evaporator superheat is an indication of an underfed evaporator coil.

8. Explain why low evaporator superheat is an indication of an overfed evaporator coil.

Charging a TVX Air-Conditioning System with Refrigerant

Name	Date	Grade

Objectives: Upon completion of this exercise, you will able to properly charge a split-type air-conditioning system that is equipped with a thermostatic expansion valve.

Introduction: As part of this lab exercise, you will install gages on a split-type air-conditioning system that is equipped with a thermostatic expansion valve. Because the equipment is field installed, the exact refrigerant charge for the system is not found on the nameplate of the system. To properly charge this system, we will rely on the condenser subcooling to ensure that the refrigerant charge is correct.

Text References: 10.1, 10.2, 10.3, and 10.4.

Tools and Materials: Gage manifold, refrigerant tank, high-quality thermocouple thermometer, temperature/pressure chart, goggles, gloves, refrigerant scale, and foam insulation tape.

Safety Precautions: When handling and working with refrigerants, be sure to protect yourself from frostbite, as escaping refrigerant is extremely cold. Wear goggles as well as gloves and other pieces of personal protection equipment. Make certain that the charging process is performed slowly and carefully to ensure that excessive system pressures do not result. Always follow all of your teacher's instructions.

PROCEDURES

1. Make certain that the system you are working on has been properly evacuated to the required vacuum level determined by your instructor. At this point, the gages should be on the system, with the low-side hose connected to the suction-line service port on the condensing unit and the high-side hose connected to the liquid-line service port on the condensing unit. Both valves on the gage manifold should be in the closed position.

2. Inspect the system and make certain that the thermal bulb on the thermostatic expansion valve is properly secured to the suction line and that the bulb is properly insulated. If the insulation is missing or if the bulb is loosely installed, add insulation and resecure the bulb as needed.

3. Determine the correct refrigerant for the system. This information is provided on the nameplate of the unit.

4. Obtain the proper tank of refrigerant.

5. Place the refrigerant tank on the charging scale and turn the scale on.

6. Connect the center hose from your gage manifold to the refrigerant tank.

7. Open the valve on the refrigerant tank.

8. Very briefly, loosen the center hose connection on the gage manifold to purge any air that might be in the center hose.

9. Retighten the center hose connection on the gage manifold.

10. On refrigerant scales equipped with a zero-adjust feature, reset the scale so that the display reads "0 LBS, 0 OZ." If your scale does not have this feature, write down the weight of the cylinder in pounds and ounces.

11. Make certain that the air-conditioning system is off.

12. With the refrigerant tank positioned so that vapor refrigerant will leave the tank, open both the high-side and the low-side valves on the gage manifold. Refrigerant will flow from the tank through the gage manifold and into the system. **NOTE:** On systems that operate with blended refrigerants, the refrigerant must leave the tank as a liquid. In this case, position the tank so that liquid refrigerant will leave the tank and introduce refrigerant only to the high-pressure side of the system.

13. Continue to allow refrigerant to flow into the system until the system pressure equalizes with the pressure in the refrigerant tank.

14. Close the valves on the gage manifold.

15. Secure the sensor of your thermocouple thermometer to the liquid line at the outlet of the condenser coil. Make certain that the sensor is well insulated by wrapping it, along with the liquid line, with foam insulation tape. Poor thermal contact between the sensor and the liquid line will result in improper system charging, as the liquid-line temperature reading will be different from the actual liquid-line temperature.

16. Inspect the system and make certain that there are no restrictions or obstructions in the air paths through the evaporator and condenser coils. This means checking for dirty air filters, dirty coils, closed supply registers, blocked returns, and other possible restrictions to airflow that might be present.

17. Energize the air-conditioning system and allow it to operate for a minimum of 5 minutes.

18. Obtain and record the following information:
    - Ambient temperature surrounding the condensing unit _____ °F
    - Indoor ambient temperature (return air to the evaporator coil) _____ °F
    - Supply air temperature coming off the evaporator coil _____ °F
    - Evaporator saturation pressure _____ psig
    - Evaporator saturation temperature _____ °F
    - Condenser saturation pressure _____ psig
    - Condenser saturation temperature _____ °F
    - Condenser outlet temperature _____ °F
    - Compressor amperage _____ A
    - Temperature differential between return and supply air temperatures _____ °F
    - Temperature differential between outside ambient and condenser saturation temperatures _____ °F

19. Calculate the condenser subcooling:

	Condenser saturation temp:	_____ °F
−	Condenser outlet temp:	_____ °F
	Condenser subcooling:	_____ °F

20. If the condenser subcooling is lower than the 15°F–20°F range, add 4 oz of refrigerant to the system, allow it to operate for at least 5 minutes, and return to step 19. If the condenser subcooling is greater than the 15°F–20°F range, you will need to recover some refrigerant from the system, allow it operate for at least 5 minutes, and return to step 19. If the condenser subcooling is within the 15°F–20°F range, the system is properly charged. Proceed to step 21.

21. Now that the system is charged, obtain and record the following information:

- Ambient temperature surrounding the condensing unit _____ °F
- Indoor ambient temperature (return air to the evaporator coil) _____ °F
- Supply air temperature coming off the evaporator coil _____ °F
- Evaporator saturation pressure _____ psig
- Evaporator saturation temperature _____ °F
- Condenser saturation pressure _____ psig
- Condenser saturation temperature _____ °F
- Condenser outlet temperature _____ °F
- Condenser subcooling _____ °F
- Compressor amperage _____ A
- Temperature differential between return and supply air temperatures _____ °F
- Temperature differential between outside ambient and condenser saturation temperatures _____ °F
- Total amount of refrigerant added to the system _____ lb _____ oz

22. Compare the recorded values from step 18 to those recorded in step 21. For each of the listed parameters, determine whether it increased, decreased, or remained the same.

23. Provide explanations for the changes noted in step 22.

Maintenance of Workstation and Tools: Make certain that the gage manifold is properly removed from the system and that the hoses are connected to the blank ports on the manifold to prevent hose contamination. Make certain that the service port caps are replaced on the system service valves. Return all tools to their proper places. Make certain that refrigerant tanks are closed and properly stored.

Summary Statement: Explain how condenser subcooling changes with respect to the refrigerant charge in an operating air-conditioning system. Explain the effects of an overcharge/undercharge on system operation and compressor amperage.

QUESTIONS

1. Explain the function of the thermostatic expansion valve.

2. Explain, in plain terms, what condenser subcooling is.

3. Explain how condenser subcooling is calculated.

4. Explain why we use condenser subcooling to charge air-conditioning and refrigeration systems that are equipped with thermostatic expansion valves.

5. Why is it important to secure the loose ends of the service hoses to the blank ports on the gage manifold?

6. Why does the sensor on the thermocouple thermometer need to be insulated?

7. Why do you have to wait at least 5 minutes before adding additional refrigerant to the system?

8. Explain why low condenser subcooling is an indication of an undercharged system.

9. Explain why high condenser subcooling is an indication of an overcharged system.

Calibrating Instruments

Unit Overview

Instruments used by the air-conditioning and refrigeration technician must be accurate and reliable. They should be checked periodically to ensure that they read or correspond to a standard or correct reading.

Temperature-measuring instruments can be checked with known temperatures such as ice and water, boiling water, and body temperature. Pressure gages can be checked using the temperature/pressure relationship chart and cylinders with clean, fresh refrigerant. Vacuum gages and instruments can be checked using a vacuum pump. The voltage and amperage readings of electrical test equipment can be compared with high-quality bench meters. The ohmmeter can be used to read the resistance of several high-quality resistors at various resistance ranges. Flue-gas analysis instruments should be maintained as recommended by the manufacturer.

Key Terms

Calibration or Calibrate
Clamp-on Ammeter
Flue-Gas Analysis Instruments
Leak Detection Devices
Temperature/Pressure Relationship Chart

Temperature-Measuring Instruments
 (Thermometer)
Temperature Reference Points
Temperature-Sensing Element
Volt-Ohm-Milliammeter (VOM)

REVIEW TEST

Name	Date	Grade

Circle the letter that indicates the correct answer.

1. Three reference points for checking temperature-measuring instruments are:
   A. 0°F, 100°F, 110°F.
   B. −460°F, 0°F, 32°F.
   C. 32°F, 98.6°F, 212°F.
   D. 32°F, 100°F, 250°F.

2. To check the temperature leads of an electronic thermometer for low temperatures, they should be stirred in a container of:
   A. crushed ice and water with half a teaspoon of salt.
   B. crushed ice and water with a pinch of sugar.
   C. crushed ice and pure water.
   D. cubed ice and water.

3. To check the temperature leads of an electronic thermometer for high temperatures, they should be:
   A. placed on the bottom of a pan of boiling water.
   B. suspended in a pan of boiling water.
   C. suspended in the vapor just above a pan of boiling water.
   D. suspended in a pan of boiling water when above sea level at least 1000 feet.

4. Pressure gages can be checked by:
   A. connecting them to cylinders of refrigerant with a known temperature and comparing the pressure reading with that on a temperature/pressure relationship chart.
   B. connecting them to an operating refrigeration system and comparing the pressure to the superheat.
   C. connecting them to an operating refrigeration system and comparing the pressure to that indicated by the manufacturer for that system.
   D. connecting them one at a time to a small cylinder of nitrogen under low pressure.

5. The ohmmeter feature of a volt-ohm-milliammeter can be checked by:
   A. comparing it with an ohmmeter at a supply store.
   B. checking the manufacturer's specifications.
   C. using Ohm's law.
   D. making a zero adjustment and then taking and comparing resistance readings of several high-quality resistors with known resistances.

6. **The volt scale on a volt-ohm-milliammeter may be checked by:**
   A. taking a line voltage reading at the disconnect box.
   B. comparing a voltage reading from the meter being checked with that of a high-quality bench voltmeter.
   C. taking a current reading with a clamp-on ammeter and using Ohm's law to determine the voltage.
   D. using known resistances to check the voltage.

7. **The clamp-on ammeter can be checked by:**
   A. comparing a reading on the meter being checked with a reading taken at the same time with a high-quality bench meter.
   B. taking a current reading at the disconnect box of an operating condensing unit.
   C. using Ohm's law in a 208 V compressor circuit.
   D. both A and B.

8. **A common refrigerant leak detection device is the:**
   A. micron gage.
   B. halide torch.
   C. U-tube manometer.
   D. all of the above.

9. **To determine whether a leak detection device is operating properly, spray pure refrigerant into the sensing element.**
   A. true
   B. false

10. **A flue-gas analysis kit contains a:**
    A. CO_2 tester, draft gage, and thermometer.
    B. micron gage and refrigerant reference leak vial.
    C. smoke tester and halide torch.
    D. both A and C.

11. **The most accurate thermometer is the:**
    A. glass stem thermometer.
    B. electronic thermometer.
    C. dial-type thermometer.
    D. pocket dial-type thermometer.

12. **Which of the following gages is considered to be the most accurate gage for checking very low vacuum levels when evacuating a system?**
    A. U-tube manometer
    B. compound manifold gage
    C. micron gage
    D. none of the above

13. **Which of the following leak detectors will find the smallest leak?**
    A. soap bubbles
    B. electronic leak detector
    C. halide torch
    D. listening for the leak

14. **The element in a halide torch is made of:**
    A. copper.
    B. steel.
    C. silver.
    D. cast iron.

Calibrating Thermometers

Name	Date	Grade

Objectives: Upon completion of this exercise, you will be able to check the accuracy of some typical thermometers and calibrate them if they are designed to be calibrated.

Introduction: You will use a variety of thermometers and check their calibration using different temperature standards.

Text References: Paragraphs 11.1, 11.2, and 11.3.

Tools and Materials: One each of the following if available to you in your lab: a dial-type thermometer with immersion stem, a glass stem thermometer, an electronic thermometer, a room thermostat cover with a thermometer in it, a heat source, ice, and water.

Safety Precautions: You will be working with hot water (212°F) and a heat source. Do not spill the water on yourself or get burned from the heat source.

PROCEDURES
For Immersion-Type Thermometers

1. From several different thermometers, list the type and the range (high and low reading). Example: glass stem, mercury-type thermometer with a high of 250°F and a low of −20°F:

    A. Type _____ High mark _____ Low mark _____

    B. Type _____ High mark _____ Low mark _____

    C. Type _____ High mark _____ Low mark _____

    D. Type _____ High mark _____ Low mark _____

    Notice that some of these thermometers may indicate as low as 32°F and may be checked in ice and water. Some of them may indicate as high as 212°F and may be checked in boiling water. Later we will deal with thermometers that may not be immersed in water, such as the one on a thermostat cover. All thermometers should be checked at two reference points and somewhere close to the range in which they will be used.

2. In a mixture of ice and water, submerge all of the thermometers that are submersible and that indicate down to freezing, and record their actual readings below. Make sure that both the ice and water reach to the bottom of the container, stir the mixture, and allow time for the thermometers to indicate.

    A. Type _____ Reading _____

    B. Type _____ Reading _____

    C. Type _____ Reading _____

    D. Type _____ Reading _____

3. Use a container of rapidly boiling water for the thermometers that are submersible and that indicate as high as 212°F and submerge them in the boiling water. DO NOT LET THEM TOUCH THE BOTTOM OF THE PAN. Stir them in the water and allow time for them to indicate. Record the readings below:

A. Type _____ Reading _____

B. Type _____ Reading _____

C. Type _____ Reading _____

D. Type _____ Reading _____

4. If the range of the thermometer does not reach a low of 32°F or a high of 212°F, record its high and low indications. For example, if it does not go to 212°F but only to 120°F, record 120°F.

A. Low reading _____°F High reading _____°F

B. Low reading _____°F High reading _____°F

C. Low reading _____°F High reading _____°F

D. Low reading _____°F High reading _____°F

5. Examine each of the thermometers checked in step 2 and step 3 for its method of calibration. Submerge each again and change the calibration to the correct reading of 32°F or 212°F. AFTER A THERMOMETER HAS BEEN CALIBRATED, CHECK IT AGAIN. Your body heat may have caused an error and you may have to change the calibration again. If a thermometer cannot be calibrated, a note may be placed with it to indicate its error, for example, "This thermometer reads 2°F high."

PROCEDURES

For Dial-Type Thermometers Such as in the Room Thermostat Cover

1. Place the thermostat cover in a moving air stream at room temperature, such as in a return air grille or a fan inlet.

2. Place one of the calibrated thermometers next to the room thermostat cover. The moving air stream will quickly cause both temperature indicators to reach the same temperature.

3. Change the calibration of the room thermostat cover to read the same as the calibrated thermometer.

4. Allow the thermometers to remain in the air stream for another short period of time because your body heat may have caused an error.

5. Recalibrate if necessary.

Maintenance of Workstation and Tools: Return all tools and equipment to their respective places and clean the work area.

Summary Statement: Describe the purpose of three different temperature-measuring instruments and include how their sensing elements work.

QUESTIONS

1. Why should the thermometer element not touch the bottom of the pan when checking it in boiling water?

2. Why should the thermometer be stirred when checking it in ice and water?

3. How do you calibrate a mercury thermometer when the numbers are etched on the stem?

4. What can you do if a thermometer is not correct and cannot be calibrated?

5. What is the purpose of calibrating a thermometer?

6. Name three ready reference temperatures that may be used to check a thermometer.

7. Why were the thermostat cover and the temperature lead placed in a moving air stream in step 1 of the procedures for dial-type thermometers?

8. What tool was used to adjust the calibration in the thermostat cover?

9. Why is it important to keep the calibrated thermometer in the cover of the thermostat when calibrating it?

10. Why should you recheck a thermometer after calibration?

Checking the Accuracy or Calibration of Electrical Instruments

Name	Date	Grade

Objectives: Upon completion of this exercise, you will be able to check the accuracy of and calibrate an ohmmeter, an ammeter, and a voltmeter.

Introduction: You will use field and bench techniques to verify the accuracy of and calibrate an ohmmeter, a voltmeter, and an ammeter so that you will have confidence when using them. Reference points that should be readily available to you will be used.

Text References: Paragraphs 11.1, 11.2, and 11.5.

Tools and Materials: A calculator; a resistance heater (an old duct heater will do); several resistors of known value (gold band electronic resistors); an ammeter; a voltmeter; a volt-ohm-milliameter (all meters with insulated alligator clip leads); 24-volt, 115-volt, and 230-volt power sources; and various live components in a system available in your lab.

Safety Precautions: You will be working with live electrical circuits. Caution must be used while taking readings. Do not touch the system while the power is on. Working around live electrical circuits will probably be the most hazardous part of your job as an air-conditioning, heating, and refrigeration technician. You should perform the following procedures only after you have had proper instruction and only under the close personal supervision of your instructor. You must follow the procedures as they are written here, as well as the procedures given by your instructor.

PROCEDURES

Ohmmeter Test

1. Turn the ohmmeter selector switch to the correct range for one of the known resistors. For example, if the resistor is 1500 ohms, the R × 100 scale will be correct for most meters because it will cause the needle to read about mid scale. Your instructor will show you the proper scale for the resistors he gives you to test. Place the alligator clip ends together and adjust the meter to 0 using the 0-adjust knob.

Using several different resistors of known value, record the resistances:

Meter range setting, R × _____ Resistor value _____ ohms
Meter reading _____ ohms   Difference in reading _____ ohms

Meter range setting, R × _____ Resistor value _____ ohms
Meter reading _____ ohms   Difference in reading _____ ohms

Meter range setting, R × _____ Resistor value _____ ohms
Meter reading _____ ohms   Difference in reading _____ ohms

Meter range setting, R × _____ Resistor value _____ ohms
Meter reading _____ ohms   Difference in reading _____ ohms

Meter range setting, R × _____ Resistor value _____ ohms
Meter reading _____ ohms   Difference in reading _____ ohms

Voltmeter Test

In this test, you will use the best voltmeter you have and compare it to others. Your instructor will make provisions for you to take voltage readings across typical low-voltage and high-voltage components. You must perform these tests in the following manner and follow any additional safety precautions from your instructor.

Use only meter leads with insulated alligator clips.

You and your instructor should ensure that all power to the unit you are working on is turned off.

Fasten alligator clips to terminals indicated by your instructor.

Make sure that all range selections have been set properly on your meter.

When your instructor has approved all connections, turn the power on and record the measurement.

1. Take a voltage reading of both a high- and low-voltage source using your best meter (we will call it a STANDARD). Your instructor will show you the appropriate terminals.

   Record the voltages here:

   Low-voltage reading _____ V   High-voltage reading _____ V

2. Use your other meters such as the VOM and the ammeter's voltmeter feature and compare their readings to the standard meter. Record the following:

   STANDARD: Low V reading _____ V   High V reading _____ V

   VOM: Low V reading _____ V   High V reading _____ V

   Ammeter: Low V reading _____ V   High V reading _____ V

   Record any differences in the readings here:

   VOM reads _____ V high, or _____ V low

   Ammeter reads _____ V high, or _____ V low

Ammeter Test

(Follow all procedures indicated in the voltmeter test.)

1. WITH THE POWER OFF, connect the electric resistance heater to the power supply.

2. Fasten the leads of the most accurate voltmeter to the power supply directly at the heater.

3. Take an ohm reading WITH THE POWER OFF: _____ ohms

4. Clamp the ammeter around one of the conductors leading to the heater.

5. Change the selector switch to read amps. HAVE YOUR INSTRUCTOR CHECK THE CONNECTIONS, and then turn the power on long enough to record the following. (The reason for not leaving the power on for a long time is that the resistance of the heater will change as it heats, and this will change the ampere reading.)

   Volt reading _____ V   Ampere reading _____ A

   Volt reading _____ V ÷ Ohm reading _____ Ω = Ampere reading _____ A

6. Use a bench meter, if possible, and compare to the clamp-on ammeter.

   Bench meter reading _____ A   Clamp-on ammeter reading _____ A

   Difference in the two _____ A

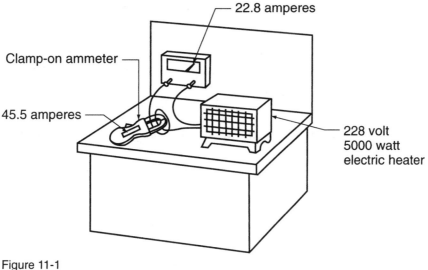

22.8 amperes

Clamp-on ammeter

45.5 amperes

228 volt
5000 watt
electric heater

Figure 11-1

7. Double the conductor through the ammeter and watch the reading double, Figure 11-1.

Maintenance of Workstation and Tools: Put each meter back in its case with any instruments that it may have. Turn all electrical switches off. Return all tools and equipment to their respective places.

Summary Statement: Describe the accuracy of each meter that you tested.

QUESTIONS

1. Describe an analog meter.

2. Describe a digital meter.

3. Which of the above meters will usually take the most physical abuse?

4. Why does the amperage change on an electrical heater when it becomes hot?

5. What does the gold band mean on a high-quality carbon resistor used for electronics?

6. What should the resistance be in a set of ohmmeter leads when the probes are touched together?

7. Why must you adjust an ohmmeter to 0 with the leads touching each time you change scales?

8. What should be done with a meter with a very slight error?

9. What should be done with a meter that has a large error and cannot be calibrated?

10. Why must you be sure that your meters and instruments are reading correctly and are in calibration?

Basic Electricity and Magnetism

Unit Overview

Matter is made up of atoms. Atoms are made up of protons, neutrons, and electrons. Protons have a positive charge, neutrons have no charge, and electrons have a negative charge. An atom with an excess of electrons has a negative charge. An atom with an excess of protons has a positive charge. An electron in an atom with a surplus of electrons will be attracted to an atom with a shortage of electrons.

Electrons travel easily in good conductors, such as copper and silver. They do not travel easily in insulators, such as glass, rubber, and plastic.

An electrical current produced to flow in one direction is called direct current (DC). Current that continually reverses itself is called alternating current (AC).

Voltage is an electrical measurement that indicates the difference in potential in two charges. The ampere is the unit used to measure the quantity of electrons moving past a given point in a specific period of time, and an ohm is the unit of measurement of resistance to electron flow. An electrical circuit must have a power source, a conductor to carry the electron flow, and a device to use the current.

Ohm's law: Voltage (E) = Amperage (I) × Resistance (R).

Components in a circuit may be wired in series or in parallel.

A coil with an electric current flowing through the loops of wire will cause an iron bar to be attracted into it. Switching devices designed to use this action are solenoids, relays, and contactors.

A multimeter is a common electrical measuring instrument. One that is often used is the volt-ohm-milliammeter (VOM).

Semiconductors have been developed that provide many alternatives with regard to the design of control circuits in air-conditioning, heating, and refrigeration equipment.

Key Terms

Alternating Current	Direct Current	Multimeter	Semiconductors
Ammeter	Electrical Power	Negative Charge	Series Circuit
Ampere	Electron	Ohm	Sine Wave
Atom	Fuse	Ohmmeter	Solenoid
Capacitance	Impedance	Ohm's Law	Transformer
Capacitor	Inductance	Parallel Circuit	Voltage
Circuit Breaker	Insulator	Positive Charge	Voltmeter
Clamp-on Ammeter	Magnetism	Proton	
Conductor	Microfarad	Root-Mean-Square (RMS)	

REVIEW TEST

Name	Date	Grade

Circle the letter that indicates the correct answer.

1. What type of charge does an electron have?
A. negative
B. positive
C. neutral
D. negative-positive

2. What part of an atom freely moves from one atom to another in a good conductor?
A. electron
B. proton
C. neutron

3. Typical insulators are:
A. copper, silver, gold.
B. steel, bronze, brass.
C. glass, rubber, plastic.
D. aluminum, nickel, magnesium.

4. Electricity can be produced from:
A. magnetism.
B. chemicals.
C. heat.
D. all of the above.

5. **The unit of measurement for resistance is the:**
   A. volt.
   C. ohm.
   B. ampere.
   D. milliampere.

6. **An ampere is:**
   A. the difference in potential between two charges.
   B. a measure of the quantity of electrons flowing past a point in a given period of time.
   C. the unit of measure for resistance.
   D. the unit of measure for capacitance.

7. **Switches in electrical circuits are usually wired:**
   A. in parallel/series.
   B. near the power source.
   C. in parallel.
   D. in series.

8. **Loads or power-consuming devices are usually wired:**
   A. in parallel/series.
   B. near the power source.
   C. in parallel.
   D. in series.

9. **If there were a current flowing of 6 A in a 120 V circuit, what would the resistance be?**
   A. 10 ohms
   C. 17 ohms
   B. 720 ohms
   D. 20 ohms

10. **If the resistance in a 120 V electrical circuit were 30 ohms, what would the current be in amperes?**
    A. 100 A
    C. 4 A
    B. 3600 A
    D. 6 A

11. **What would the voltage be in a DC circuit if there were a current flow of 6 A and a load resistance of 8 ohms?**
    A. 120 V
    C. 115 V
    B. 14 V
    D. 48 V

12. **If a conductor carrying an alternating current were formed into a coil, the magnetic field would:**
    A. be decreased.
    B. be increased.
    C. remain the same.

13. **A magnetic field can be used to:**
    A. generate electricity.
    B. operate relays, contactors, and solenoids.
    C. cause electric motors to operate.
    D. all of the above.

14. **The primary and secondary windings in a transformer are:**
    A. wired in parallel.
    B. wired in series.
    C. not connected electrically.
    D. connected with a solenoid switching device.

15. **The unit of measurement for the charge a capacitor can store is the:**
    A. inductive reactance.
    B. microfarad.
    C. ohm.
    D. joule.

Making Voltage and Amperage Readings with a VOM

Name	Date	Grade

Objectives: Upon completion of this exercise, you will be able to make voltage and amperage readings on actual operating equipment using a VOM. You will be able to do this under the supervision of your instructor.

Introduction: You will be using a VOM to make voltage readings and a clamp-on ammeter to make AC amperage readings. Your ammeter may be an attachment to your VOM, or it may be designed to use independently.

Text References: Paragraphs 12.1, 12.2, 12.7, 12.8, 12.9, 12.10, 12.11, 12.12, 12.13, 12.14, 12.15, 12.17, and 12.20.

Tools and Materials: A VOM with insulated alligator clip test leads, a clamp-on ammeter, $1/4$" and $5/16$" nut drivers, and a straight-blade screwdriver.

Safety Precautions: Working around live electricity can be very hazardous. Use a meter with insulated alligator clips on the ends of the leads. Make all meter connections with the POWER OFF. Have your instructor inspect and approve all connections before turning the power on. Your instructor should have given you thorough instruction in using the VOM and clamp-on ammeter. The meters will vary in design, and you will need specific instruction in the use of each. The scales on your meter may differ from those indicated in this exercise.

PROCEDURES

Making Voltage Readings

1. With the power off, locate the control voltage transformer in an air-conditioning or heating unit.

2. Determine the output (secondary 24 V) terminals.

3. Turn the function switch to AC and the range switch to 50 V on your VOM. The black test lead should be in the common (−) jack, and the red test lead in the (+) jack in your meter.

4. Connect the insulated alligator clips to the transformer output (24 V) terminals. Either lead can go on either terminal as there is no + or − polarity for AC voltage.

5. Have your instructor check the connections. Turn the power on. (You may have to adjust the thermostat to call for heating or cooling.) Record the AC reading: _____V

6. Turn the power off.

7. Remove the leads.

8. You will now measure the voltage at the input (primary) side of the transformer. The meter function switch should be set at AC. Set the range switch at 250 V.

9. With the power off, fasten the meter test leads with alligator clips to the two terminals at the input side of the control transformer.

10. Ask your instructor to check the connections. With approval, turn the power on. Record the AC voltage: _____ V

11. Turn the power off.

12. Remove the leads.

13. Ask your instructor to show you the location of the 230 V terminals on the lead side of the contactor of an air-conditioning unit.

14. Be sure the power to the unit is turned off.

15. Set the meter function switch at AC and the range switch at 500 V. (Always select a setting higher than the anticipated voltage.)

16. Connect the meter lead alligator clips to the terminals.

17. Have your instructor inspect the connections and with approval turn the power on. Record the voltage reading: _____V

18. Turn the power off.

19. Remove the leads.

Making Amperage Readings

1. With the power off, place the clamp-on ammeter jaws around a wire to the fan motor.

2. Have your instructor check your setup, including any necessary settings on your meter.

3. With approval, turn the power on. Record the amperage reading: _____ A

4. Turn the power off.

5. Remove the meter.

6. Have your instructor show you a wire leading to an air-conditioning compressor or another component that has a 208 V or 230 V power source.

7. With the power off, place the ammeter jaws around this wire.

8. Have your instructor inspect your meter, and with approval turn the power on. Record the amperage reading: _____ A

9. Turn the power off.

10. Remove the meter.

11. With the power off, disconnect one lead from the output of the 24 V control transformer. Wrap a small wire ten times around one jaw of the ammeter. Connect one end of this wire to the terminal where you just removed the wire. Connect the other end to the wire you removed.

12. Ask your instructor to check your setup, including any settings you may need to make on your meter. With the instructor's approval, turn the power on. Record your amperage reading: _____ A Your reading will actually be ten times the actual amperage because of the ten loops of wire around the jaws of the ammeter. Divide your reading by ten, and record the actual amperage: _____ A

13. Turn the power off. Remove the wire and ammeter. Connect the wire back to the transformer terminal as you originally found it.

Maintenance of Workstation and Tools: Replace all panels on equipment with correct fasteners. Return all meters and tools to their proper places.

Summary Statement: Describe, in your own words, how you would make a voltage reading in the 230 V range. Describe how you would use a clamp-on ammeter to make an AC amperage reading.

QUESTIONS

1. Why is alternating current rather than direct current normally used in most installations?

2. Describe why the voltage is different from the input to the output of a transformer.

3. Is a typical control transformer a step-up or step-down transformer?

4. Describe a VOM.

5. What are typical settings for the function switch on a VOM?

6. What is the range switch used for on a VOM?

7. Explain how the magnetic field can be increased around an iron core.

8. Explain how a solenoid switch works.

9. Describe inductance.

10. When measuring voltage, is the meter connected in series or parallel?

Measuring Resistance and Using Ohm's Law

Name	Date	Grade

Objectives: Upon completion of this exercise, you will be able to make current, voltage, and resistance readings. You will also determine the current, voltage, and resistance of a circuit using Ohm's law.

Introduction: You will be using a VOM to make resistance readings and a VOM to check your calculations using Ohm's law.

Text References: Paragraphs 12.7, 12.8, 12.9, 12.10, 12.11, 12.12, 12.13, 12.17, and 12.20.

Tools and Materials: A VOM with insulated alligator clip test leads, a clamp-on-type ammeter, $1/4$" and $5/16$" nut drivers, and a straight-blade screwdriver.

Safety Precautions: Use a meter with insulated alligator clips on the ends of the leads. Make all meter connections with the power off. Have your instructor inspect all connections before turning the power on. Your instructor should have given you thorough instruction in using all features of your meter.

PROCEDURES

1. Make the 0-ohms adjustment on the VOM in the following manner: Set the function switch to either $-$DC or $+$DC. Turn the range switch to the desired ohms range. Probably R $\times$ 100 will be satisfactory for this exercise. Connect the two test leads together and rotate the 0 ohms control until the pointer indicates 0 ohms. (Paragraph 12.20 in your text will indicate the ohm range for each range selection.)

2. With the power off, disconnect one lead to an electric furnace heating element.

3. Clip one lead of the meter to each terminal of the element. Do not turn the power on. Read and record the ohms resistance: _____ Disconnect the meter leads.

4. Connect the heating element back into its original circuit. Place the clamp-on ammeter jaws around one lead to the heating element.

5. Have your instructor inspect and approve your setup. Turn the power on and quickly read the amperage. Record the amperage: _____ A You must take the reading quickly because as the element heats up, the resistance increases. Turn the power off.

6. Turn the function switch on the VOM meter to AC and set the range switch to 500 V.

7. With the power off, connect a meter lead to each terminal on the heating element.

8. Have your instructor inspect and approve your setup. Turn the power on. Record the voltage: _____V Turn the power off.

9. Check your readings by using Ohm's law. Paragraph 12.11 in your text explains Ohm's law. To check the amperage reading, use the following formula: $I = E/R$

10. Substitute the voltage and resistance reading from your notes. Divide the voltage by the resistance. Your answer should be close to your amperage reading.

Maintenance of Workstation and Tools: Remove test leads from your meter and coil them neatly or follow your instructor's directions. Replace any panels. Return all tools and equipment to their proper places.

Summary Statement: Describe the process of zeroing the ohm scale on a meter and measuring the resistance of a component.

QUESTIONS

1. Describe the difference between a step-up and step-down transformer.

2. What do the letters VOM stand for? What does a typical VOM measure?

3. Describe how to make the 0 ohms adjustment on a VOM.

4. What is inductance?

5. Describe how you would use a typical clamp-on ammeter to measure the amperage in a circuit.

6. Explain how a large diameter wire can safely carry more current than a smaller size.

7. What can happen if a wire too small in diameter is used for a particular installation?

8. How and why should circuits be protected from current overloads?

9. What are two methods used in most circuit breakers to protect electrical circuits?

10. What is a ground fault circuit interrupter?

Introduction to Automatic Controls

Unit Overview

The air-conditioning, heating, and refrigeration industry requires many types of automatic controls to stop, start, and modulate equipment. These controls primarily detect and control temperature and components to control and protect equipment that provides safe and comfortable conditions for people.

The bimetal device is probably the most commonly used to detect thermal change.

Fluid expansion is another method of sensing temperature change.

A thermocouple is also used to detect and control thermal change. A thermopile consists of many thermocouples ganged together.

The thermistor is an electronic solid-state component that varies its resistance to current flow based on temperature. The thermistor is usually very small, can respond to very small temperature changes, and with other electronic components can start, stop, or modulate system components.

Key Terms

Bimetal Device
Bourdon Tube
Fluid Expansion Device
Modulating Thermistor
Rod and Tube
Positive Temperature Coefficient (PTC)

Snap Disc
Thermistor
Thermocouple
Thermopile
Thermostat

REVIEW TEST

Name	Date	Grade

Circle the letter that indicates the correct answer.

1. **Modulating controls are used to:**
   A. start motors.
   B. stop motors.
   C. vary the speed of a motor to control liquid flow.
   D. stop the flow of a liquid.

2. **A thermostat is used:**
   A. to control space temperature.
   B. to control product temperature.
   C. as a safety device.
   D. all of the above.

3. **A bimetal device is:**
   A. constructed of two unlike metals attached back to back.
   B. used to control water pressure.
   C. used to control refrigerant temperature/pressure relationship.
   D. all of the above.

4. **The movement or travel of a bimetal device can be lengthened by:**
   A. increasing the voltage.
   B. increasing the length and forming it into a coil or helix.
   C. decreasing the temperature.
   D. all of the above.

5. An example of a bimetal device is a:
   A. snap-disc.
   B. mercury-filled tube and diaphragm.
   C. thermistor.
   D. Bourdon tube.

6. The Bourdon tube:
   A. tends to straighten out with an increase in voltage.
   B. tends to curl inward with an increase in pressure.
   C. tends to straighten outward with an increase in electrical current.
   D. tends to straighten out with an increase in vapor pressure.

7. The thermocouple:
   A. is constructed of several different metals.
   B. is constructed of two different metals fastened on one end.
   C. varies its resistance with temperature.
   D. uses pressure to generate an electrical current.

8. The thermocouple:
   A. generates an electrical current when heated at the hot junction.
   B. varies its resistance with temperature.
   C. produces an increase in pressure when heated at the hot junction.
   D. produces an increase in pressure when heated at the cold junction.

9. The thermopile:
   A. is constructed with several thermocouples ganged together.
   B. varies its resistance with temperature.
   C. tends to straighten out with an increase in vapor pressure.
   D. produces a pressure to activate a Bourdon tube.

10. The thermistor is an electronic solid-state semiconductor that:
   A. produces an electrical current when heated.
   B. shows an increase in pressure when heated.
   C. causes fluid expansion when heated.
   D. varies its resistance to current flow based on its temperature.

11. A common use for a thermocouple is:
   A. a household thermostat.
   B. to check pressure.
   C. to check water level.
   D. to detect a pilot light flame.

12. The thermistor may be used to:
   A. detect flame.
   B. determine water level.
   C. measure temperature.
   D. check voltage.

13. The liquid-filled mercury tube is used for:
   A. measuring voltage.
   B. checking water level.
   C. room thermostats for heating.
   D. detecting pilot light flames.

14. Why is it important to prove that the pilot flame is lit on a gas furnace before turning on the main gas burner?

15. What is an application for a positive temperature coefficient (PTC) thermistor device?

Automatic Controls

Name	Date	Grade

Objectives: Upon completion of this exercise, you will be able to identify several temperature-sensing elements and state why they respond to temperature changes.

Introduction: You will use several types of temperature-sensing devices, subjecting them to temperature changes and recording their responses.

Text References: Paragraphs 13.1, 13.2, 13.3, 13.4, 13.5, and 13.6.

Tools and Materials: A straight-blade and a Phillips screwdriver, a millivolt meter if available (a digital meter should read down to the millivolt range), a VOM for measuring ohms, a source of low heat (the sun shining in an area or heat from a warm air furnace), an air acetylene unit, ice and water, a fan-limit control, a remote-bulb thermostat (one that closes on a rise in temperature refrigeration) with a temperature range close to room temperature, a thermocouple from a gas furnace, and a thermistor temperature tester.

Safety Precautions: Care should be used with the open flame. DO NOT APPLY THE FLAME HEAT TO THE REMOTE-BULB THERMOSTAT.

PROCEDURES

Checking Bimetal and Remote-Bulb Devices

1. Lightly fasten the fan-limit control straight up in a vise.

2. Connect the two ohmmeter leads together, turn the meter to the R × 1 scale and zero the meter. If the meter is a digital meter, it is not necessary to do the zero adjustment.

3. Remove the cover of the fan-limit control and fasten the ohmmeter leads to the terminals for the fan operation. Your instructor may need to show you which terminals, Figure 13-1. The meter should read infinity. This indicates an open circuit and is normal with the control at room temperature.

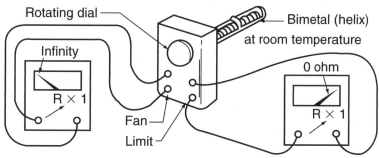

Figure 13-1

4. Light the air acetylene torch and pass the flame close to, but not touching, the fan control bimetal element. Pass the flame back and forth. Notice the dial on the front of the control start to turn as the element is heated. Soon the control contacts will close and the ohmmeter will read 0 ohm resistance. The control may give an audible click at this time. A high-wattage light bulb may also be used.

5. Remove the heat source as soon as the meter reads 0 ohm and allow the control to cool. In a few moments, the control contacts will open and the ohmmeter will again show an open circuit by reading infinity. This closing and opening of these contacts would start and stop a fan if in a furnace circuit.

6. Fasten the ohmmeter leads to the terminals that control the high-limit action. Again, your instructor may need to help you. The meter should show 0 ohm resistance. This circuit is normally closed at room temperature.

7. Again, move the torch slowly back and forth close to the sensing element. You may hear the audible click as the contacts close to start the fan. This is normal. With a little more heat applied, you should then hear the audible click of the high-limit contacts opening to stop the heat source. This will cause the meter to read infinity, indicating an open circuit.

8. Allow the control to cool to room temperature. The limit contacts will close first, then the fan contacts will open.

9. Remove the cover from a remote-bulb thermostat.

10. Fasten the leads from the ohmmeter to the terminals on the remote-bulb thermostat, Figure 13-2.

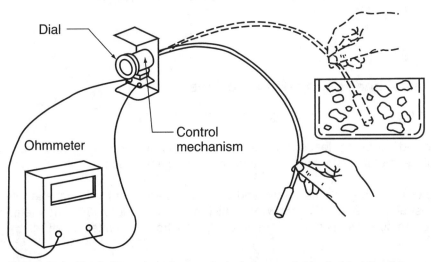

Figure 13-2 Check the action of a remote-bulb thermostat by first holding it in your hand and then placing it in ice and water.

If the ohmmeter shows that the circuit is open, turn the adjustment dial until the contacts just barely close.

11. Position the control so that you can see the action inside as the temperature begins to move its mechanism. Now grasp the control bulb in your hand and observe the control levers start to move. Soon you should hear the audible click of the control contacts closing. If not, you have the adjustment dial turned so high that hand temperature will not cause the control to function. Let the bulb cool and change the adjustment dial downward and start again.

12. Now set the control at 40°F.

13. Place the control bulb in the ice-and-water mixture and observe the contacts and their action.

14. When the contacts open, remove the bulb from the water so that it will be exposed to normal room temperature. The contacts should close again in a few minutes.

Refrigeration & Air Conditioning Technology

Checking a Thermistor and a Thermocouple Temperature Device

15. Leave the thermistor device at room temperature until it has stabilized. Record the temperature: _____ °F

16. Place the lead in the palm of your hand and hold it tightly until it no longer changes. Record the temperature: _____ °F

17. Place the lead in an ice-and-water mixture and stir it until it stops changing. Record the temperature: _____ °F

18. Place the lead next to a heat source, not greater than the range of the meter, such as close to a light bulb or in the sun. Record the temperature: _____ °F

19. Fasten a thermocouple (such as a gas furnace thermocouple) lightly in a vise. Fasten one millivolt meter lead to the end of the thermocouple and the other lead to the thermocouple housing, Figure 13-3.

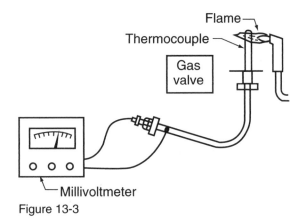
Figure 13-3

20. Light the torch and pass the flame lightly over the end of the thermocouple. If the meter needle moves below 0 volts, reverse the leads. This is direct current and the meter may be connected wrong. DO NOT OVERHEAT THE THERMOCOUPLE LEAD. Record the thermocouple voltage: _____ mV

21. Allow the thermocouple to cool and watch the meter needle move towards 0 millivolts.

Maintenance of Workstation and Tools: Turn off the torch at the tank valve. Bleed the torch hose. Return all tools, materials, and equipment to their respective places.

Summary Statements: Describe the action of the four temperature-sensitive devices that you checked and observed. Tell how each functions with a temperature change.

A. Bimetal device:

B. Liquid-filled bulb device:

C. Thermocouple device:

D. Thermistor device:

QUESTIONS

1. State three methods used to extend the length of a bimetal device for more accuracy.

2. How does the action of a room thermostat make and break an electrical circuit to control temperature?

3. What causes a bimetal element to change with a temperature change?

4. Is a bimetal-sensing element used to control liquids by submerging it in the liquid?

5. Name two applications for a bimetal-sensing element.

6. What causes a mechanical action in a liquid-filled bulb with a temperature change?

7. Name an application for a liquid-filled bulb-sensing device.

8. What is the tube connecting the bulb to the control called on a liquid-filled remote-bulb control?

9. Describe how a thermocouple reacts to temperature change.

Automatic Control Components and Applications

Unit Overview

Being able to recognize a control and understand its function is important. A control may be an operating control, such as a room thermostat, or it may be a safety control to shut down the system, such as a motor-winding thermostat.

Space-temperature thermostats are either low voltage or line voltage. A heat anticipator cuts off the heating equipment before the set point is reached. A cold anticipator starts the air-conditioning system a little early. Line-voltage thermostats may be used in window-unit air conditioners, household refrigerators, reach-in coolers, and free-standing package air conditioners.

Measuring the temperature of fluids is also very important. The term fluid applies to both liquid and vapor states of matter. Sensing temperature in fast-moving airstreams, such as in ductwork and furnace heat exchangers, is usually done by inserting a sensor into the airstream.

Pressure-sensitive devices are normally used with refrigerant, air, gas, or water. The high- and low-pressure controls in refrigeration and air-conditioning equipment are two of the most widely used pressure controls in this industry.

Most pressure and temperature controls have a cut-out, cut-in, and differential adjustment. The cut-out of a control is where the control interrupts or opens the electrical circuit. The cut-in is where the electrical circuit closes. The differential is the difference between the cut-in and cut-out points.

Pressure transducers are devices that measure pressure by first sensing the pressure in some manner, and then converting the sensed pressure signal to an electronic signal to be processed by a microprocessor.

Oil-pressure controls are used on larger compressors to ensure adequate lubricating oil pressure. Air-pressure controls are often used with heat pumps to sense air-pressure drop due to ice buildup. Gas-pressure switches are used with gas-burning equipment to ensure the proper gas pressure before the burners are allowed to ignite.

Key Terms

Air-Pressure Control
Ambient Temperature
Boiler Relief Valve
Cold Anticipator
Crank-Case Heat
Differential
Dual-Pressure Controls
Electromechanical Controls
Fluid
Gas-Pressure Regulator
Gas-Pressure Switch
Heat Anticipator
Helix Coil

Inert Gas
Line-Voltage Thermostat
Low-Voltage Thermostat
Mechanical Controls
Mercury Bulb
Motor-Temperature Cutout
Motor-Temperature Sensing
    Thermostat
Net Oil Pressure
Oil-Pressure Safety Control
Pressure Relief Valve
Pressure Transducer
Range

Sail Switch
Snap or Detent Action
Snap-Disc
Space-Cooling Thermostat
Space-Heating Thermostat
Subbase
Temperature Sensor
Valve Seat
Water Pressure Regulating
    Valve

REVIEW TEST

Name	Date	Grade

Circle the letter that indicates the correct answer.

1. The heat anticipator is:
   A. an adjustable resistor in a space-heating thermostat used to stop a furnace prematurely.
   B. a fixed resistor in a space-cooling thermostat used to stop an air-conditioning system prematurely.
   C. a fixed resistor in a space-cooling thermostat used to start an air-conditioning system prematurely.
   D. an adjustable resistor in a space-heating thermostat used to start a furnace prematurely.

2. A cold anticipator is:
   A. an adjustable resistor in a space-heating thermostat used to stop a furnace prematurely.
   B. a fixed resistor in a space-cooling thermostat used to stop an air-conditioning system prematurely.
   C. an adjustable resistor in a space-heating thermostat used to start a furnace prematurely.
   D. a fixed resistor in a space-cooling thermostat used to start an air-conditioning system prematurely.

3. The heating thermostat _____ on a rise in temperature.
   A. opens
   B. closes
   C. stays the same
   D. fluctuates

4. Detent is:
   A. a slow action to ensure that electrical contact is made.
   B. a snap action.
   C. a part of the wiring in some electric motors.
   D. a type of gas used in mercury bulbs.

5. The low-voltage thermostat is energized by:
   A. a relay.
   B. a transformer.
   C. the inert gas.
   D. the line voltage.

6. The cold anticipator is:
   A. wired in series with the cooling contacts.
   B. wired in series with the bimetal.
   C. wired in parallel with the cooling contacts.
   D. wired in parallel with the compressor relay.

7. What is meant by the term "temperature of the ambient air"?
   A. temperature of the coldest air
   B. temperature of very warm air
   C. temperature of air within an electric motor
   D. temperature of surrounding air

8. What are two types of motor-temperature protection devices?
   A. resistors and capacitors
   B. diodes and heat sinks
   C. bimetal and thermistor
   D. silicon and germanium

9. Most hermetic compressors are cooled by:
   A. refrigerant from the high-pressure line.
   B. refrigerant from the compressor discharge line.
   C. ambient air.
   D. suction gas.

10. A thermistor motor-temperature protection device operates by:
   A. reversing the polarity when the motor overheats.
   B. causing a circuit to open when the motor overheats.
   C. producing excessive voltage when the motor overheats.
   D. causing a bimetal to warp when the motor overheats.

11. The term fluid pertains to:
   A. a liquid.
   B. a liquid or a vapor.
   C. steam.
   D. all of the above.

12. Which of the following is used in some applications to sense temperature in a moving airstream?
   A. helix coil
   B. silicon diode
   C. heat exchanger
   D. heat sink

13. The pressure control can be an operating control or:
   A. a safety-type control.
   B. used to measure current in a refrigeration system.
   C. used to detect excessive voltage.
   D. used instead of an electrical circuit breaker.

14. **A switch that detects air movement in a duct with electric heat is a:**
    A. pressure relief valve.
    B. sail switch.
    C. P&T valve.
    D. positive displacement valve.

15. **The oil-pressure safety control is:**
    A. a pressure relief valve.
    B. a P&T valve.
    C. a pressure differential control.
    D. an electrically controlled valve.

16. **The difference between the suction pressure and the compressor oil pump outlet pressure is:**
    A. the net oil pressure.
    B. the refrigeration system high- and low-pressure differential.
    C. the ambient air-pressure differential.
    D. the suction- and liquid-line pressure differential.

17. **The air-pressure control on a heat pump outdoor coil:**
    A. measures the pressure-temperature difference across the coil.
    B. measures the high- and low-pressure refrigerant differential.
    C. senses the air-pressure drop across the coil to determine when it should go into defrost.
    D. indicates the pressure at the heat pump thermostat.

18. **Gas-pressure switches:**
    A. detect the presence of gas pressure before gas burners are allowed to ignite.
    B. are considered sail switches.
    C. operate only when an interruption of gas transmission is anticipated.
    D. are considered a pressure relief valve.

19. **An example of a mechanical control is a:**
    A. high-pressure switch.
    B. low-pressure control.
    C. water pressure regulating valve.
    D. gas furnace limit control.

20. **An example of an electromechanical control is a:**
    A. high-pressure switch.
    B. water pressure regulating valve.
    C. pressure relief valve.
    D. boiler valve.

21. **The difference between the cut-out and the cut-in of a control is the:**
    A. span.
    B. dead band.
    C. differential.
    D. sum.

22. **A control adjustment that provides for the correct minimum and maximum pressures or temperatures is the:**
    A. dead band.
    B. range.
    C. cut-in.
    D. all of the above.

23. **A device that senses the pressure in a system and then converts the sensed pressure signal to an electronic signal usually to be processed by a microprocessor is a:**
    A. pressure transducer.
    B. pressure regulator.
    C. pressure adjuster.
    D. none of the above.

Controlling Temperature

Name	Date	Grade

Objectives: Upon completion of this exercise, you will be able to use a temperature-measuring device to determine the accuracy of thermostats while observing the control of the temperature of product and space.

Introduction: You will use a thermometer with its element located at a temperature-control device to determine the accuracy of a thermostat used in a refrigerated box and one used to control space temperature.

Text References: Paragraphs 14.1, 14.2, 14.3, and 14.4.

Tools and Materials: A quality thermometer (electronic type with a thermistor or thermocouple is preferred because of its response time), a VOM, a small screwdriver, a medium-temperature refrigerated case that uses a remote-bulb temperature control, and a temperature thermostat.

Safety Precautions: You will be using the screwdriver to make corrections in the temperature control setting. Remove the control's cover prior to starting the test. Your instructor must give you instruction in operating the particular refrigerated box and room thermostat before you begin. Do not touch the electrical terminals.

PROCEDURES

Checking Product Temperature Thermostat

1. Locate the remote bulb for the temperature control in a refrigerated box and fasten the thermometer sensor next to it.

2. Place a second sensor in the refrigerated space, suspended in the air.

3. Place a third sensor in the product to be cooled, for instance, a glass of water on the food shelf, Figure 14-1.

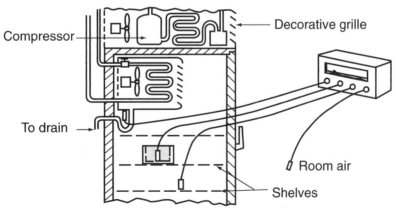

Figure 14-1 Location of temperature tester sensors. SHUT THE DOOR IF THE CASE IS NOT AN OPEN CASE.

4. After 5 minutes, record the following: Probe A, by the remote bulb _____ °F, Probe B, in the air _____ °F, Probe C, in water _____ °F

5. Listen for the compressor to stop and record the same three temperatures when it stops.

A. _____ °F   B. _____ °F   C. _____ °F

Record every 5 minutes:

A. _____ °F   B. _____ °F   C. _____ °F

A. _____ °F   B. _____ °F   C. _____ °F

A. _____ °F   B. _____ °F   C. _____ °F

A. _____ °F   B. _____ °F   C. _____ °F

A. _____ °F   B. _____ °F   C. _____ °F

A. _____ °F   B. _____ °F   C. _____ °F

A. _____ °F   B. _____ °F   C. _____ °F

6. Set the control for 5 degrees higher.

Record every 5 minutes:

A. _____ °F   B. _____ °F   C. _____ °F

A. _____ °F   B. _____ °F   C. _____ °F

A. _____ °F   B. _____ °F   C. _____ °F

A. _____ °F   B. _____ °F   C. _____ °F

A. _____ °F   B. _____ °F   C. _____ °F

A. _____ °F   B. _____ °F   C. _____ °F

A. _____ °F   B. _____ °F   C. _____ °F

Checking a Space Temperature Thermostat

7. Remove the cover of a room thermostat so the mercury contacts may be observed.

8. Place the lead of a thermometer next to the thermostat, at the same level as the bimetal element, Figure 14-2.

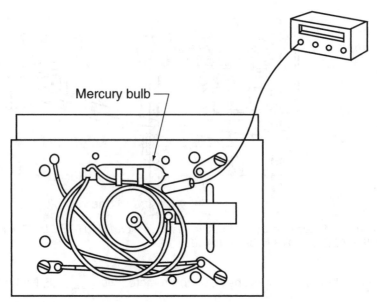

Mercury bulb

Figure 14-2

9. Set the VOM for measuring resistance using R × 1 scale. Switch the thermostat to the cooling mode and place one meter lead on "R" and the other on "Y." Set the thermostat so that the meter shows an open circuit. When the space around the thermostat warms, the contacts will close. Stand back from the thermostat (your body heat will affect its response) and observe the action of the contacts. If the room temperature is varying, the contacts should open and close from time to time. If they do not function in about 5 minutes, you may move closer to the thermostat so your body heat will cause a response. This can be verified by the use of an ohmmeter if the thermostat is on the workbench.

10. Record the time and temperature changes.

Time _____	Temperature _____ °F
Time _____	Temperature _____ °F
Time _____	Temperature _____ °F
Time _____	Temperature _____ °F
Time _____	Temperature _____ °F
Time _____	Temperature _____ °F
Time _____	Temperature _____ °F
Time _____	Temperature _____ °F

Maintenance of Workstation and Tools: Clean the workstation and return all tools and equipment to their respective places.

Summary Statements: Describe the action of the two thermostats in relation to the thermometer that you used. Did the thermostats need calibrating? Describe how this calibrating was done or could have been done.

QUESTIONS

1. Which type of thermometer responds the fastest, a mercury bulb or an electronic thermistor?

2. How does a bimetal sensing element make and break an electrical circuit?

3. What type of action does a bimetal element produce with a temperature change?

4. What type of action does a thermocouple produce with a change in temperature?

5. What type of action does a thermistor produce with a change in temperature?

6. Which of the following, a thermistor or a bimetal sensing element, is used more commonly in room thermostats?

7. Why does the product temperature lag behind the air temperature in a refrigerated box?

8. Explain the differential setting on a temperature control.

9. Describe a good location for a space temperature thermostat.

10. Describe the action of a typical bimetal thermostat.

Pressure-Sensing Devices

Name	Date	Grade

Objectives: Upon completion of this exercise, you will be able to determine the setting of a refrigerant-type low-pressure control and reset it for low-charge protection.

Introduction: You will use dry nitrogen as a refrigerant substitute and determine the existing setting of a low-pressure control, set that control to protect the system from operating under a low-charge condition, and use an ohmmeter to prove the control will open and close the circuit.

Text References: Paragraphs 14.9 and 14.11.

Tools and Materials: An adjustable type of low-pressure control (such as in Figure 14-3) screwdriver, $1/4$" flare union, goggles, gloves, ohmmeter, gage manifold, and a cylinder of dry nitrogen.

Safety Precautions: You will be working with dry nitrogen under pressure, and care must be taken while fastening and loosening connections. Wear goggles and gloves.

Figure 14-3 Typical low-pressure control. *Photo by Bill Johnson*

PROCEDURES

1. Connect the low-pressure control to the gage manifold, Figure 14-4.

2. Typically for an R-134a system, the lowest allowable pressure for a medium temperature system is 7 psig. A typical high setting may be 38 psig. Set the control at these approximate settings.

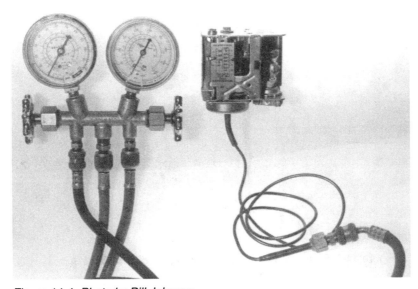

Figure 14-4 *Photo by Bill Johnson*

3. Turn both valves to the off position in the gage manifold.

4. Turn the cylinder valve on, allowing pressure to enter the center line of the gage manifold.

5. Attach the ohmmeter leads to the terminals on the low-pressure control. The meter should read ∞ infinity, showing that the control contacts are open. If the contacts are closed, the control is set for operation below atmosphere and must be reset for a higher value.

6. Slowly open the gage manifold valve to the low-side gage line and record the pressure at which the control contacts close. **NOTE:** IF THE CONTROL MAKES CONTACT SO FAST THAT YOU CANNOT FOLLOW THE ACTION, ALLOW A SMALL AMOUNT OF GAS TO BLEED OUT AROUND THE GAGE LINE AT THE FLARE UNION WHILE SLOWLY OPENING THE VALVE. You may have to repeat this step several times in order to determine the exact pressure at which the control contacts close. Record the pressure reading at which the contacts close: _____ psig (This is the cut-in pressure of the control.)

7. With the control contacts closed and pressure on the control line, shut the gage manifold valves. Pressure is now trapped in the gage line and control.

8. Slowly bleed the pressure out of the control by slightly loosening the connection at the flare union. Record the pressure at which the control contacts open: _____ psig (This is the cut-out pressure of the control.) You may have to perform this operation several times to get the feel of the operation. YOU NOW KNOW WHAT THE CONTROL IS SET AT TO CUT IN AND OUT. Record here:

- Cut-in _____ psig
- Cut-out _____ psig
- Differential (cut-in pressure − cut-out pressure) _____ psig

Procedures for Resetting the Pressure Control

9. Determine the setting you want to use and record here:

- Cut-in _____ psig
- Cut-out _____ psig
- Differential _____ psig

10. Set the control range indicator at the desired cut-in point, Figure 14-5.

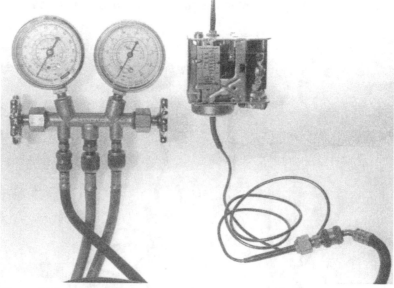

Figure 14-5 Setting the range adjustment. *Photo by Bill Johnson*

11. Set the control's differential indicator to the desired differential, Figure 14-6.

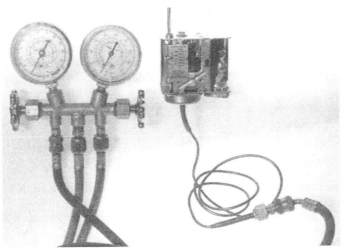

Figure 14-6 Setting the differential adjustment.
Photo by Bill Johnson

12. Recheck the control pressure settings. Record your results here:

  • Cut-in _____ psig
  • Cut-out _____ psig
  • Differential _____ psig

You may notice that the pointer and the actual cut-in and differential are not the same. The pointer is only an indicator. The only way to properly set a control is to do just as you have done, then make changes until the control responds as desired. If the control differential is not correct, make this adjustment first. For more differential, adjust the control for a higher value. For example, if the differential is 15 psig and you want 20 psig, turn the differential dial to a higher value.

13. If the control needs further setting, make the changes and recheck the control using pressure. Record the new values here.

  • Cut-in _____ psig
  • Cut-out _____ psig
  • Differential _____ psig

Maintenance of Workstation and Tools: Return all tools and materials to their respective places. Clean your work area.

Summary Statement: Describe the range and differential functions of this control.

QUESTIONS

1. Why should a unit have low-pressure protection when a low charge is experienced?

2. What typical pressure-sensing device may be used in a low-pressure control?

3. What is the function of the small tube connecting the pressure-control-sensing device to the system?

4. If the cut-out point of a pressure control were set correctly but the differential set too high, what would the symptoms be?

5. If a control were cutting in at the correct setting and the differential were set too high, what would the symptoms be?

Controlling the Head Pressure in a Water-Cooled Refrigeration System

Name	Date	Grade

Objectives: Upon completion of this exercise, you will be able to set a water pressure regulating valve for a water-cooled condenser and set the cut-out and cut-in point for a high-pressure control in the same system.

Introduction: You will control the head pressure in a water-cooled refrigeration (or air-conditioning) system. This system will have a wastewater condenser and a water pressure regulating valve to control the head pressure. You will set the high-pressure control for protection against a water shutoff.

Text References: Paragraphs 14.9 and 14.16.

Tools and Materials: A gage manifold, valve wrench, goggles, gloves, and a water-cooled, wastewater refrigeration system with a water-regulating valve.

Safety Precautions: You will be working with refrigerant, so care should be taken. Wear your goggles and gloves. Never use a jumper to take a high-pressure control out of a system because excessive high pressure may occur.

PROCEDURES

1. See text, Figure 14-62, for an example of a water pressure regulating valve.

2. Fasten the high- and low-pressure gages on the system.

3. Open the service valves to allow pressure to the gages.

4. Purge a small amount of refrigerant from the center line to clean the gage lines.

5. Start the system.

6. After 5 minutes of running time, record the following:
   • Suction pressure _____ psig
   • Discharge pressure _____ psig

7. Turn the water-regulating valve adjustment clockwise 2 turns, wait 5 minutes, and record the following:
   • Suction pressure _____ psig
   • Discharge pressure _____ psig

8. Turn the water-regulating valve adjustment counterclockwise 4 turns, wait 5 minutes, and record the following:
   • Suction pressure _____ psig
   • Discharge pressure _____ psig

9. After completion of the above, slowly turn the adjustment stem of the water pressure regulating valve in the direction that causes the head pressure to rise until the unit stops because of head pressure. Caution: Do not let the pressure rise above the following:
   - R-12 147 psig or 115°F condensing temperature
   - R-22 243 psig or 115°F condensing temperature
   - R-502 263 psig or 115°F condensing temperature
   - R-134a 159 psig or 115°F condensing temperature
   - R-404A 291 psig or 115°F condensing temperature
   - R-410A 391 psig or 115°F condensing temperature

10. If the compressor stops before the above settings, it is set correctly.

11. If the high-pressure control does not stop the compressor before these pressures, slowly turn the range adjustment on the high-pressure control to stop the compressor.

12. After completion of checking and setting the high-pressure control, reset the water pressure regulator to maintain the following pressures:
   - R-12 127 psig or 105°F condensing temperature
   - R-22 211 psig or 105°F condensing temperature
   - R-502 232 psig or 105°F condensing temperature
   - R-134a 135 psig or 105°F condensing temperature
   - R-404A 253 psig or 105°F condensing temperature
   - R-410A 340 psig or 105°F condensing temperature

13. Remove the gage manifold.

Maintenance of Workstation and Tools: Return all tools to their respective places. Place the system back in the manner in which you found it.

Summary Statement: Describe what would happen to the water flow if the condenser tubes became dirty.

QUESTIONS

1. Why is a refrigerant cylinder not used to set a high-pressure control?

2. What is the purpose of the high-pressure control?

3. What is the purpose of the water pressure regulating valve on a wastewater system?

4. What senses the pressure of the water pressure regulating valve?

5. Where is the typical water pressure regulating valve connected to the system?

6. Where is the high-pressure control typically connected to the refrigeration system?

Troubleshooting Basic Controls

Unit Overview

Controls are either electrical, electronic, mechanical, or a combination of these.

Electrical devices can be considered as power-consuming or nonpower-consuming (power-passing) as far as their function in a circuit is concerned.

Power-consuming devices, also called loads, can be motors, heaters, or any other system component that uses power. Incandescent light bulbs and strip heaters are resistive loads. Loads that have magnetic coils in them are called inductive loads. Power-consuming devices are wired in parallel with the power source. The power-consuming device must have a potential voltage. It must be wired between two legs of a power source.

Non-power-consuming devices pass power from the power source to the load and are normally wired in series with the power-consuming devices in a circuit. Non-power-consuming devices are more commonly known as switches. Switches can be manual devices or automatic devices. Automatic switches can open or close their contacts based on temperature, pressure, fluid movement, fluid level, or a number of other physical properties.

To troubleshoot many electrical circuits, start with a voltmeter. The voltmeter range selector switch must be turned to a voltage setting higher than the voltage supply. Use extreme caution. Keep your hands well back on the probes from the meter. Be sure that the probe tips touch only the terminals intended. Before beginning any troubleshooting procedures, read the text material. Then work under the supervision of an experienced technician or instructor. Unit 15 in the text explains many troubleshooting procedures that will not be repeated here.

To troubleshoot a low-voltage circuit using a clamp-on ammeter, you may need to use a 10-wrap multiplier to achieve an accurate reading.

If there is proper voltage to the power-consuming device, then the continuity should be checked with the power off. The ohmmeter function of the multimeter may be used for this. One side of the coil should be removed from the circuit and it can be checked for continuity or proper resistance.

When electrically troubleshooting switches and power-consuming devices using a voltmeter, technicians must ask themselves where Line 1 and Line 2 are not whether the switch is open or closed. Measuring across the same line with a voltmeter, whether it is Line 1 or Line 2, will always give 0 volts.

The service technician must have a good electrical diagram to work from when troubleshooting. There are two types of diagrams furnished with equipment: the pictorial, and the line diagram.

Key Terms

Electrical Heat Relay	**Line Side of the Switch**	**Potential Voltage**
Fan Relay Coil	**Load Side of the Switch**	**Power-Consuming Devices**
Fixed Resistor	**Non-power-Consuming Devices**	**Variable Resistor**
Insulated Jumper	**Package Air Conditioner**	

REVIEW TEST

Name	Date	Grade

Circle the letter that indicates the correct answer.

1. An electrical power-consuming component:
   A. is normally wired in series in an electrical circuit.
   B. is normally used to pass power to another component.
   C. is often a basic switch.
   D. uses power and is normally wired in parallel.

2. Potential voltage is a term that indicates:
   A. the voltage across a switch.
   B. the voltage across a switching device.
   C. the voltage between two power legs of a power supply.
   D. the pressure drop across a cooling coil.

3. The range selector switch on a voltmeter that is to be used to measure the voltage at an AC circuit should be:
   A. set at 500 mA.
   B. R × 100.
   C. 10 V.
   D. a voltage higher than the supply voltage.

4. The line side of a switch is the:
   A. side the power-consuming device is connected to.
   B. neutral side of the circuit.
   C. side nearest the power source.
   D. side with the white wire.

5. The electric heat relay is:
   A. a magnetic coil used to stop and start the current through an electric heating element.
   B. a type of circuit breaker.
   C. another term for the heat anticipator in a thermostat.
   D. a type of variable resistor.

6. The continuity through a heat relay can be checked with:
   A. an ohmmeter.
   B. a voltmeter.
   C. a clamp-on ammeter.
   D. a micron gage.

7. The heat anticipator in a thermostat is often:
   A. a variable resistor.
   B. a fixed resistor.
   C. a capacitor.
   D. a variable relay coil.

8. The cool anticipator in a thermostat is often:
   A. a variable resistor.
   B. a fixed resistor.
   C. a capacitor.
   D. a variable relay coil.

9. A 10-wrap amperage multiplier is used with a:
   A. voltmeter.
   B. clamp-on ammeter.
   C. ohmmeter.
   D. megger.

10. Transformers are normally rated:
    A. by AC voltage.
    B. by volt-amperes.
    C. by capacitance.
    D. with microfarads.

11. When using a 10-wrap multiplier, the reading on the meter must be:
    A. multiplied by 10.
    B. divided by 10.
    C. adjusted by subtracting 10.
    D. adjusted by adding 10.

12. The electrical line wiring diagram is often called:
    A. a pictorial diagram.
    B. a ladder diagram.
    C. an orthographic projection.
    D. an oblique drawing.

13. The electrical wiring diagram that shows the approximate location of components is:
    A. a pictorial diagram.
    B. a ladder diagram.
    C. an orthographic projection.
    D. an oblique drawing.

14. It is important to know whether the electrons flow from negative to positive or positive to negative in an AC electrical circuit.
    A. true
    B. false

15. An insulated jumper wire may be used to temporarily bypass components in an electrical circuit.
    A. true
    B. false

Air-Conditioning Circuit Evaluation

Name	Date	Grade

Objectives: Upon completion of this exercise, you will be able to follow the circuit in a typical electric air-conditioning system and check the amperage in a low-voltage circuit.

Introduction: You will use a volt-ohmmeter and a clamp-on ammeter to follow the circuit characteristics of a high-voltage and a low-voltage circuit.

Text References: Paragraphs 15.1, 15.2, 15.3, 15.4, 15.5, 15.6, and 15.7.

Tools and Materials: A VOM, goggles, screwdriver, clamp-on ammeter and some low-voltage wire (about 3 feet of 1 strand, 18 gage), a typical packaged or split air-conditioning system, and panel lock and tag.

Safety Precautions: You will be taking voltage readings on a live circuit. Do not perform this function until your instructor has given you instruction on the use of the meter. The power must be off, locked, and tagged at all times until the instructor has approved your setup and tells you to turn it on. Wear goggles, and do not touch any of the bare terminals. Do not perform this exercise if your feet are wet because this may subject you to electrical hazard.

PROCEDURES

1. Turn the power to the air-conditioning system off. Lock and tag the panel. Remove the door to the control compartment.

2. Locate the following:

   - The line power coming into the unit
   - The line and load side of the compressor contactor
   - The condenser fan motor leads
   - The compressor "run," "start," and "common" leads
   - The low-voltage terminal block

3. Turn the voltmeter selector switch to the correct voltage range. With the instructor's approval and supervision, turn the power on and start the unit. CAUTION: BE SURE THAT THE CONTROL COMPARTMENT IS ISOLATED FROM THE AIRFLOW PATTERN SO AIR IS NOT BYPASSING THE CONDENSER DURING THIS TEST.

   Check the voltage at the following points and record:

   - Line voltage _____ V
   - Line side of the contractor _____ V
   - Load side of the contractor _____ V
   - Compressor, Common to Run terminal _____ V
   - Compressor, Common to Start terminals _____ V
   - Low voltage, C to R terminals _____ V
   - Low voltage, C to Y terminals _____ V

4. Turn the power off at the disconnect, lock, and tag.

5. Remove one wire from the magnetic holding coil of the compressor contactor and the indoor fan relay and record the following:

- Resistance of the fan relay coil _____ ohms
- Resistance of the compressor contactor _____ ohms

6. Check the following resistances at the compressor terminals:

- Resistance from Common to Run _____ ohms
- Resistance from Common to Start _____ ohms
- Resistance from Run to Start _____ ohms

7. Remove the wire from the R terminal and make a 10-wrap coil around the jaw of the ammeter for reading the amperage in the low-voltage circuit. See the text, Figure 15-15, for an example of how this can be done.

8. After the instructor has inspected and approved your setup, start the unit and record the following amperages:

- Indoor fan relay _____ A
- Compressor contactor and fan relay _____ A

Maintenance of Workstation and Tools: Replace all wires and panels and return the tools and materials to their respective places. Clean your work area.

Summary Statement: Describe how the low-voltage circuit controls the power to the high-voltage circuit.

QUESTIONS

1. Why is low voltage used to control high voltage?

2. What is the typical amperage draw in the low-voltage circuit?

3. What does the term VA mean when applied to a control transformer?

4. How much amperage can a 60 VA transformer put out?

5. How much amperage can a typical thermostat handle?

Troubleshooting Simple Electric Circuits

Name	Date	Grade

Objectives: Upon completion of this exercise, using a voltmeter and an ammeter, you will be able to properly evaluate a simple series circuit and determine whether the circuit is operational. As part of this exercise, you will determine whether system components are power-consuming or power-passing devices.

Introduction: A major part of the troubleshooting process involves locating the problem in a circuit as quickly and efficiently as possible. By understanding how electric circuits operate and how voltage and amperage readings indicate the condition of various circuit components, the troubleshooting process can be mastered with a fair amount of practice and dedication.

Text References: 15.1, 15.2, 15.3, 15.4, 15.5, 15.6, and 15.7.

Tools and Materials: A power cord (length of SJ cable with a male plug attached to one end), a manual single-pole–single-throw switch, fuses (both good and defective, if possible), a line voltage thermostat, 75-watt light bulbs (good and defective, if possible), a light socket, wire, wire nuts, electrical tape, goggles, wire cutters, wire strippers, screwdrivers, a voltmeter, and an ammeter.

Safety Precautions: When working on or around electric circuits, make every attempt to avoid becoming part of the electric circuit. Avoid wearing metallic jewelry. Wear rubber-soled work boots. Whenever possible, work on circuits that are deenergized to avoid receiving an electric shock. When working on live circuits, be sure that you do not touch bare conductors. Always follow all of your teacher's instructions.

PROCEDURES

1. Wire up the circuit as shown in Figure 15-1 (Figure 15-3 in your textbook). When wiring the circuit, use a good fuse and a good light bulb.

2. Identify each of the circuit components as either a power-passing or power-consuming device and circle the appropriate term here:

   Switch: power-passing   power-consuming

   Fuse: power-passing   power-consuming

   Thermostat: power-passing   power-consuming

   Light Bulb: power-passing   power-consuming

3. Have your instructor inspect the circuit to make certain that all connections are correct.

4. Place the switch in the open (off) position and the thermostat in the closed position.

5. Energize the circuit.

6. Take a voltage reading between the neutral line and point 1 in the circuit. Point 1 is the line side of the manual switch. Record the voltage here: _____ V

7. Take a voltage reading between the neutral line and point 2 in the circuit. Point 2 is the load side of the manual switch. Record the voltage here: _____ V

8. Take a voltage reading between point 1 and point 2 and record the voltage reading here: _____ V

9. From the readings taken in steps 6, 7, and 8 you should be able to verify that the switch is in the open (off) position.

10. Take a voltage reading across the light bulb and record it here: _____ V

11. The reading obtained in step 10 should indicate why the bulb is not on.

12. Turn the switch to the closed (on) position. The bulb should light up.

13. Repeat the voltage readings in steps 6 and 7 and record them here:
    - Voltage between neutral and point 1: _____ V
    - Voltage between neutral and point 2: _____ V

    These readings should indicate that there is power passing through the switch because there is power between neutral and point 2.

14. Take a voltage reading between neutral and point 4 and record it here: _____ V

15. Take a voltage reading between neutral and point 6 and record it here: _____ V
    A positive voltage reading between neutral and point 6 indicates that all power-passing devices in the circuit are in the closed position and power is being passed to the circuit load.

16. Take a voltage reading across the bulb and record it here: _____ V

17. Take an amperage reading of the circuit and record it here: _____ A
    A positive voltage reading across the load indicates that the light bulb is being supplied power.
    A positive amperage reading indicates not only that the load is being supplied power, but the load is good as it is consuming power.

18. Disconnect power to the circuit.

19. Remove the good fuse and replace it with a defective one.

20. With the manual switch in the open (off) position and the thermostat in the closed position, reenergize the circuit.

21. Turn the switch to the closed (on) position.

22. Take a voltage reading between neutral and point 1 and record it here: _____ V

23. Take a voltage reading between neutral and point 2 and record it here: _____ V

24. Take a voltage reading between neutral and point 3 and record it here: _____ V

25. Take a voltage reading between neutral and point 4 and record it here: _____ V

26. Take a voltage reading between point 3 and point 4 and record it here: _____ V
    A positive voltage reading between point 3 and point 4 indicates that the fuse is no good and is not passing power on to the rest of the circuit.

27. Take an amperage reading of the circuit and record it here: _____ A

    An amperage reading of 0 indicates that no current is flowing in the circuit and that no power is being consumed. A positive amperage reading indicates that there is a complete circuit, current is flowing, and that power is being consumed.

28. Based on the previous steps in this exercise, make a prediction about how much voltage is being supplied to the light bulb and record it here: _____ V

29. Take an actual voltage reading across the light bulb and record it here: _____ V

30. If the recorded values in steps 28 and 29 are different, reevaluate the circuit and repeat this portion of the exercise.

31. Deenergize the circuit and replace the defective fuse with the good one.

32. With the circuit still deenergized, replace the good light bulb with a defective one.

33. With the manual switch in the open (off) position and the thermostat in the closed position, restore power to the circuit.

34. Turn the manual switch to the closed (on) position.

35. Take a voltage reading between neutral and point 1 and record it here: _____ V

36. Take a voltage reading between neutral and point 2 and record it here: _____ V

37. Take a voltage reading between neutral and point 3 and record it here: _____ V

38. Take a voltage reading between neutral and point 4 and record it here: _____ V

39. Take a voltage reading between neutral and point 5 and record it here: _____ V

40. Take a voltage reading between neutral and point 6 and record it here: _____ V

41. Take a voltage reading across the bulb and record it here: _____ V

42. Take an amperage reading of the circuit and record it here: _____ A

43. What conclusions can be drawn from the information recorded in steps 35–42?

44. Now, open the thermostat contacts and take the following voltage measurements:

    • Voltage between neutral and point 5: _____ V
    • Voltage between neutral and point 6: _____ V

45. What conclusion can be drawn about the thermostat based solely on the information obtained in step 44?

Maintenance of Workstation and Tools: Make certain that all circuits are deenergized before disconnecting the components. Return all tools to their proper places. Leave the work area neat and clean.

Summary Statement: Voltmeters and ammeters serve different functions when used to evaluate an electric circuit. When used together, these pieces of test equipment allow the service technician to quickly identify the circuit component that is preventing the system from operating properly. It is important for the technician to identify whether a particular circuit component is a load or a switch, as the measured voltage across it will need to be interpreted differently in each case.

QUESTIONS

1. Why is it that we often see multiple switches wired in series with a load?

2. If a load is wired in a circuit that has multiple switches in series with it, what conditions must be present in order for the load to be energized?

3. If a circuit has a current draw of 0 amps, is the load energized? Explain your answer?

4. If a circuit is comprised of a load and multiple switches wired in series with the load and the load is operating, is it possible that one or more of the switches is in the open position?

5. If voltage is being supplied to a load and the load is not operating, what can be said about the switches in the circuit?

6. If voltage is being supplied to a load and the load is not operating, what can be said about the load in the circuit?

7. If voltage is being supplied to a load and the load is not operating, what can be said about the current flow in the circuit?

8. If voltage is being supplied to a load and the load is not operating, what further testing can be done to check the condition of the load in the circuit?

Advanced Automatic Controls—Direct Digital Controls (DDCs) and Pneumatics

Unit Overview

Heating, air-conditioning, and refrigeration systems were manually operated for many years until controls were developed to take the place of manual operation. An example would be a coal heating system in a large building. Originally, an operator (stoker) manually shoveled coal into the boiler or furnace until an automatic conveyer-belt stoker with a timer on it was developed. The modern office building today may be operated by a computer in a distant city that is scanning the weather, even the long-term predicted weather, to decide how to react to keep the building comfortable. The building tenants may even have a clause in their lease specifying prescribed space conditions. The computer can also act as a diagnostic tool to tell the technician what the likely problem is when dispatched

to one of the building systems. There are even home systems that have some of these features built in.

The first large system controls were pneumatic (air) controls. These systems were simple enough that a person with a basic understanding of mechanics could maintain them. The latest systems of today are electronic. Many of the older pneumatic systems have been successfully merged to operate with electronic systems.

These system controls consist of many components that are arranged in control loops for each zone, simplifying the overall picture into smaller steps. The control technician of today must become oriented to electronic controls. Proper training is available for anyone who seeks it.

Key Terms

Analog Signal
Control Loop
Direct Digital Control (DDC)
Electronic-Control
Energy Management

Error Codes
Gain
Load Shed
Modulating Flow
Pilot Positioner

Pneumatic Control
Sensitivity
Smart Recovery

REVIEW TEST

Name	Date	Grade

Circle the letter that indicates the correct answer.

1. A control loop has _____ components.
   A. twelve
   B. three
   C. seven
   D. nine

2. Pneumatic controls use which of the following as the power to operate?
   A. water
   B. electricity
   C. magnetic
   D. air

3. Pneumatic controls can be used in an explosive atmosphere.
   A. true   B. false

4. Modulating flow means:
   A. extra fast.
   B. extra slow.
   C. variable.
   D. no flow.

5. **The sensitivity of a control has to do with:**
   A. how it feels.
   B. time of day.
   C. how it responds to a change in conditions.
   D. its valve.

6. **Electronic control components are often mounted:**
   A. underwater.
   B. on a circuit board.
   C. without wires.
   D. in series with pressure.

7. **The typical voltage that DDC controls use as their signal is:**
   A. 24 volts.
   B. 230 volts.
   C. 115 volts.
   D. 0 to 10 volts.

8. **An analog signal has an infinite number of steps.**
   A. true   B. false

9. **A pilot positioner is used with which type of controls?**
   A. pneumatic
   B. electronic
   C. electromechanical
   D. DDC

10. **Electronic controls are only used in commercial buildings.**
    A. true   B. false

Types of Electric Motors

Unit Overview

Electric motors provide the power for the prime movers of air, water, refrigerant, and fuels. Some motors must start under heavy loads and still develop their rated horsepower under running conditions. Others run in a refrigerant atmosphere.

Motors are designed with different starting and running characteristics. Some compressor motors, for instance, must be able to start under heavy starting loads. A small fan does not need as much starting torque. There are single-phase motors for light loads and domestic use and three-phase motors for heavier commercial or industrial use. Power supplies for motors for homes and light commercial use are normally 115 V, 230 V, or 208 V.

A capacitor-start motor is a split-phase motor with a capacitor wired in series with the start windings to provide more starting torque. A capacitor-start, capacitor-run motor also has a capicitor wired into the run winding circuit to provide a more efficent operation. A permanent split-capacitor (PSC) motor has windings similar to the split-phase and a run capacitor but no start capacitor. It is an efficient motor, but it has a low starting torque. It can be a multispeed motor when used for fans. The shaded-pole motor has very little starting torque and is used in light-duty applications. The three-phase motor is used in heavy commercial or industrial installations. A special three-phase power must be available. Single-phase hermetic motors are designed to operate in a refrigerant vapor atmosphere. Two-speed motors are used by some manufacturers to control the capacity required from small compressors. Variable-speed motors are used in many applications because most motors do not need to operate at full speed and load capacity except during the peak temperature of the season.

Key Terms

Capacitor-Start, Capacitor-Run Motor
Capacitor-Start Motor
Centrifugal Switch
Current Relay
DC Converter
DC Motors
ECM Motor
Electronic Relay
End Bells
Frequency
Full-Load Amps
Hertz

Induction Motor
Inverter
Locked Rotor Amps
Permanent Split-Capacitor Motor (PSC)
Positive-Temperature-Coefficient Start Device
Potential Relay
Pulse Width Modulator
Rotor
Run Load Amps
Run Windings
Single-Phase Hermetic Motor

Slip
Squirrel Cage Rotor
Start Windings
Stator
Three-Phase Motor
Torque
Transistor
Two-Speed Compressor Motor
Two-Pole, Split-Phase Motor
Variable Speed Motor
Winding Thermostat

REVIEW TEST

Name	Date	Grade

Circle the letter that indicates the correct answer.

1. **When an alternating current is applied to the running windings of a split-phase electric motor, an alternating current:**
   A. with the same polarity is induced in the rotor.
   B. with reverse polarity is induced in the rotor.
   C. with the same polarity is induced in the start winding.
   D. circuit to the start winding is opened by the electronic relay.

2. **A two-pole split-phase motor operates at a speed just under:**
   A. 1800 rpm.   C. 4800 rpm.
   B. 3600 rpm.   D. 5200 rpm.

3. **Start windings are placed _____ the run windings.**
   A. outside   C. between
   B. inside   D. in series with

4. **The start winding has:**
   A. fewer turns than the run winding.
   B. more turns than the run winding.
   C. the same number of turns as the run winding.

5. **A four-pole split-phase motor runs _____ a two-pole split-phase motor.**
   A. faster than
   B. at the same speed as
   C. slower than

6. **When a typical split-phase motor reaches approximately 75% of its operating speed, the start winding circuit is removed from the circuit by a:**
   A. circuit breaker.
   B. capacitor.
   C. centrifugal switch.
   D. dual-voltage switch.

7. **A three-phase power supply can be found in most:**
   A. homes.
   B. factories.
   C. mobile homes.

8. **The electronic relay is used with some electric motors to:**
   A. provide current to the start windings.
   B. relay current to the start capacitor.
   C. provide current to the run capacitor.
   D. open the circuit to the start windings after the motor has started.

9. **The amount by which the current leads or lags the voltage in an alternating-current circuit is known as the:**
   A. phase angle.
   B. electronic relay.
   C. inductive angle.
   D. starting torque.

10. **In an alternating current circuit in which there is a capacitor, the current _____ the voltage.**
    A. leads
    B. lags
    C. is in phase with

11. **Capacitors wired in series with the start windings provide:**
    A. a lag in starting time.
    B. a greater starting torque.
    C. an aid in starting three-phase motors.
    D. more efficient operation after the motor has reached its operating speed.

12. **The shaded-pole motor:**
    A. has little starting torque.
    B. has excessive starting torque.
    C. has greater efficiency than most split-phase motors.
    D. is designed for three-phase operation.

13. **Three-phase motors have:**
    A. start windings.
    B. a start capacitor.
    C. no start windings or start capacitors.
    D. both start windings and a start capacitor.

14. **The term "slip" is used to indicate:**
    A. the amount of horsepower lost due to the motor resistance.
    B. the amount of efficiency lost when there is no run capacitor.
    C. the difference in motor rpm between the time the start winding circuit is opened and when the motor reaches full operational speed.
    D. the difference between the motor full operational rpm rating and the actual speed under a load.

15. **The hermetic motor must run in a:**
    A. warm ambient temperature.
    B. refrigerant vapor atmosphere.
    C. cool ambient temperature to keep from overheating.
    D. low-temperature refrigeration system.

16. **The start winding circuit in a hermetic motor is interrupted by a potential relay or a current relay because _____ mechanical switch.**
    A. of the electric arc produced by a
    B. of the capacitive circuit produced by a
    C. of the lubrication needed by a
    D. of poor reliability of the

17. **The positive-temperature-coefficient start device is a type of:**
    A. capacitor.
    B. thermistor.
    C. motor winding.

18. **A two-speed compressor motor may operate as a:**
    A. three-phase motor.
    B. two- or four-pole motor.
    C. potential relay type of motor.

19. **Three-phase motors have a:**
    A. low starting torque.
    B. high starting torque.
    C. poor operating efficiency.

20. **Most small hermetic motors are cooled by:**
    A. water.
    B. air.
    C. refrigerant gas.
    D. cold air from the evaporator fan.

Identification of Electric Motors

Name	Date	Grade

Objectives: Upon completion of this exercise, you will be able to identify different types of electric motors, state their typical applications, and sketch a diagram showing how they are wired into circuits.

Introduction: You will be provided an air-conditioning unit and a refrigeration unit. You will remove enough panels on each system for you to identify the motors. You will draw a diagram showing how the motor is wired into the circuit.

Text References: Paragraphs 17.8, 17.9, 17.12, 17.13, 17.14, 17.15, 17.16, 17.17, 17.21, and 17.24.

Tools and Materials: A VOM, straight-blade and Phillips screwdrivers, $1/4$" and $5/16$" nut drivers, and a flashlight.

Safety Precautions: Turn the power off before removing any panels. Lock and tag the distribution box where you turned the power off and keep the single key in your possession. Properly discharge any capacitors in the circuit.

PROCEDURES

Air-Conditioning Unit

(Turn the power off. Lock and tag the box. Keep the key in your possession.)

1. Remove the panel to the evaporator section, exposing the indoor fan motor.

2. Using a good light, describe the following:
   - Number of wires entering the fan motor terminal box _____
   - Color of the wires _____
   - Is there a run capacitor? _____
   - Is there a start capacitor? _____
   - Motor voltage _____
   - Type of motor (capacitor-start or other type) _____

3. Replace the panels with the correct fasteners.

4. Remove the panel to the condenser section, exposing the outdoor fan motor.

5. Using a good light, describe the following:
   - Number of wires entering the fan motor terminal box _____
   - Color of the wires _____
   - Is there a run capacitor? _____
   - Is there a start capacitor? _____
   - Motor voltage _____
   - Type of motor (capacitor-start or other type) _____

6. Follow all wires going into the compressor and describe the following:

- Number of wires entering the box _____
- Is there a run capacitor? _____
- Is there a start capacitor? _____
- Is there a start assist (relay or a PTC device)? _____
- Type of compressor motor (capacitor-start or other type) _____

7. Replace the panels with the correct fasteners.

Refrigeration Unit

(Turn the power off. Lock and tag the box. Keep the key in your possession.)

8. Remove the panel to the evaporator section, exposing the indoor fan motor.

9. Using a good light, describe the following:

- Number of wires entering the fan motor terminal box _____
- Color of the wires _____
- Is there a run capacitor? _____
- Is there a start capacitor? _____
- Motor voltage _____
- Type of motor (capacitor-start or other type) _____

10. Replace the panels with the correct fasteners.

11. Remove the panel to the condenser section, exposing the outdoor fan motor.

12. Using a good light, describe the following:

- Number of wires entering the fan motor terminal box _____
- Color of the wires _____
- Is there a run capacitor? _____
- Is there a start capacitor? _____
- Motor voltage _____
- Type of motor (capacitor-start or other type) _____

13. Replace the panels using the correct fasteners.

14. Follow all wires going into the compressor and describe the following:

- Number of wires entering the fan motor terminal box _____
- Color of the wires _____
- Is there a run capacitor? _____
- Is there a start capacitor? _____
- Motor voltage _____
- Type of motor (capacitor-start or other type) _____

15. Replace the panels with the correct fasteners.

16. Draw a wiring diagram showing how the fan motors and compressor are wired into the circuit in the air-conditioning unit.

17. Draw a wiring diagram of a typical shaded-pole motor.

18. Draw a wiring diagram of a typical three-speed permanent split-capacitor motor.

19. Draw a wiring diagram of a single-phase compressor motor with a start relay and a start and run capacitor.

Maintenance of Workstation and Tools: Return all tools to their respective places.

Summary Statements: Describe the difference between a shaded-pole and a permanent split-capacitor fan motor. Explain which is the most efficient.

QUESTIONS

1. What is the reason for a start capacitor on a motor?

2. Must a motor with a start capacitor have a start relay? Why?

3. What is the purpose of a centrifugal switch in the end of an open motor?

4. Why you can't have a centrifugal switch in a hermetic compressor motor?

5. Name two methods used in hermetic compressor motors to take the start winding out of the circuit when the motor is up to speed.

6. How is a start capacitor constructed?

7. How is a run capacitor constructed?

8. How are capacitors rated?

9. Can a 230 V capacitor be used for a 440 V application?

10. Can a 440 V capacitor be used for a 230 V application?

11. Describe a PTC device.

12. How is a typical open electric motor cooled?

13. How is a typical hermetic compressor motor cooled?

Compressor Winding Identification

Name	Date	Grade

Objectives: Upon completion of this exercise, you will be able to identify the terminals to the start and run windings in a single-phase hermetic compressor.

Introduction: You will use a VOM and learn the difference in the resistance of the start and run windings of a single-phase hermetic motor.

Text References: Paragraphs 17.4, 17.6, 17.8, 17.9, 17.10, 17.11, 17.12, 17.13, 17.14, 17.15, 17.16, 17.17, and 17.18.

Tools and Materials: A VOM, straight-blade and Phillips screwdrivers, and $1/4$" and $5/16$" nut drivers.

Safety Precautions: Turn the power off on any piece of equipment you are servicing. Lock the distribution box where you turned the power off and keep the single key in your possession. Properly discharge all capacitors.

PROCEDURES

1. You will be provided a single-phase compressor, either in a piece of equipment or from stock.

2. Turn the power off if the compressor is in a piece of equipment. Lock and tag the panel or box, keeping the single key on your person. Properly discharge all capacitors.

3. Remove the compressor.

4. If the compressor is wired into the circuit, turn the power on with the supervision of your instructor, and check the voltage at the terminals, from terminal to terminal and terminal to ground.

5. With the power off, remove the terminal wiring one at a time and label the wires so that you can replace them correctly. **NOTE:** THE CROSSING OF THE WIRES CAN DO PERMANENT DAMAGE BEFORE YOU CAN SHUT THE COMPRESSOR OFF. DO NOT MAKE A MISTAKE.

6. Set the ohmmeter selector switch to R $\times$ 1 and zero the meter needle if applicable.

7. Draw a diagram of the terminal layout, Figure 17-1.

   Record the following:

   1 to 2 _____ ohms

   1 to 3 _____ ohms

   2 to 3 _____ ohms

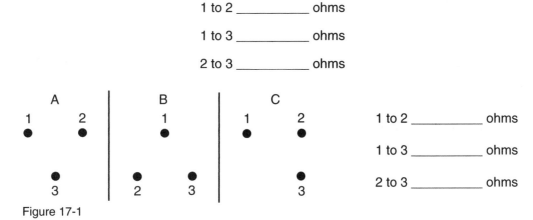

Figure 17-1

8. Based on your resistance readings, identify the start winding: terminal _____ to _____

9. Based on your resistance readings, identify the run winding: terminal _____ to _____

Maintenance of Workstation and Tools: Return all equipment and tools to their respective places. Replace all panels with correct fasteners if appropriate.

Summary Statement: Describe the function of the start and run winding in a single-phase compressor.

QUESTIONS

1. Name the three terminals on a single-phase compressor.

2. Name two types of relays used to start a single-phase compressor motor.

3. Why does a hermetic compressor not use a centrifugal switch like an open motor?

4. Behind which terminal is the internal thermal overload located in a hermetic compressor?

5. What must be done to repair a compressor with a defective internal overload?

6. Which winding has the most resistance, the run or the start?

7. What cools the motor in a hermetic compressor?

8. Does oil ever touch the windings of a hermetic compressor?

9. Is there any difference in the materials in a hermetic compressor motor and an open motor?

10. What is the approximate loaded speed of a 3600 rpm compressor?

Application of Motors

Unit Overview

A technician will often have to determine an appropriate substitute for an electric motor when an exact replacement is not available. Consequently, the technician should understand factors in the design and application of these motors. Some design factors are power supply, work requirements, motor insulation type or class, bearing types, and mounting characteristics.

Motor drives connect a motor such as a fan or pump to the driven load. These are normally belt or direct drive. The direct drive may involve a coupling, or the load may be mounted on the motor shaft.

Key Terms

Belly-Band Mount Motor
Cradle Mount Motor
Delta Transformer Connection
Direct Drive Motor
Drip-Proof Motor
End Mount
Explosion-Proof Motor
Frequency
Full-Load Amperage (FLA)

Hertz
Locked-Rotor Amperage (LRA)
Motor Service Factor
Phase
Rated or Run-Load Amperage (RLA)
Resilient Mount Motor
Rigid Mount Motor
Wye Transformer Connection

REVIEW TEST

Name	Date	Grade

Circle the letter that indicates the correct answer.

1. **The alternating current frequency used in the United States is:**
   A. 50 cps.
   B. 60 cps.
   C. 120 cps.
   D. 208 cps.

2. **If the voltage supplied to the motor is well below the motor's rated voltage, the current draw will be _____ it should be.**
   A. lower than
   B. the same as
   C. higher than

3. **The current motor will draw at startup is known as the:**
   A. full-load amperage.
   B. run-load amperage.
   C. locked-rotor amperage.

4. **A sleeve bearing would typically be used in a motor that:**
   A. operates with a light load and where noise is a factor.
   B. operates with a heavy load and where noise is not a factor.
   C. operates with a heavy load in a dirty atmosphere.
   D. operates with a heavy load in a moist atmosphere.

5. **A bearing used on the bottom of a motor where the shaft is positioned vertically is called a _____ bearing.**
   A. sleeve
   B. ball
   C. thrust
   D. bronze

6. **Roller bearings are normally used with:**
   A. small light-duty motors.
   B. small motors where there is a noise factor.
   C. large motors operating with a heavy load.

7. A direct-drive motor may be connected to the load with a:
   A. cradle mount.
   B. rigid-base mount.
   C. pulley.
   D. coupling.

8. The width of a drive belt is classified as:
   A. large or small.
   B. narrow or wide.
   C. A or B.
   D. 1 or 2.

9. An instrument often used to check the alignment of two shafts is the:
   A. micrometer.
   B. dial indicator.
   C. manometer.
   D. sling psychrometer.

10. Single-phase motors can be wired to operate on three-phase power.
   A. true
   B. false

11. The approximate full-load amperage for a single-phase, 2 HP motor operating on 230 V is:
   A. 9 amps.
   B. 12 amps.
   C. 24 amps.
   D. 4.4 amps.

12. The approximate full-load amperage for a 3-phase, 5 HP motor operating on 460 V is:
   A. 7.6 amps.
   B. 2.7 amps.
   C. 11.0 amps.
   D. 1.1 amps.

13. A motor is rated at 230 V. Which of the following is the upper and lower allowable voltage for this motor?
   A. 210 to 245 V
   B. 200 to 400 V
   C. 187.2 to 253 V
   D. 207 to 253 V

14. Using Figure 18-9 in the text, a 1 HP motor with an RPM of 1800 and a voltage rating of 200–208 V has an amperage rating of 3.80 amps. Using the service factor for this motor, what is the maximum allowable operating amperage?
   A. 4.10 amps
   B. 4.37 amps
   C. 5.12 amps
   D. 3.80 amps

15. A loose-fitting belt on a belt-driven device will cause:
   A. bearing wear.
   B. overheating motor.
   C. pulley wobble.
   D. worn pulley grooves.

16. A flexible coupling between the drive and driven device:
   A. corrects minor misalignment.
   B. prevents liquid from entering the bearings.
   C. reduces suction gas noise.
   D. prevents airflow problems.

Motor Application and Features

Name	Date	Grade

Objectives: Upon completion of this exercise, you will be able to fully describe a basic motor used for a fan in an air handler.

Introduction: You will remove a motor from an air handler, inspect it, describe the type of motor, and describe its features such as the speeds, insulation class, shaft size, rotation, mount characteristics, voltage, amperage, phase, and bearing type.

Text References: Paragraphs 18.1, 18.2, 18.3, 18.4, 18.5, 18.6, and 18.7.

Tools and Materials: A VOM, adjustable wrench, straight-blade and Phillips screwdrivers, an Allen head wrench set, and $1/4$" and $5/16$" nut drivers.

Safety Precautions: Turn the power off to any unit you are servicing. Lock and tag the power distribution panel where you turned the power off and keep the single key in your possession. Do not force any parts that you are disassembling or assembling, because they are made to work together and should not be bent or damaged in any way. Properly discharge all capacitors.

PROCEDURES

1. Make sure that the power is off to the air handler to which you have been assigned, the power panel is locked and tagged, and you have the key. Remove the panels. Properly discharge all capacitors.

2. Carefully remove the fan section, fan, and motor as an assembly. Label all wires you disconnect.

3. Remove the motor from the mounting and fill in the following information:

   - Number of motor speeds _____
   - Insulation class of motor _____
   - Diameter of the motor shaft _____ in.
   - Rotation looking at the shaft _____
     NOTE: If the motor has two shafts, look at the end where the motor leads enter.
   - Type of motor mount _____
   - Voltage rating _____ V
   - Full-load amperage _____ A
   - Locked-rotor amperage _____ A
   - Phase of motor _____
   - Type of motor (PSC, shaded-pole) _____
   - Type of bearings (sleeve or ball) _____
   - Number of motor leads _____

4. Replace the motor in its mount.

5. Replace the fan section back into its unit. If any wires were removed, replace them as they were or as they should be.

6. Replace panels with the correct fasteners.

Maintenance of Workstation and Tools: Return all tools to their respective places. Make sure that your work area is clean.

Summary Statements: Describe how the motor was used in the exercise above. Explain whether it is a belt- or direct-drive motor, and state which is better for the application.

QUESTIONS

1. Why are sleeve bearings preferable for some fan applications?

2. What are the advantages of a ball-bearing motor?

3. What is the name of the plug on the bottom of a bearing with a grease connection?

4. Do all bearings have to be lubricated each year?

5. What does "resilient mount" mean when applied to a motor?

6. Why must resilient-mount motors have a ground strap?

7. What are the upper and lower voltages at which a motor rated at 230 V may be safely operated? _____ lower, _____ upper

8. What are the upper and lower voltages at which a motor rated at 208 V may be safely operated? _____ lower, _____ upper

9. If a motor were to be operated below its rated voltage for long periods of time, what would happen?

10. Why do some motors have a different insulation class than others?

Motor Controls

Unit Overview

The switching devices used to close or open the power-supply circuit to the motor are relays, contactors, and starters. The size of the motor and the job it does usually determine which of the switching devices are used. Both the FLA and the LRA must be considered when choosing the type of switching device.

The relay has a magnetic coil that closes one or more sets of contacts. It is designed for light duty and cannot be repaired. The contactor is a larger version of the relay. It can be repaired or rebuilt. A motor starter is different from a contactor because it has motor overload protection built into it.

Most motors need motor overload protection in addition to the circuit protection provided by fuses or circuit breakers. Inherent protection is provided by internal thermal overload devices in the motor windings, such as the thermally activated bimetal. External protection is provided by current overload devices that break the circuit to the contactor coil.

When a motor has been stopped for safety reasons, it should not be restarted immediately. Some manufacturers design their control circuits with a manual reset or a time delay. Others will not let the motor restart until the thermostat has been reset.

Key Terms

Contactor
External Motor Protection
Inherent Motor Protection
Magnetic Overload Protection
Manual Reset
Motor Starters

National Electrical Code® (NEC)
Pilot Duty Relay
Relay
Motor Service Factor
Solder Pot
Time-Delay Relay

REVIEW TEST

Name	Date	Grade

Circle the letter that indicates the correct answer.

1. **The locked-rotor amperage for a motor is approximately _____ times that of the full-load amperage.**
   A. two
   B. four
   C. five
   D. ten

2. **The pilot duty relay is used primarily to switch on and off:**
   A. small motors.
   B. large motors.
   C. contactors and motor starters that in turn will switch the motor on or off.
   D. condenser fan motors.

3. **If the relay does not operate properly,**
   A. it can be rebuilt.
   B. the circuit breaker will perform its job.
   C. it must be replaced.
   D. the contacts may be cleaned with sandpaper.

4. **The contactor is a _____ with electrical contacts.**
   A. circuit breaker
   B. magnetic coil
   C. solder pot
   D. bimetal device

5. **If a contactor fails, it generally is:**
   A. discarded.
   B. repaired or rebuilt.
   C. replaced with a start capacitor.
   D. replaced with a potential relay.

6. **The motor starter has:**
   A. a belt and pulley.
   B. one or more overload protectors, magnetic coil, and electrical contacts.
   C. a start capacitor.
   D. been replaced with a potential relay.

7. **Inherent motor protection is provided by a:**
   A. circuit breaker.
   B. solder pot.
   C. line fuse.
   D. thermally activated device in the motor windings.

8. **External motor protection:**
   A. breaks the circuit to the contactor or motor starter.
   B. breaks the circuit to the motor windings.
   C. is a thermally activated device in the motor windings.
   D. is controlled by the full-load amperage rating.

9. **The service factor of an electric motor is determined by the:**
   A. National Electrical Code.
   B. motor's reserve capacity.
   C. National Electrical Manufacturer's Association.
   D. wire size leading to the motor.

10. **A solder pot is an overload-sensing device:**
    A. made of a low-melting solder that normally can be reset.
    B. made of a low-melting solder that normally cannot be reset.
    C. used to repair circuits that have been damaged due to a current overload.
    D. used as an inherent motor protection.

11. **Crankcase heat for compressors is often controlled by using:**
    A. thermostats.
    B. thermisters.
    C. timers.
    D. auxiliary contacts.

12. **The _____ in a contactor keep(s) equal tension on all contact surfaces:**
    A. stationary contacts
    B. holding coil
    C. springs
    D. moveable contacts

13. **Dirty and pitted contacts in a contactor:**
    A. cause resistance and heat.
    B. cause the motor to run faster.
    C. have no effect.
    D. none of the above.

14. **The moving armature in a contactor is actuated by:**
    A. the springs.
    B. the solenoid coil.
    C. the contacts.
    D. the stationary electromagnet.

Identification of a Relay, Contactor, and Starter

Name	Date	Grade

Objectives: Upon completion of this exercise, you will be able to state the differences between a relay, a contactor, and a starter, and describe the individual parts of each.

Introduction: You will disassemble, inspect the parts, and reassemble a relay, a contactor, and a starter. You will compare the features of each.

Text References: Paragraphs 19.3, 19.4, and 19.5.

Tools and Materials: Straight-blade and Phillips screwdrivers, VOM, relay, contactor, and starter.

Safety Precautions: Disassemble these devices carefully and reassemble them correctly.

PROCEDURES

1. Disassemble all parts of the relay. Lay the parts out in order on the bench. You may only be able to remove the cover.

2. Disassemble the contactor. Lay the parts out in order on the workbench. Remove the coil, the contacts (both movable and stationary), and the springs. See the text, Figure 19-9, for a photo of a disassembled contactor.

3. Disassemble the starter. Lay the parts out on the bench. Remove the contacts, both the movable and the stationary, the coil, and the overloads. See the text, Figure 19-13, for a photo of a disassembled starter.

4. Assemble the relay back together.

5. Assemble the contactor. Check to make sure that all contacts are functioning, that they close when the armature is pushed up through the coil. Make sure that all screws holding the stationary contacts are tight.

6. Assemble the starter. Check to make sure that all contacts are functioning, that they close when the armature is pushed through the coil. Make sure that all screws holding the stationary contacts and the overloads are tight.

Maintenance of Workstation and Tools: Return all tools and parts to their proper places. Make sure that your work area is left clean.

Summary Statements:

A. State an application suitable for a relay where a contactor is not needed, and describe why.

B. State an application suitable for a contactor where a relay cannot be used, but a starter is not necessary. Tell why.

QUESTIONS

1. What is the difference between a relay and a contactor?

2. What is the difference between a contactor and a starter?

3. What should be done with a defective relay?

4. Are magnetic overloads fastened to a starter?

5. What effect would a loose connection have on a solder-pot-type overload?

6. What effect would a loose connection have on a bimetal overload?

7. What effect would a loose connection have on a magnetic-type overload?

8. Which of the following overloads is not affected by ambient temperature: solder-pot, bimetal, or magnetic?

9. What is the difference between a relay and a potential relay?

10. What is the contact surface of a typical contactor made of?

Troubleshooting Electric Motors

Unit Overview

Mechanical motor problems normally occur in the bearings of the shaft where the drive is attached. Bearings may be worn due to lack of lubrication or to grit that has gotten into the bearings and caused them to wear. Roller and ball-bearing failure can often be determined by the bearing noise. When sleeve bearings fail, they normally lock up or sag so that the motor will not start. Most bearings can be replaced. It may be easier and less expensive to replace a small motor than to replace the bearings.

Electrical problems may consist of an open winding, a short circuit from the winding to ground, or a short circuit from winding to winding. The best way to check a motor for electrical soundness is to know what the measurable resistance is for each winding and check it with an ohmmeter.

A short circuit from winding to ground or the frame of the motor may be detected with a good ohmmeter. There should be no circuit from winding to ground. A megger, which has the capacity to detect very high resistances, may be used to determine high-resistance ground circuits.

Motor capacitors may be checked to determine whether they are defective. Be sure to follow safety precautions such as those stated in the text when checking capacitors.

Be sure to check the wiring supplying power to the motor for a loose connection, frayed wiring, or oxidation.

Key Terms

Belt Tension
Belt Tension Gage
Ground, Earth
High Limit Control
Keyway
Megger, Megohmmeter
Open Winding

Oxidation
Pulley Alignment
Pulley Puller
Set Screw
Single Phasing
Winding Short Circuit
Winding-to-Ground Short Circuit

REVIEW TEST

Name	Date	Grade

Circle the letter that indicates the correct answer.

1. Motor bearings _____ by the air-conditioning and refrigeration technician.
A. should never be replaced
B. are occasionally replaced

2. An open winding in an electric motor means that:
A. the winding is making contact with the motor frame.
B. one winding is making contact with another winding.
C. a wire in one winding is broken.
D. the centrifugal switch to the start winding is open.

3. If there is a decrease in resistance in the run winding:
A. the motor probably will not start.
B. the motor may draw too much current while it is running.
C. the centrifugal start winding switch will stop the motor.
D. the load on the motor is probably too great.

4. A megger:
A. is an ammeter that can measure very small amounts of current.
B. is a digital voltmeter that measures very high voltages.
C. measures high-voltage direct current.
D. is a meter with the capacity to measure very high resistances.

5. **The recommended way to discharge a motor start capacitor is to:**
   A. place a screwdriver across the two terminals.
   B. assume it will discharge itself over a period of time.
   C. use an ohmmeter.
   D. short the capacitor from pole to pole with a 20,000-ohm, 5-watt resistor, holding it with insulated pliers.

6. **A run capacitor:**
   A. has an appearance similar to a resistor.
   B. is usually contained in a metal can and is oil filled.
   C. is usually enclosed inside the motor housing.
   D. is a dry type in a shell of paper or plastic.

7. **The electrical components for a fully welded hermetic compressor are checked:**
   A. by disassembling the compressor shell and checking them, the same as with an open motor.
   B. at the terminal box outside the motor.
   C. at the disconnect box for the compressor unit.
   D. at the thermostat.

8. **If the evaporator fan motor has an open-run winding and is not operating but the compressor continues to run, the:**
   A. condenser will ice up.
   B. evaporator will ice up.
   C. condenser fan motor will overheat.
   D. metering device will increase the flow of refrigerant.

9. **The high-limit control on a gas furnace:**
   A. keeps an excessive amount of fuel oil from entering the furnace.
   B. shuts the furnace down before the temperature exceeds the thermostat setting.
   C. shuts the furnace down when the temperature in the furnace or flue becomes excessive.
   D. keeps the conditioned space from becoming too hot should there be a malfunction in the thermostat.

10. **If a condenser fan motor winding is grounded, it will usually:**
    A. trip a circuit breaker to shut the motor off.
    B. keep running because the current is only going to the earth ground.
    C. run a little faster because the current flow will be greater.
    D. cause the evaporator to ice over.

Belt-Drive Motors

Name	Date	Grade

Objectives: Upon completion of this exercise, you will be able to disassemble and reassemble a belt-drive fan and motor and align the pulleys.

Introduction: You will remove a belt-drive fan section from an air handler, take the motor and pulley off, and remove the pulley from the fan shaft. You will then reassemble the fan section, align the pulleys, and set the belt tension.

Text References: Paragraphs 20.3, 20.4, and 20.5.

Tools and Materials: A VOM, pulley puller, two adjustable wrenches, $^3/_8$"–$^1/_2$" and $^9/_{16}$"–$^5/_8$" open or box-end wrenches, oil can with light oil, straight-blade and Phillips screwdrivers, straightedge, Allen wrench set, emery cloth (300 grit), soft face hammer, and $^1/_4$" and $^5/_{16}$" nut drivers.

Safety Precautions: Turn the power off to the unit and lock and tag the panel before you start to disassemble it. Properly discharge all capacitors.

PROCEDURES

1. You will be assigned an air handler or furnace with a belt-drive fan motor. Turn the power off. Lock and tag the panel and keep the single key in your possession. Check the voltage with a meter to ensure that the power is off. Properly discharge all capacitors.

2. Take off enough panels to remove the fan section.

3. Remove the fan section. Label any wiring that is disconnected.

4. Release the tension on the belt by loosening the motor tension adjustment.

5. Remove the belt and then the pulley from the motor shaft. DO NOT FORCE IT OR HAMMER THE SHAFT.

6. Remove the pulley from the fan shaft. See Figure 20-3 and Figure 20-5 in the text. DO NOT DISTORT THE SHAFT BY FORCING IT OFF.

7. Clean the fan shaft and motor shaft with fine emery cloth, about 300 grit.

8. Wipe, clean, and oil slightly.

9. Assemble the fan pulley to the fan shaft and the motor pulley to the motor shaft.

10. Place the motor back in its bracket and tighten it finger tight.

11. Align the pulleys as per Figure 20-9 in the text.

12. Tighten the motor adjustment for tightening the belt, as per the text. Then tighten the motor onto its base.

13. Place the fan section back into the air handler and connect the motor wires.

14. Give the compartment a visual inspection and turn the fan motor over by hand. Make sure no wires or obstructions are in the way.

15. After your instructor has inspected and approved your work, shut the fan compartment door, turn the power on, and start the fan. Listen for any problems.

Maintenance of Workstation and Tools: Wipe up any oil, clean the workstation, and return all tools to their places.

Summary Statement: Describe your experience of removing the fan shaft pulley from its shaft.

QUESTIONS

1. Describe the effects of a belt that is too tight.

2. Why must pulleys be aligned accurately?

3. Why must a fan or motor shaft not be hammered?

4. State the advantages of a belt-drive motor over a direct-drive motor.

5. Do pulleys ever wear to the point of needing replacement?

6. How do you adjust a pulley for a faster-driven pulley speed?

7. Are all belt and pulley grooves alike? Explain your answer.

8. What is a matched set of belts?

9. What are fan shafts made of?

10. What causes fan motors to turn at different design speeds?

11. What is the speed of a blower if the motor turns at 1000 rpm and the motor's drive pulley is twice the diameter of the pulley on the blower shaft?

Examination of a
Hermetic Compressor Electric Motor

Name	Date	Grade

Objectives: Upon completion of this exercise, you will be able to evaluate a motor in a hermetic compressor to determine whether it is electrically sound and safe to start.

Introduction: You will use an ohmmeter to check a hermetic compressor from terminal to terminal and from terminal to ground to determine whether the motor is safe to operate.

Text References: Paragraphs 20.6, 20.7, 20.8, 20.9, 20.10, and 20.14.

Tools and Materials: A VOM, straight-blade and Phillips screwdrivers, and $1/4$" and $5/16$" nut drivers.

Safety Precautions: Make sure the power is turned off before beginning this exercise. Lock and tag the panel where the power is turned off and keep the single key on your person. Properly discharge all capacitors.

PROCEDURES

1. You will be assigned a hermetic compressor to use for this exercise. If the compressor is part of a system, turn the power off, and lock and tag the panel. Properly discharge all capacitors.

2. Label the terminal wires and remove them from the terminals.

3. Turn the range selector switch on the VOM to the R × 1 scale. Fasten the two leads together and zero the meter.

4. Determine which terminals are Common, Run, and Start, either by using the resistance method or by the identification in the terminal box.

5. Fasten one lead on the VOM to the Common terminal.

6. Touch the other lead to the Run terminal and record the reading here: _____ ohms

7. Now, touch the lead to the Start terminal and record the reading here: _____ ohms

8. If you have a book that lists this compressor and gives the winding resistance, compare the ones you took with those in the book.

9. Turn the ohmmeter selector switch to the R × 1000 and zero the meter again.

10. Fasten one lead to a good frame ground on the compressor; the copper process tube or discharge line will do. Touch the other lead to another part of the compressor frame. The meter should read 0 resistance.

11. With one lead on the ground, touch the other lead to each of the three terminals. The reading should be the same for each terminal. Record here: _____ ohms

12. If the compressor has a winding thermostat with leads wired to external terminals, with one meter lead attached to the ground, touch the other meter lead to each of them and record the reading here: _____ ohms The winding thermostat should also be checked to the motor terminals by placing one lead on the winding thermostat and the other on each of the motor terminals. Record the readings here: _____ ohms, _____ ohms, _____ ohms

Maintenance of Workstation and Tools: Replace all panels that may have been removed and put all tools in their respective places.

Summary Statement: Describe a circuit to ground in a motor.

QUESTIONS

1. What would be the results of starting a motor that is grounded?

2. How would a motor that has an open-run winding react at starting time?

3. How would a motor that has an open-start winding react while starting?

4. How should the resistance reading compare in each winding of a three-phase motor?

5. If a motor's windings all had the correct resistance reading, the correct power were supplied to the windings, and the compressor would not start, what would the probable cause be?

6. Describe a shunted winding.

7. What voltage does a potential relay operate from?

8. Describe how a current relay operates removing the start winding from the circuit.

Checking a Capacitor with an Ohmmeter

Name	Date	Grade

Objectives: Upon completion of this exercise, you will be able to field check a start or run capacitor using an ohmmeter.

Introduction: You will use the ohmmeter feature of a VOM to check a capacitor. This will not give the actual capacitance of a capacitor but will show a short, open, or grounded capacitor. The procedures in this exercise can be demonstrated better using an ohmmeter with a needle scale.

Text References: Paragraphs 20.11 and 20.12.

Tools and Materials: A straight-blade screwdriver; VOM; insulated needle-nose pliers; 20,000-ohm, 5-watt resistor; and a start and run capacitor.

Safety Precautions: Always short a capacitor from terminal to terminal before applying an ohmmeter, or the ohmmeter may be damaged. The capacitor should be shorted with a 20,000-ohm, 5-watt resistor. Use insulated needle-nose pliers.

PROCEDURES

1. You will be given a start and run capacitor either from stock or from a unit. If you are going to remove it from a unit, TURN THE POWER OFF. Lock and tag the panel while the power is off. Keep the only key in your possession.

2. Short the capacitor from terminal to terminal. See text, Figure 20-27, for an explanation of how this procedure is done.

3. Set the range selector switch on the meter to ohms and the R × 100 scale.

4. Using the run capacitor, touch one lead on each terminal of the run capacitor while watching the meter. If you get no response, reverse the leads. The needle should rise and fall. It will finally fall back to ∞ if the capacitor is good. If the needle only falls back to a point, the capacitor has an internal short of the value showing on the meter.

5. Now change the ohmmeter range selector switch to R × 1000.

6. Place one lead on a capacitor terminal and touch the other lead to the meter can. Any needle movement at all indicates that the capacitor is shorted to the can. It is defective.

7. Did the run capacitor show "good" in this test?

8. Using the start capacitor, short between the terminals. **NOTE:** If the capacitor has a resistor between the terminals, it is there for the purpose of bleeding the capacitor charge. There should be no stored charge if the resistor is good.

9. Turn the ohmmeter range selector switch to the R × 100 scale.

10. Fasten one meter lead to one terminal and then touch the other meter lead to the other terminal. The meter needle should rise and fall back. IT WILL FALL BACK MUCH MORE SLOWLY THAN WITH THE RUN CAPACITOR BECAUSE OF THE GREATER CAPACITANCE. **NOTE:** The needle will only fall back to the value of the bleed resistor. If you need to know whether the needle will fall all the way back to infinity, the bleed resistor must be disconnected.

11. You will not perform a ground check on a start capacitor because the case is plastic and does not conduct electricity.

12. Did the start capacitor show "good" in this test?

Maintenance of Workstation and Tools: Return all parts and tools back to their respective places. Make sure that your work area is left clean.

Summary Statement: Describe how a capacitor functions in a starting circuit and in a running circuit.

QUESTIONS

1. What is a start capacitor container made of?

2. What is a run capacitor container made of?

3. What is the fill material in a start capacitor?

4. What is the fill material in a run capacitor?

5. What is the purpose of a bleed resistor in a start capacitor?

6. What would the symptoms be of an open start capacitor if the compressor were started?

7. What would the symptoms be of a shorted run capacitor if the compressor were started?

8. What is the purpose of the run capacitor in a motor circuit?

Evaporators and the Refrigeration System

Unit Overview

When heat is allowed to move naturally, it moves from a warm place to a colder place. Heat is actually absorbed into the refrigerant in the evaporator in a household refrigerator while the refrigerant is changing state from a liquid to a gas. It is then compressed and pumped outside the refrigerator and released into the kitchen. There are four important components that cause this to happen: the evaporator, the compressor, the condenser, and the expansion device.

Refrigeration used to preserve food often has evaporator temperatures below the freezing point of water. Under these conditions, water on the evaporator will freeze, causing an ice buildup. The system must be designed to melt this ice so that the evaporator can function properly. In higher-temperature systems, the defrost of this ice is called off-cycle defrost. It uses the space temperature air to cause the defrost. Lower-temperature refrigeration systems must provide heat from external sources such as electric heat to cause the defrost.

The boiling temperature of the liquid refrigerant determines the evaporator coil operating temperature.

The boiling temperature is determined by the type of refrigerant and the pressure it is under. The service technician should be able to determine the pressure and temperatures that are correct for the system being serviced.

Older evaporators were of the natural convection type. More recent designs use blowers to force air over the evaporator. This has improved the efficiency and allows evaporators to be much smaller. Many large or long evaporators are designed with multiple circuits in which the refrigerant flows. By using multiple circuits, the amount of pressure drop can be reduced.

In the operation of an evaporator, refrigerant enters the evaporator from the metering device, approximately 75% liquid and 25% vapor. The vaporized refrigerant along with the pressure drop across the expansion device begin to cool the remaining liquid. As the partial-liquid, partial-vapor mixture moves through the evaporator, more of the liquid boils to a vapor as a result of heat absorbed into the coil. Near the end of the evaporator, all of the liquid should be boiled away.

Key Terms

Boiling
Boiling Temperature
Compressor
Condenser
Defrost
Dehumidification
Direct-Expansion Evaporator (Dry Type)
Evaporator
Expansion (Metering) Device
Film Factor
Finned Tube Evaporator
Flooded Evaporator

High-Temperature Refrigeration
Hot Gas
Hot Pull Down
Low-Temperature Refrigeration
Medium-Temperature Refrigeration
Multiple Circuit Coil
Off-Cycle Defrost
Saturated Vapor
Stamped Evaporator
Starved Coil
Superheat

REVIEW TEST

Name	Date	Grade

Circle the letter that indicates the correct answer.

1. **The evaporator in a refrigeration system:**
   A. rejects heat from the conditioned space.
   B. absorbs heat from the conditioned space.
   C. feeds a partial vapor to the compressor.
   D. feeds the liquid refrigerant to the compressor.

2. **Refrigerant entering the evaporator is approximately _____ vapor.**
   A. 50%
   B. 25%
   C. 10%
   D. 75%

3. An evaporator made up of two pieces of sheet metal that has an impression of refrigerant piping is called a _____ evaporator.
   A. finned tube
   B. dry-type
   C. stamped
   D. liquid-to-liquid

4. Heat that is added to the refrigerant after it has all vaporized is called:
   A. superheat.
   B. latent heat.
   C. artificial heat.
   D. extended heat.

5. A large or long evaporator would probably have:
   A. stainless steel tubing.
   B. an excess of liquid refrigerant.
   C. an excess of vapor refrigerant.
   D. multiple refrigerant circuits.

6. The refrigerant leaving the compressor is often referred to as:
   A. hot liquid refrigerant.
   B. hot gas.
   C. a partial liquid.
   D. condensed vapor.

7. A typical evaporator temperature in an air-conditioning system is:
   A. 40°F.
   B. 55°F.
   C. 75°F.
   D. 200°F.

8. Changing state from a liquid to a gas is often called:
   A. condensing.
   B. flooding.
   C. boiling.
   D. starving.

9. If the evaporator temperature is below 32°F, there must be some provision for:
   A. defrost.
   B. adding refrigerant.
   C. increasing the rate of vaporization.
   D. flooding the evaporator.

10. The expansion device is located between the:
    A. compressor and condenser.
    B. evaporator and compressor.
    C. condenser and evaporator.

11. Off-cycle defrost occurs:
    A. with supplemental electric heat when the system is shut down.
    B. with hot gas when the system is operating.
    C. from the space temperature when the system is shut down.
    D. with hot liquid refrigerant from the compressor.

12. A saturated vapor condition exists when:
    A. there is still liquid refrigerant in the evaporator.
    B. all of the liquid has boiled to a vapor.
    C. there is excessive superheat.
    D. there is evaporator flooding.

13. Another name for expansion device is:
    A. liquid-line receiver.
    B. suction-line filter drier.
    C. high-side service valve.
    D. metering device.

14. A direct expansion-type evaporator is also called a:
    A. flooded evaporator.
    B. dry-type evaporator.
    C. multiple circuit evaporator.
    D. stamped evaporator.

15. The _____ of the air passing over the evaporator has the greatest effect on the film factor.
    A. quantity
    B. temperature
    C. humidity
    D. velocity

16. Refrigeration designed for the purpose of storing flowers or candy is considered _____ refrigeration.
    A. high-temperature
    B. medium-temperature
    C. low-temperature
    D. dry-type

17. For low-temperature refrigeration, _____ from the compressor discharge line may be used to defrost the evaporator.
    A. warm liquid
    B. hot gas
    C. moisture
    D. hot crankcase oil

18. The fin spacing on low-temperature evaporators is _____ on medium-temperature evaporators.
    A. the same as
    B. closer together than
    C. farther apart than

19. The capillary tube is a type of:
    A. safety device.
    B. pressure gage.
    C. filter.
    D. metering device.

20. The latent heat absorbed by the refrigerant while it is changing state is _____ the sensible heat added to the vapor leaving the coil.
    A. the same as
    B. less concentrated than
    C. more concentrated than

Evaporator Functions and Superheat

Name	Date	Grade

Objectives: Upon completion of this exercise, you will be able to take pressure and temperature readings and determine superheat of the refrigerant in the evaporator operating under normal and loaded conditions.

Introduction: You will be working with an evaporator in an operating system. Any typical evaporator in a refrigeration system operating in its normal environment can be used as long as there are gage ports for both high- and low-side gages. You will be taking pressure and temperature readings.

Text References: Paragraphs 21.3, 21.4, 21.5, 21.9, 21.13, and 21.14.

Tools and Materials: A gage manifold, a straight-blade screwdriver, ¼" and ⁵⁄₁₆" nut drivers, light gloves, goggles, and a thermometer that can be attached to the suction line, a small amount of insulation for the thermometer, and an evaporator in an operating system.

Safety Precautions: You will be working with refrigerant under pressure. Care should be taken not to allow the refrigerant to get in your eyes. Wear goggles at all times when working with refrigerants. Liquid refrigerant will freeze the skin. If a leak develops and liquid refrigerant escapes, DO NOT TRY TO STOP IT WITH YOUR HANDS. Light gloves are recommended when handling all refrigerants. Your instructor must give you instruction in using the gage manifold before beginning this exercise. Be careful of all moving parts and especially the evaporator fan.

PROCEDURES

1. Put on your goggles and gloves. Fasten the suction-gage line to the suction gage port. The high-side line may be attached if desired; however, it is not necessary.

2. Purge air from gage lines at manifold.

3. Securely fasten the thermometer to the suction line close to the evaporator (electrical tape works well) and insulate the lead for about 4 inches in each direction, Figure 21-1.

4. Start the unit, allow it to run for at least 15 minutes, and then record the following:

   Suction pressure: _____ psig

   Determine boiling temperature from the PT chart: _____ °F

   Suction-line temperature: _____ °F

   Determine evaporator superheat: Suction-line temperature − boiling temperature = _____ °F superheat

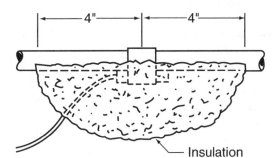

Figure 21-1   Insulation

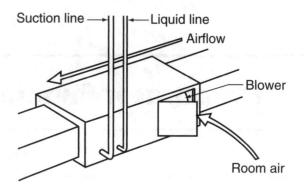

Suction line — ‖ ‖—Liquid line

Airflow

Blower

Room air

Figure 21-2

5. Now, increase the load on the evaporator. If it is a reach-in box, open the door and place some room temperature objects in the box. If it is an air conditioner, partially open the fan compartment door to increase the airflow across the evaporator. See Figure 21-2. After 15 minutes, record the following:

Suction pressure: _____ psig

Boiling temperature from the PT chart: _____ °F

Suction-line temperature: _____ °F

Determine evaporator superheat: Suction-line temperature − boiling temperature = _____ °F superheat

6. Remove the gages and thermometer and return the unit to normal running condition with all panels in place.

Maintenance of Workstation and Tools: Wipe any oil off the gages and replace all plugs and caps. Return all equipment and tools to their respective places.

Summary Statement: Describe what happened when a load was placed on the evaporator.

QUESTIONS

1. What is the typical superheat for an evaporator operating under normal conditions? _____ °F

2. Why is it important to have superheat at the end of the evaporator?

3. What is the job of the evaporator in a refrigeration system?

4. The refrigerant entering the evaporator is approximately _____ % liquid and _____ % vapor.

5. The refrigerant leaving the evaporator is _____ % vapor.

6. Where does the refrigerant go when it leaves the evaporator?

7. The refrigerant in the evaporator must be (colder or warmer) than the air passing over it for heat to transfer into the coil.

Evaporator Performance

Name	Date	Grade

Objectives: Upon completion of this exercise, you will be able to evaluate the performance of a direct-expansion evaporator.

Introduction: You will use superheat and coil-to-air temperature difference to evaluate a direct-expansion evaporator.

Text References: Paragraphs 21.3, 21.4, 21.5, 21.9, 21.12, and 21.13.

Tools and Materials: An electric thermometer, screwdriver, electrician's tape, some insulation, service valve wrench, gage manifold, goggles, and a refrigeration system with a thermostatic expansion valve and service valves.

Safety Precautions: You will be fastening gages to the refrigeration system. Wear goggles and gloves. You must have had instruction in using the gage manifold before starting this exercise.

PROCEDURES

1. Put on your goggles and gloves. Fasten the high- and low-side gage lines to the service valves. Open the service valves by turning them off the back seat. Turn them clockwise.

2. Purge the air from the gage lines by loosening the fittings slightly at the manifold.

3. Fasten a thermometer lead to the suction line of the evaporator where the line leaves the coil. This will be in the same area as the TXV (thermostatic expansion valve) sensor. Be sure the temperature lead is insulated.

4. Fasten another lead in the return air of the evaporator.

5. Start the system and allow it to pull down to within 10°F of its intended use—low, medium, or high temperature—and record the following before the thermostat shuts the compressor off:
   - Suction pressure _____ psig
   - Temperature at the suction line _____ °F
   - Refrigerant boiling temperature (converted from suction pressure) _____ °F
   - Superheat (suction-line temperature − boiling temperature) _____ °F
   - Return air temperature _____ °F
   - TD (temperature difference between return air and refrigerant boiling temperature) _____ °F
   - Discharge pressure _____ psig
   - Condensing temperature (converted from discharge pressure) _____ °F

6. Back seat the service valves and remove the gage lines.

7. Remove the temperature leads and replace them in their case.

Maintenance of Workstation and Tools: Replace all panels and return all tools to their respective places. Make sure the refrigerated box is returned to its correct condition.

Summary Statement: Describe how well the evaporator is performing in this test.

QUESTIONS

1. What is a typical superheat for a typical direct expansion evaporator performing in the correct temperature range?

2. What is the difference between a direct-expansion and a flooded evaporator?

3. What device is used to evenly feed refrigerant to an evaporator with more than one circuit?

4. What would the symptoms of a flooded evaporator be on a direct-expansion system?

5. What would the symptoms of a dirty coil be?

6. Why is it necessary to insulate the bulb on the thermometer when taking the superheat of an evaporator?

7. What is the problem when the superheat reading is too low?

8. What is the problem when the superheat reading is too high?

9. What does the term "starved evaporator" mean?

10. When the superheat is too low, what can happen to the compressor?

Checking an Evaporator Under Excessive Load

Name	Date	Grade

Objectives: Upon completion of this exercise, you will be able to check an evaporator while it is under excessive load and observe the changing conditions while the load is being reduced.

Introduction: You will be assigned a packaged commercial refrigeration system with a thermostatic expansion valve that has been turned off long enough to be warm inside. You will then turn it on with an ammeter, with gages and thermometer attached, and record the conditions while the temperature is being lowered in the box.

Text References: Paragraphs 21.3, 21.4, 21.5, 21.9, 21.12, 21.13, and 21.14.

Tools and Materials: A gage manifold, service valve wrench, four-lead thermometer, ammeter, goggles, gloves, straight-blade and Phillips screwdrivers, ¼" and ⁵⁄₁₆" nut drivers, and a commercial refrigeration system with thermostatic expansion valve.

Safety Precautions: Wear goggles and gloves while working with pressurized refrigerant.

PROCEDURES

1. Put on your goggles and gloves. Connect the high- and low-side gage lines.

2. If the system has service valves, turn the valve stems off the back seat far enough for the gages to read. If the system has Schrader valves, the gages will read when connected. Bleed air from gage lines.

3. Fasten the four temperature leads in the following locations:

   - Suction line leaving the evaporator (before a heat exchanger, if any). See the text, Figure 21-25, for an example of evaporator pressures and temperature.
   - Inlet air to the evaporator.
   - Air entering the condenser.

4. Fasten the ammeter around the common wire on the compressor.

5. Record the following information two times—TEST 1 at 10 minutes after start-up and TEST 2 when the box has reduced the temperature to near the shut-off temperature.

   - Type of refrigerant R-_____

	TEST 1	**TEST 2**	
• Suction pressure	_____	_____	psig
• Suction-line temperature	_____	_____	°F
• Evaporator boiling temperature	_____	_____	°F
• Superheat	_____	_____	°F
• Air temperature entering evaporator	_____	_____	°F
• Evaporator TD (air entering the evaporator − refrigerant boiling temperature)	_____	_____	°F
• Discharge pressure	_____	_____	psig

- Air temperature entering condenser   _____   _____°F
- Condensing temperature   _____   _____°F
- Condenser TD (condensing temperature
  − entering air temperature)   _____   _____°F
- Compressor full-load amperage   _____   _____A

6. Remove all instruments and gages.

7. Replace all panels with the correct fasteners.

8. Return the system to the condition stated by the instructor.

Maintenance of Workstation and Tools: Return all tools to their respective workstations.

Summary Statement: Describe how the amperage, suction, and discharge pressure responded as the temperature in the refrigerated box was lowered.

QUESTIONS

1. What is the function of the evaporator in a refrigeration system?

2. How does the boiling refrigerant absorb heat when it is cold?

3. What should the typical superheat be at the end of the evaporator before the heat exchanger? _____ °F

4. What is the purpose of the fins on an evaporator?

5. Where does the condensate come from that forms on the evaporator?

6. What is done with the condensate mentioned in question 5?

7. What is a dry-type evaporator?

8. What is the ratio of liquid-to-vapor refrigerant at the beginning of the evaporator? _____% liquid and _____% vapor

Condensers

Unit Overview

One of the purposes of the condenser is to reject the heat absorbed by the evaporator. The refrigerant changes from a vapor to a liquid in the condenser. While this change of state is taking place, the greatest amount of heat is rejected or given up. The condenser is operated with higher temperatures and pressures than the evaporator. It is often located outside, where the heat can be rejected without making a difference to the outside air temperature.

Water-cooled condensers are quite efficient compared to those that are air cooled and operate at much lower condensing temperatures. The tube within a tube, shell and coil, and shell and tube are common types of water-cooled condensers.

In a water-cooled condenser, the water is used to remove the heat from the refrigerant. The water then can be wasted down the drain or it can be pumped to a remote area where the heat is removed and the water reused. Normally, the condensing temperature of the refrigerant is approximately 10°F higher than the water leaving the system. In a recirculating water system, approximately 3 gallons of water per minute per ton of refrigeration is recirculated. Wastewater systems use about 1½ gallons per minute.

A cooling tower is a component in which the outside air passes over the heated water from the condenser to remove the heat that has been rejected into it. A natural draft tower utilizes natural drafts to pass the air over the water. Forced-draft cooling towers have a fan to force air over a wetted surface to remove heat by evaporation.

In an evaporative condenser, the refrigerant condenser is located inside the tower. The water is used over and over with a pump circulating it. As the water is evaporated, it is replaced with a makeup system. Freeze protection must be provided in the winter in colder climates.

Air-cooled condensers use air as the medium to remove heat from the refrigerant.

Practical methods used to help control head pressures and prevent equipment wear are fan cycling, dampers, Variable Frequency Drive (VFD) motors and condenser flooding. Fans can be turned off and on to maintain a constant head pressure. A shutter or damper can be opened or closed to help maintain a constant head pressure. Adding refrigerant to the condenser can raise the head pressure.

Heat from the condensers of refrigeration systems can be captured in the winter to heat space and/or domestic hot water.

Key Terms

Air-Cooled Condenser
Air Shutters (or Dampers)
Condenser
Condenser Flooding
Condensing Temperature
Cooling Tower
Desuperheating
Evaporative Condenser
  (Cooling Tower)

Fan Cycling
Forced-Draft Tower
Heat Reclaim
High-Efficiency Condenser
Makeup Water
Natural-Draft Tower
Recirculated Water System
Shell-and-Coil Condenser
Shell-and-Tube Condenser

Subcooling
Tube-within-a-Tube Coil
Variable Frequency
  Drive (VFD)
Wastewater System
Water-Cooled Condenser

REVIEW TEST

Name	Date	Grade

Circle the letter that indicates the correct answer.

1. **The refrigerant in a condenser in a refrigeration system:**
   A. absorbs heat.
   B. gives up heat.
   C. changes to a vapor.
   D. is under lower pressure than in an evaporator.

2. **The condenser is operated at _____ the evaporator.**
   A. a lower pressure than
   B. a higher pressure than
   C. the same pressure as
   D. a lower temperature than

3. The tube-within-a-tube condenser is _____ condenser.
   A. a water-cooled
   B. an air-cooled
   C. an evaporative
   D. a high-pressure

4. Mineral deposit acts as _____ in a water-cooled condenser.
   A. a conductor
   B. an insulator
   C. a cleansing solution
   D. a rust inhibitor

5. The coil in a shell-and-coil condenser must be cleaned:
   A. by removing the end caps.
   B. with a brush.
   C. chemically.
   D. by air pressure.

6. The shell-and-tube condenser can be cleaned:
   A. with a brush after removing the end caps.
   B. only with chemicals.
   C. with compressed air.
   D. by sending it to a specialized shop where it is immersed in a cleaning solution.

7. The varied temperatures of a wastewater-type condenser has an effect on the:
   A. liquid refrigerant in the receiver.
   B. vaporization that takes place in the condenser.
   C. head pressure in the condenser.
   D. operation of the capillary tube.

8. In a water-cooled condenser, the condensing temperature is normally _____ higher than the leaving water.
   A. 40°F   C. 25°F
   B. 35°F   D. 10°F

9. In a wastewater condenser, a regulating valve is used to control the:
   A. refrigerant flow.
   B. refrigerant temperature.
   C. water flow.
   D. water temperature.

10. The primary reason for using a recirculating water system in a condenser is to:
    A. save refrigerant.
    B. more efficiently vaporize the refrigerant.
    C. save or reuse the water.
    D. raise the water temperature.

11. The cooling tower is used to:
    A. cool the refrigerant.
    B. raise the head pressure in the condenser.
    C. remove heat from the condenser water.
    D. increase the superheat in the suction line.

12. A natural-draft cooling tower should be located to take advantage of the:
    A. afternoon sun.
    B. prevailing wind.
    C. prevailing rain.
    D. longer summer days.

13. Evaporative condenser is another term for cooling tower.
    A. true   B. false

14. The hot refrigerant gas normally enters the air-cooled condenser from the:
    A. top.
    B. side.
    C. bottom.
    D. metering device.

15. The refrigerant entering the condenser:
    A. is superheated.
    B. is subcooled.
    C. is in the form of a liquid.
    D. comes directly from the receiver.

16. The condensing temperature of the refrigerant in a standard air-cooled condenser is approximately _____ higher than the ambient air.
    A. 10°F
    B. 30°F
    C. 55°F
    D. 100°F

17. The refrigerant used in a condenser flooding application comes from the:
    A. metering device.
    B. receiver.
    C. evaporator.
    D. cooling tower.

18. A motor control device used to gradually change the speed of a condenser fan motor with changing outdoor ambient temperatures without causing large head pressure swings is the:
    A. modulator.
    B. oscillator.
    C. variable frequency drive (VFD).
    D. none of the above.

Condenser Functions and Superheat

Name	Date	Grade

Objectives: Upon completion of this exercise, you will be able to take refrigerant and air temperature readings at the condenser. You will be able to determine condensing temperatures from the temperature/pressure chart using the compressor discharge pressure and determine the condensing superheat under normal and loaded conditions.

Introduction: You will apply a gage manifold and thermometers to an air-cooled condenser of a typical refrigeration system to take pressure and temperature readings. You will also determine the superheat at the condenser under normal and loaded conditions.

Text References: Paragraphs 22.1, 22.18, and 22.22.

Tools and Materials: A thermometer that reads up to 200°F and that can be strapped to the compressor discharge line, small piece of insulation, thermometer for recording the condenser air inlet temperature, goggles, light gloves, manifold gage, piece of cardboard, straight-blade screwdriver, ¼" and ⁵⁄₁₆" nut driver, service valve wrench, and a temperature/pressure conversion chart.

Safety Precautions: The discharge line may be hot, so give it time to cool with the compressor off before fastening the thermometer. Be very careful when working around the turning fan. Use care in fastening the refrigerant gages. Wear goggles and light gloves. Do not proceed with this exercise until your instructor has given you instruction in the use of the gage manifold.

PROCEDURES

1. With the compressor off and after the discharge line has cooled, fasten a thermometer to the discharge line, insulate it, and place a thermometer in the air entering the condenser, Figure 22-1.

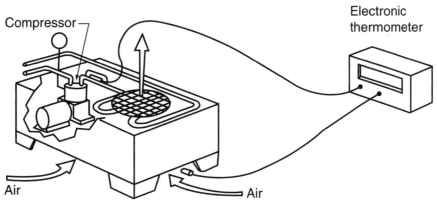

Figure 22-1 Thermometer leads to compressor discharge line where airflow enters condenser.

2. Put on your goggles and gloves. Fasten the high-pressure gage to the high-pressure side of the system. You may also fasten a low-side gage if you want, although you will not need it.

3. Start the unit and record the following after 15 minutes:

Refrigerant type R-_____

Discharge pressure _____ psig

Discharge-line temperature _____°F

Condensing temperature (converted from pressure on the PT chart) _____°F

Discharge-line superheat (discharge-line temperature − condensing temperature) _____°F

Air temperature entering the condenser _____°F

Temperature difference (condensing temperature − air temperature) _____°F

4. By blocking the condenser airflow, Figure 22-2, increase the head (discharge) pressure by 25 psig and record the following:

Discharge pressure _____ psig

Discharge-line temperature _____°F

Condensing temperature (converted from pressure on the PT chart) _____°F

Discharge-line superheat (discharge-line temperature − condensing temperature) _____°F

Air temperature entering the condenser _____°F

Temperature difference (condensing temperature − air temperature) _____°F

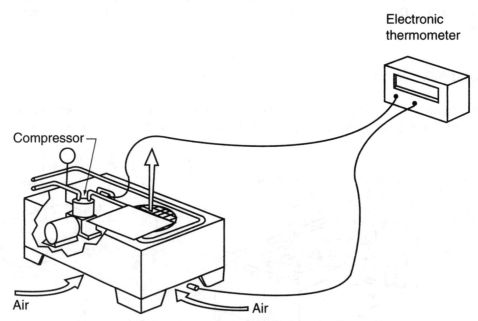

Figure 22-2 Increasing the head pressure by partially blocking the condenser.

Maintenance of Workstation and Tools: Stop the unit and remove the thermometer from the discharge line after it has cooled, remove the gages, and wipe off any oil that may have accumulated on the unit. Replace all tools. Replace all panels and leave the unit as you found it.

Summary Statement: Describe why the compressor discharge line is so much hotter than the condensing temperature and why it does not follow the temperature and pressure relationship.

QUESTIONS

1. What must be done to the hot gas leaving the compressor before it may be condensed?

2. The refrigerant leaving the condenser must be a _____. (liquid, vapor, or solid)

3. What is the typical difference in temperature between the condensing refrigerant and the air entering the condenser?

4. What happens to the temperature of the discharge line when the head pressure is increased?

5. Where does the condensing take place in a typical condenser?

6. What would the head pressure do if the fan motor on an air-cooled condenser stopped working?

7. What job does the condenser do in the refrigeration system?

8. Where is the condenser located in relation to the evaporator on a typical refrigeration system?

9. The head pressure of the refrigerant in a running refrigeration machine is determined by the condensing temperature. True or false?

10. The condensing temperature is determined by the condensing medium minus air or water. True or false?

Air-Cooled Condenser Performance

Name	Date	Grade

Objectives: Upon completion of this exercise, you will be able to perform basic evaluation procedures on an air-cooled condenser.

Introduction: You will use a thermometer and a gage manifold to evaluate an air-cooled condenser to determine its performance.

Text References: Paragraphs 22.1, 22.15, 22.16, 22.17, and 22.21.

Tools and Materials: A gage manifold, thermometer, goggles, light gloves, adjustable wrench, valve wrench, and a refrigeration system with an air-cooled condenser and service valves.

Safety Precautions: Wear goggles and gloves while connecting and disconnecting gage lines. Make sure you have had instruction in using a gage manifold.

PROCEDURES

1. Put on your goggles and gloves. Fasten the high- and low-side gage manifold lines to the service valves.

2. Turn the service valves off the back seat (clockwise) enough to obtain a gage reading.

3. Purge the gage lines to clean them by loosening the fittings slightly at the manifold.

4. Place the thermometer in the air stream entering the condenser.

5. Start the system and wait until it is within 10°F of the correct temperature range inside the box.

6. Record the following:

   - Type of box (low, medium, or high temperature) _____
   - Type of condenser (with or without fan) _____
   - Suction pressure _____ psig
   - Head pressure _____ psig
   - Condensing temperature (converted from head pressure) _____°F
   - Condensing entering air temperature _____°F
   - Temperature difference (condensing temperature − air-in temperature) _____°F

7. Is this condenser running at less than 30°F temperature difference? If the answer is yes, it is probably running efficiently.

8. If the temperature difference is more than 30°F, check the condenser to see whether it is clean.

9. Remove the gages.

Maintenance of Workstation and Tools: Return the refrigeration system to its proper condition. Clean the workstation and return all tools to their respective places.

Summary Statement: Describe the condenser-to-entering-air relationship of a typical air-cooled condenser.

QUESTIONS

1. Name the three functions that a condenser performs.

2. What is considered the highest temperature that the discharge gas should reach upon leaving the compressor?

3. What may be used on an air-cooled condenser with a fan to prevent the head pressure from becoming too low in mild weather?

4. What are the system results of an air-cooled condenser operating with a head pressure that is too low?

5. Why must an air-cooled condenser be located where the air will not recirculate through it?

6. What does the compressor amperage do when the head pressure rises?

7. What would the symptoms be if a condenser had two fan motors and one stopped running?

8. What is the purpose of the fins on the condenser?

9. What materials are typical refrigeration condensers made of?

Evaluation of a Water-Cooled Condenser

Name	Date	Grade

Objectives: Upon completion of this exercise, you will be able to evaluate a simple water-cooled refrigeration system.

Introduction: You will use gages and a thermometer to evaluate a simple water-cooled system. You will do this by evaluating the heat exchange in the condenser.

Text References: Paragraphs 22.1, 22.2, 22.8, 22.9, and 22.10.

Tools and Materials: A gage manifold, goggles, light gloves, adjustable wrench, valve wrench, thermometer, and a water-cooled refrigeration system.

Safety Precautions: Wear goggles and gloves while working with refrigerant under pressure.

PROCEDURES

1. Put on your goggles and gloves. Connect the high- and low-side gages.

2. Slightly open the service valves from their back seat, far enough for the gages to register the pressure.

3. Fasten one thermometer lead to the discharge line of the compressor.

4. Place one other lead in the entering water and one in the leaving water of the condenser in such a manner that the lead will give accurate water readings. If there are no thermometer wells, you may use the method in the text, Figure 22-12.

5. Start the system and allow it to run until it is operating at a steady state, that is, until the readings are not fluctuating. Record the following:
   - Type of refrigerant _____
   - Suction pressure _____ psig
   - Discharge pressure _____ psig
   - Discharge-line temperature _____°F
   - Condensing temperature (converted from discharge pressure) _____°F
   - Condenser leaving water temperature _____°F
   - Condenser entering water temperature _____°F
   - TD (temperature difference between the leaving water and the condensing refrigerant) _____°F

6. Disconnect the gages and thermometer leads. Replace all panels that may have been removed with the correct fasteners.

Maintenance of Workstation and Tools: Return all tools to their respective places.

Summary Statements: Describe the above system's ability to transfer heat into the water. Was the system transferring the heat at the correct rate?

QUESTIONS

1. Is a water-cooled condenser more efficient than an air-cooled condenser?

2. Why are water-cooled condensers not as popular as air-cooled condensers in small equipment?

3. What is the typical difference in temperature between the condensing refrigerant and the leaving water for a condenser that is operating efficiently?

4. Name two places that the heat-laden water from a water-cooled condenser may be piped to.

5. Why is the discharge line hotter than the condensing refrigerant?

6. Name the three functions that take place in the condenser.

7. What controls the waterflow through a water regulating valve?

8. What does the temperature difference across a condenser do if the water flow is reduced?

9. What does the temperature difference between the condensing refrigerant and the leaving water do if the condenser tubes are dirty?

10. What is the low form of plant life that grows in condenser water?

Determining the Correct Head Pressure

Name	Date	Grade

Objectives: Upon completion of this exercise, you will be able to determine the correct head pressure for a system under the existing operating conditions for air-cooled condensers.

Introduction: You will use pressure and temperature measurements to determine the correct head pressure for an air-cooled condenser under typical operating conditions. If a unit has an abnormal load on the evaporator, the head pressure will not conform exactly to the ambient temperature relationship.

Text References: Paragraphs 22.1, 22.2, 22.8, 22.9, 22.15, 22.16, 22.17, and 22.18.

Tools and Materials: A gage manifold, goggles, light gloves, adjustable wrench, service valve wrench, straight-blade and Phillips screwdrivers, ¼" and ⁵⁄₁₆" nut drivers, thermometers, and an air-cooled refrigeration system with service valves or Schrader valve ports.

Safety Precautions: Wear goggles and gloves while working with refrigerant under pressure.

PROCEDURES

1. Your refrigeration system should have been running for a period of time so that it is operating under typical conditions. Using the thermometer, determine the temperature of the air entering the condenser. Record here: _____°F

2. Determine from the air entering the condenser what you think the maximum condensing temperature should be. Air entering plus 30°F maximum equals _____°F.

3. Convert the above condensing temperature to head pressure using the temperature/pressure relationship chart in the text, Figure 3-15: _____ psig

4. Put on your goggles and gloves. Now that you have determined what the head pressure should be, connect the high- and low-side gage lines and record the following:

   • Actual head pressure _____ psig

   • Actual condensing temperature _____°F

   • Air temperature entering condenser _____°F

   • TD (condensing temperature − temperature of the air entering the condenser) _____°F

   If the last item is over 30°F, is the system under excess load? Is the evaporator absorbing excess heat because it is operating above its design application?

5. Remove the gages and replace all panels with the correct fasteners.

Maintenance of Workstation and Tools: Return all tools to their respective places.

Summary Statements: Describe the action in the condenser that you have been working with. Was it performing the way it should under the existing conditions?

QUESTIONS

1. State the three functions of a condenser.

2. Why does the head pressure rise when the evaporator has an excess load?

3. What would the TD be for a high-efficiency condenser?

4. What two temperatures are used for TD on a water-cooled condenser?

5. What TD is used to determine the condenser performance of a water-cooled condenser?

6. Why must a minimum head pressure be maintained for equipment, particularly if it is located outside?

7. Name three methods used to control head pressure on air-cooled equipment.

8. What method is used to control head pressure on water-cooled equipment that wastes the water?

9. Why would the entering water temperature vary on a wastewater system?

10. What would be the symptoms for an air-cooled condenser where the air leaving the condenser is hitting a barrier and recirculating?

Compressors

Unit Overview

The compressor compresses the refrigerant vapor from the evaporator and pumps the refrigerant throughout the system.

Refrigerant vapor enters the compressor through the suction valve and fills the cylinder(s). This refrigerant is cool but has absorbed heat in the evaporator. Most of this heat is latent heat absorbed while it was changing state from a liquid to a vapor. A smaller amount may be sensible heat, called superheat. The compressor compresses this vapor causing it to become very warm, maybe as high as 200°F, and pumps it to the condenser.

There are five major types of compressors used in the refrigeration and air-conditioning industry. These are the reciprocating, screw, rotary, scroll, and centrifugal compressors.

Reciprocating compressors are categorized by their housings and by their drive mechanisms. The housing categories are open and hermetic. The drive mechanisms are direct or belt drive. In hermetic compressors, the motor and compressor are both contained in a single shell or housing and operate in a refrigerant atmosphere in this shell. Hermetic compressors may be of the fully welded type where the compressor and motor are fully enclosed inside a welded shell, or they may be of the serviceable type where the shell is bolted together. Open-drive compressors may be belt or direct drive. Any compressor with the drive on the outside of the housing must have a shaft seal to keep the refrigerant from escaping.

Some compressors use thermistors in or near the motor windings to sense motor heat. Thermistors are variable resistors that change in resistance when exposed to different temperatures. The thermistors are wired to a solid-state electronic compressor protection module that will sense when the compressor should be cycled off because of excessive heat.

Key Terms

Centrifugal Compressor
Clearance Volume
Compression Ratio
Compressor
Compressor Head
Connecting Rod
Crankshaft
Direct-Drive Compressor
Discharge Pressure
Discharge Valve

Discus Valve
Hermetic Compressor
Internal Motor Overload
Liquid Slugging
Muffler
Open Compressor
Piston
Reciprocating Compressor
Refrigerant Cylinder Valves
Rotary Compressor

Screw Compressor
Scroll Compressor
Serviceable Hermetic
 Compressor
Suction Pressure
Suction Valve
Thermistor
Valve Plate
Vapor Pump
Welded Hermetic Compressor

REVIEW TEST

Name	Date	Grade

Circle the letter that indicates the correct answer.

1. The temperature of the suction gas entering the compressor can best be described as:
A. cool.   B. warm.

2. The suction gas contains heat absorbed while it was in the:
A. metering device.   C. condenser.
B. evaporator.   D. compressor.

3. Most of the heat absorbed by the suction gas is:
A. heat of compression.   C. latent heat.
B. sensible heat.   D. defrost heat.

4. The temperature of the compressed gas leaving the compressor can best be described as:
A. cold.   C. warm.
B. cool.   D. very warm.

5. The reciprocating compressor is used most frequently in:
A. large industrial refrigeration systems.
B. air-conditioning systems in large buildings.
C. small- and medium-size refrigeration systems and residential and light commercial air conditioning.

6. **A hermetic compressor can be described as _____ compressor.**
   A. an open-type
   B. a belt-drive
   C. a fully welded or serviceable
   D. a centrifugal

7. **The screw-type compressor is used to a great extent in:**
   A. domestic refrigerators.
   B. residential air conditioning.
   C. small commercial refrigeration units.
   D. large commercial and industrial systems.

8. **In a hermetic compressor:**
   A. the motor and compressor operate in a refrigerant liquid atmosphere.
   B. the motor and compressor operate in a refrigerant vapor atmosphere.
   C. only the compressor operates in a refrigerant atmosphere.
   D. the belt drive must be adjusted precisely.

9. **The crankshaft in a reciprocating compressor:**
   A. serves as the crankcase oil heater.
   B. changes the rotary motion to a back and forth or reciprocating motion.
   C. operates the centrifugal fan.
   D. operates the ring valves.

10. **The flapper or reed valve is used in the:**
    A. scroll compressor.
    B. condenser.
    C. reciprocating compressor.
    D. oil pump.

11. **A bimetal hermetic motor internal overload is:**
    A. located in the motor rotor.
    B. embedded in the motor windings.
    C. located in the terminal box.
    D. fastened to the outside of the motor housing.

12. **The shaft seal on an open-drive compressor:**
    A. is normally constructed of brass.
    B. is the bearing surface for the shaft.
    C. provides the lubrication for the shaft.
    D. keeps the refrigerant vapor in and the atmosphere out of the compressor.

13. **The clearance volume is the:**
    A. amount of vapor refrigerant the compressor cylinder will hold during each stroke.
    B. volume by which the compressor is rated.
    C. space between the piston at top-dead center and the bottom of the valve plate of the compressor.
    D. space between the seat and the ring valve.

14. **The discus valve design is more efficient than a flapper valve design because:**
    A. it allows a cylinder to be designed with less clearance volume.
    B. it will operate with less spring tension.
    C. the reexpanded refrigerant in the cylinder is much greater.
    D. it will not allow liquid into the cylinder.

15. **Liquid slugging will likely damage the reciprocating compressor because the:**
    A. liquid does not mix well with the crankcase oil.
    B. compressor cannot compress liquid.
    C. electrical components are not designed to operate with a liquid refrigerant.
    D. arcing from the bimetal will deteriorate the refrigerant.

16. **Overheated oil may turn to carbon and create _____ in the system.**
    A. moisture
    B. air
    C. dirt
    D. acid

17. **The rotary and scroll compressors are used primarily in:**
    A. large industrial installations.
    B. large refrigeration systems.
    C. air-conditioning systems in large buildings.
    D. residential and light commercial air conditioning and refrigeration.

18. **The screw compressor utilizes _____ to compress the refrigerant gas.**
    A. matching tapered machined gears
    B. a piston and cylinder
    C. a type of centrifugal fan
    D. a type of rotor

19. **Connecting rods connect the:**
    A. motor to the compressor.
    B. compressor head to the housing.
    C. oil pump to the crankshaft.
    D. crankshaft to the piston.

20. **Discus, ring, and flapper values are all used in _____ compressors.**
    A. rotary
    B. reciprocating
    C. screw
    D. scroll

Compressor Operating Temperatures

Name	Date	Grade

Objectives: Upon completion of this exercise, you will be able to evaluate a suction gas-cooled compressor for the correct operating temperatures while it is operating.

Introduction: You will use gages and a thermometer to become familiar with how a suction gas-cooled compressor should feel while operating and where it should be hot, warm, and cool.

Text References: Paragraphs 23.1, 23.2, and 23.10.

Tools and Materials: A gage manifold, straight-blade and Phillips head screwdrivers, service valve wrench, $1/4$" and $5/16$" nut drivers, goggles, gloves, thermometer, and a refrigeration system with a gas-cooled compressor.

Safety Precautions: Wear goggles and gloves while working with refrigerant under pressure.

PROCEDURES

1. The refrigeration system should have been off for some period of time. It should have a welded hermetic compressor and a suction-cooled motor.

2. Put on your goggles and gloves. Connect the high- and low-side gages and obtain a gage reading. There must either be service valves or Schrader valve ports.

3. Fasten one temperature sensor to the suction line entering the compressor, Figure 23-1.

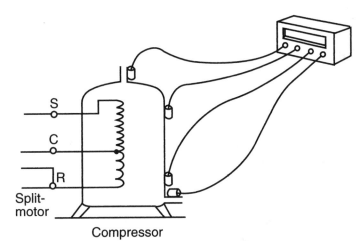

Compressor

Figure 23-1 The correct location for the temperature sensors. The sensors may be fastened with electrical tape and should have some insulation.

4. Fasten a second temperature lead to the compressor shell about 4" from where the suction line enters the shell.

5. Fasten a third temperature sensor to the bottom of the compressor shell where the oil would be.

6. Fasten the fourth temperature sensor to the discharge line leaving the compressor.

7. Record the following data before starting the system:

   - Type of refrigerant in the system _____
   - Suction pressure _____ psig
   - Refrigerant boiling temperature _____ °F
   - Discharge pressure _____ psig
   - Refrigerant condensing temperature _____ °F
   - Suction-line temperature _____ °F
   - Upper shell temperature _____ °F
   - Lower shell temperature _____ °F
   - Discharge-line temperature _____ °F

8. Start the system and let run for 30 minutes and record the following information:

   - Suction pressure _____ psig
   - Refrigerant boiling temperature _____ °F
   - Discharge pressure _____ psig
   - Refrigerant condensing temperature _____ °F
   - Suction-line temperature _____ °F
   - Upper shell temperature _____ °F
   - Lower shell temperature _____ °F
   - Discharge-line temperature _____ °F

9. Block the condenser airflow until the refrigerant is condensing at 125°F. This would be one of the following:

   - R-12   169 psig
   - R-22   278 psig
   - R-502   301 psig
   - R-134a   185 psig
   - R-404A   332 psig
   - R-410A   446 psig

10. Let the system run for 30 minutes at this condition. NOTE: YOU MAY HAVE TO TURN THE THERMO-STAT CONTROLLING THE UNIT DOWN TO KEEP IT FROM SHUTTING OFF DURING THIS RUNNING TIME. Record the following information:

   - Suction pressure _____ psig
   - Refrigerant boiling temperature _____ °F
   - Discharge pressure _____ psig
   - Refrigerant condensing temperature _____ °F
   - Suction-line temperature _____ °F
   - Upper shell temperature _____ °F
   - Lower shell temperature _____ °F
   - Discharge-line temperature _____ °F

11. Shut the system off for 15 minutes and record the following conditions again:
    - Suction pressure _____ psig
    - Refrigerant boiling temperature _____ °F
    - Discharge pressure _____ psig
    - Refrigerant condensing temperature _____ °F
    - Suction-line temperature _____ °F
    - Upper shell temperature _____ °F
    - Lower shell temperature _____ °F
    - Discharge-line temperature _____ °F

12. Reset the thermostat to the correct setting.

13. Remove the gages and thermometer sensors.

Maintenance of Workstation and Tools: Return all tools to their respective places. Replace all panels. Leave the system in the manner in which you found it unless instructed otherwise.

Summary Statements: Describe how the different temperatures of the compressor responded to a rise in head pressure and why.

Suction line:

Upper shell:

Lower shell:

Discharge line:

QUESTIONS

1. What would happen to the discharge-line temperature with an increase in superheat at the suction line?

2. What would happen to the discharge-line temperature with a decrease in suction-line temperature?

3. What would the result of an internal discharge-line leak be on a compressor's temperature?

4. What is the purpose of a crankcase heater on a compressor?

5. How is a typical hermetic compressor mounted inside the shell?

6. Is the compressor shell thought of as being on the high- or low-pressure side of the refrigeration system?

7. Is the suction or discharge line on the compressor typically larger?

8. What do high discharge gas temperatures do to the oil in a compressor?

9. How can a hot hermetic compressor be cooled with water?

10. Which of the following terminal combinations would show an open circuit if a hermetic compressor with an internal overload were to be open due to a hot compressor?

    A. common to run

    B. common to start

    C. run to start

Compressor Efficiency

Name	Date	Grade

Objectives: Upon completion of this exercise, you will be able to evaluate the performance of a hermetic compressor.

Introduction: You will use a gage manifold to measure the suction and head pressure at design conditions. You will compare this with the full-load amperage of the compressor to determine operating characteristics. To achieve design conditions, the system must be operated until the temperature in the conditioned space is lowered to near the design level.

Text References: Paragraphs 23.1, 23.2, and 23.10.

Tools and Materials: A gage manifold, goggles, gloves, adjustable wrench, voltmeter, clamp-on ammeter, valve wrench, and a refrigeration system with known full-load amperage for the compressor motor.

Safety Precautions: Wear goggles and gloves while working with refrigerant under pressure.

PROCEDURES

1. Put on your goggles and gloves. With the power off, fasten the gages to the valve ports or service valves and obtain a gage reading.

2. Fasten the ammeter to the common or run wire leading to the compressor.

3. Start the system.

4. Allow the system to run until the conditioned space is near the design temperature, just before the thermostat would shut it off.

5. Lower the thermostat so it will not stop the compressor.

6. Reduce the airflow across the condenser until the refrigerant is condensing at 125°F. The head pressure will match one of the following:
   - 169 psig for an R-12 system
   - 278 psig for an R-22 system
   - 301 psig for an R-502 system
   - 185 psig for an R-134a system
   - R-404A 332 psig
   - R-410A 446 psig

7. Record the following information:
   - Suction pressure _____ psig
   - Boiling temperature _____ °F
   - Discharge pressure _____ psig
   - Condensing temperature _____ °F
   - Compressor rated full-load amperage _____ A

- Compressor actual amperage _____ A
- Compressor rated voltage _____ V
- Actual voltage at the unit _____ V

8. If the compressor will pump the pressure differential and the full-load amperage is within 10% of the rated amperage, the compressor is pumping to near capacity. It is doing all it can.

9. Remove gages.

Maintenance of Workstation and Tools: Return all tools to their respective places. Replace all panels. Return the unit to its normal running condition.

Summary Statement: How did the compressor in the above test compare to the rated pressures and rated amperage?

QUESTIONS

1. Name three types of compressors.

2. Name two different types of compressor drives.

3. Name two ways that small compressor motors are cooled.

4. Why should you never front seat the discharge service valve while a compressor is running?

5. What does the amperage of a compressor do when the head pressure rises?

6. What would the symptoms be for a compressor that has two cylinders and only one is pumping?

7. Is the full-load amperage printed on all compressors?

8. What lubricates the compressor's moving parts?

Hermetic Compressor Changeout

Name	Date	Grade

Objectives: Upon completion of this exercise, you will be able to change a hermetic compressor in a refrigeration system.

Introduction: You will remove the refrigerant charge, completely remove a compressor from a refrigeration system, and then replace it in the system. You will then leak check, evacuate, and charge the system as though the compressor were a new one.

Text References: Paragraphs 8.4, 8.5, 8.6, 8.7, 8.11, 9.16, 9.16, 9.17, 10.1, 10.2, 10.4, 10.5, and 23.1.

Tools and Materials: A torch arrangement, leak detector (soap bubbles or halide will do), service valve wrench, tubing cutters, tube reamer, gage manifold, goggles, gloves, line tap valves (if needed, such as for a small system), straight-blade and Phillips screwdrivers, small socket set, refrigerant recovery system, scales or charging cylinder (may be needed in order to add the correct charge), vacuum pump, and a cylinder of refrigerant.

Safety Precautions: Wear goggles and gloves while transferring liquid refrigerant. DO NOT VENT REFRIGERANT TO THE ATMOSPHERE.

PROCEDURES

1. Put on your goggles and gloves. Fasten gages to the system and purge the gage lines.

2. Remove the refrigerant from the system to the appropriate standard using a refrigerant recovery system.

3. Turn off and lock out the power, remove the wiring from the compressor terminals, and mark each wire. Discharge all capacitors.

4. Remove the compressor hold-down nuts or bolts.

5. Use tubing cutters to cut the compressor out of the system. Cut the compressor suction and discharge lines at least 2" from the shell so you have a stub of piping to work with when replacing the compressor.

6. Remove the compressor from the appliance or frame and set it aside.

7. Ream the compressor stubs and the piping in the system of the burrs. DO NOT ALLOW ANY MATERIAL TO FALL DOWN THE PIPE INTO THE SYSTEM OR THE COMPRESSOR.

8. Now replace the compressor onto its mountings as though it were a new one. **NOTE:** REPLACE THE COMPRESSOR AS FAST AS POSSIBLE AND YOU WILL NOT NEED TO REPLACE THE LIQUID-LINE DRIER.

9. Use sweat tubing connectors to fasten the suction and the discharge lines back together. A good grade of low-temperature solder may be what your instructor would recommend so this process may be performed again on the same unit.

10. Pressurize the system with R-22 and nitrogen pressure to 150 psi and leak check the connections you worked with.

11. Triple evacuate the system. The motor terminal wiring and compressor mounting bolts may be refastened at this time.

12. Charge the system with the correct amount of refrigerant. Scales or a charging cylinder are recommended for critically charged systems.

13. Start the system and check for normal operation.

14. Return the system to the condition specified by your instructor.

Maintenance of Workstation and Tools: Return all tools to their correct locations.

Summary Statement: Describe when a new filter-drier would be vital for this system.

QUESTIONS

1. How long did it take you to evacuate the system?

2. What type of service ports did you use on this system?

3. How long did it take to remove the refrigerant from the system?

4. Why is it not recommended that the refrigerant be exhausted to the atmosphere?

5. Draw the terminal designation for the compressor showing common, run, and start.

6. Was this compressor mounted on external or internal springs?

7. How much refrigerant charge was used to recharge the unit?

Expansion Devices

Unit Overview

The expansion device is often called the metering device, and certain types may be referred to as expansion valves. This is a small component but one of the four necessary for the refrigeration cycle to function. It can be a valve or a fixed-bore type and is one of the dividing points between the high- and low-pressure sides of the system. The compressor is the other component separating the high- and low-pressure sides.

The expansion device is located in the liquid line between the condenser and the evaporator, usually at the inlet to the evaporator. Its function is to meter the liquid refrigerant into the evaporator. The three types of devices used on modern equipment are the thermostatic expansion valve, the automatic expansion valve, and the fixed-bore type, such as a capillary tube.

The thermostatic expansion valve meters the refrigerant to the evaporator using a thermal sensor to monitor the superheat at the suction line. Electronic expansion valves (EEVs) use a solid-state controller to instruct a stepper motor to control the modulation of the valve's port. Thermistors act as one of the inputs to the controller for evaporator superheat control. The stepper motor can act in any direction at 200 steps per second. The automatic expansion valve meters refrigerant to the evaporator by using a pressure-sensing device that maintains a constant pressure in the evaporator. The capillary tube is a fixed-bore metering device that has no moving parts. It controls refrigerant flow by pressure drop. This may be in the form of a small diameter tube, which in some applications can be quite long.

Key Terms

Algorithms
Automatic Expansion Valve
Balanced Port TXV
Capillary Tube
Cross Liquid Charge Bulb
Cross Vapor Charge Bulb
Dual Port TXV
Electronic Expansion Valve
Error

Expansion (Metering) Device
External Equalizer
Feedback Loops
Fixed-Bore Device
Liquid Charge Bulb
Offset
Pressure Limiting TXV
Proportional, Integral, and
 Derivative (PID) Controller

Refrigerant Receiver
Step Motor Electronic
 Expansion Valves
Thermostatic Expansion Valve
 (TXV)
Vapor Charge Bulb

REVIEW TEST

Name	Date	Grade

Circle the letter that indicates the correct answer.

1. The thermostatic expansion valve regulates refrigerant flow to the evaporator based on:
   A. heat of compression.
   B. suction-line superheat.
   C. external equalizer pressure.
   D. evaporator pressure.

2. The capillary tube is a type of:
   A. external equalizer.
   B. metering device.
   C. TXV sensor.
   D. device that senses superheat.

3. The TXV sensing bulb is mounted:
   A. on the liquid line between the condenser and the TXV.
   B. at the inlet to the condenser.
   C. on the suction line between the evaporator and compressor.
   D. on the compressor discharge line.

4. An external equalizer is used when:
   A. the pressure drop across the evaporator is excessive.
   B. the liquid-line pressure becomes too great.
   C. the receiver is too full of refrigerant.
   D. the pressure drop at the automatic expansion valve becomes too great.

5. When the load on the evaporator increases, the TXV:
   A. increases the flow of refrigerant.
   B. decreases the flow of refrigerant.
   C. maintains the flow of refrigerant.
   D. decreases the pressure across the evaporator.

6. The liquid charge bulb is:
   A. a sensing bulb on an automatic expansion valve.
   B. a bulb attached to one end of a capillary tube.
   C. a sensing bulb at the inlet of the expansion valve.
   D. a TXV-sensing bulb charged with a fluid similar to that of the system refrigerant.

7. Superheat is:
   A. the sensible heat absorbed by the refrigerant after it has boiled to a vapor.
   B. the heat of compression at the compressor.
   C. heat used to boil the liquid in the evaporator.
   D. latent heat given off by the condenser.

8. The sensing bulb and transmission tube are a part of the:
   A. automatic expansion valve.
   B. capillary tube.
   C. king valve.
   D. TXV.

9. A small amount of superheat in the suction line is desirable with a TXV to:
   A. keep the proper amount of refrigerant in the receiver.
   B. ensure the proper head pressure at the condenser.
   C. keep from starving the evaporator.
   D. ensure that no liquid refrigerant enters the compressor.

10. The thin metal disc connected to the needle in the TXV is called the:
   A. spring.   C. seat.
   B. diaphragm.   D. sensing bulb.

11. Liquid refrigerant should be completely boiled to a vapor:
   A. close to the inlet of the evaporator.
   B. approximately halfway through the evaporator.
   C. near the outlet of the evaporator.
   D. halfway through the suction line.

12. The cross liquid charge bulb at the TXV contains fluid that:
   A. has similar characteristics to the system fluid to prevent liquid refrigerant from entering the suction line.
   B. has different characteristics from the system fluid and helps prevent liquid refrigerant from flooding over to the compressor.
   C. causes the valve to stay open during the off cycle.
   D. follows the typical temperature/pressure relationship.

13. Balanced port TXVs are used:
   A. to feed refrigerant at a normal rate with low ambient temperatures.
   B. to feed refrigerant at a normal rate with high ambient temperatures.
   C. to stop refrigerant flow during low ambient temperatures.
   D. to stop refrigerant flow during high ambient temperatures.

14. The pressure-limiting TXV:
   A. allows the compressor to operate at low temperatures.
   B. shuts off the condenser fan during a hot pulldown.
   C. continues to feed refrigerant to the evaporator during a hot pulldown.
   D. shuts off the refrigerant flow to the evaporator when it has reached a predetermined pressure.

15. The automatic expansion valve was designed to:
   A. maintain a constant pressure in the evaporator.
   B. meter refrigerant to the evaporator with an electronic sensing device.
   C. reduce refrigerant flow when the evaporator pressure drops.
   D. vary the pressure in the evaporator under low ambient conditions.

16. **What force is cancelled out when using a balanced port TXV?**
    A. liquid force
    B. evaporator force
    C. bulb force
    D. none of the above

17. **What type of TXV uses a large port for pulldown capacities and a smaller port for holding loads?**
    A. balanced port
    B. liquid charged
    C. cross charged
    D. dual port

18. **Which type of valve uses a control board, transistors, thermistors, and a stepper motor controlled by an algorithm?**
    A. electronic expansion valve
    B. balanced port valve
    C. dual port valve
    D. none of the above

19. **The difference between the superheat set point of an electronic expansion valve and the actual superheat is called a(n):**
    A. error
    B. algorithm
    C. range
    D. differential

20. **The terms "offset" and "error" are one and the same.**
    A. true   B. false

Evaluating a Thermostatic Expansion Device

Name	Date	Grade

Objective: Upon completion of this exercise, you will be able to describe the characteristics of a thermostatic expansion valve and adjust it for more or less superheat.

Introduction: You will use a system with an adjustable thermostatic expansion valve and check the superheat when the valve is set correctly. You will then make adjustments to produce more and less superheat and watch the valve respond to adjustment.

Text and References: Paragraphs 24.1, 24.2, 24.3, 24.4, 24.5, 24.6, 24.7, and 24.8.

Tools and Materials: A gage manifold, thermometer, goggles, gloves, adjustable wrench, valve wrench, and a refrigeration system with a thermostatic expansion valve.

Safety Precautions: Wear goggles and gloves while connecting gages to the system.

PROCEDURES

1. Put on your goggles and gloves. Connect the gage manifold lines to the high- and low-sides of the system.

2. Fasten the thermometer sensor to the suction line just as it leaves the evaporator. See text, Figure 24-20. If there is a heat exchanger, make sure that the sensor is fastened between the evaporator and the heat exchanger.

3. Inspect the thermostatic expansion valve sensing bulb and be sure that it is fastened correctly to the line. See text, Figure 24-33. It should be insulated.

4. Turn the thermostat to a low setting where it will not satisfy and shut the system off.

5. Start the system and wait until it is operating at very close to design conditions (high, medium, or low temperature).

6. Record the following data when the system is operating in a stable manner:
   - Discharge pressure _____ psig
   - Condensing temperature _____ °F
   - Suction-line temperature _____ °F
   - Suction pressure _____ psig
   - Boiling temperature _____ °F
   - Superheat _____ °F

7. Turn the thermostatic expansion valve adjustment two full turns for an increase in superheat and wait for the system to stabilize. Record the following data:

- Discharge pressure _____ psig
- Condensing temperature _____ °F
- Suction-line temperature _____ °F
- Suction pressure _____ psig
- Boiling temperature _____ °F
- Superheat _____ °F

8. Turn the valve adjustment back to where it was and then two full turns toward a decrease in superheat and wait for the system to stabilize. Record the following:

- Discharge pressure _____ psig
- Condensing temperature _____ °F
- Suction line temperature _____ °F
- Suction pressure _____ psig
- Boiling temperature _____ °F
- Superheat _____ °F

9. Turn the valve adjustment back to the mid-position where it was in the beginning, and verify that the superheat is maintaining between 8°F and 12°F.

10. Remove the gages and thermometer lead.

Maintenance of Workstation and Tools: Return all tools and equipment to their respective places. Replace all panels and fasteners in the proper manner. Leave the system in the manner you found it. Be sure to turn the thermostat back to normal.

Summary Statements: Describe the change in superheat when adjusted. Indicate how much change per turn was experienced.

QUESTIONS

1. When a coil is starved of refrigerant, the superheat is _____ .

2. When a coil is flooded with refrigerant, the superheat is _____ .

3. The thermostatic expansion valve maintains a constant _____ .

4. The automatic expansion valve maintains a constant _____ .

5. The thermostatic expansion valve feeds (more or less) refrigerant with an increase in load.

6. The automatic expansion valve feeds (more or less) refrigerant with an increase in load.

Troubleshooting Exercise: Refrigerated Cooler with TXV

Name	Date	Grade

Objectives: Upon completion of this exercise, you will be able to evaluate a refrigeration system and make a repair on a no-cooling problem.

Introduction: This refrigeration unit will have a specified problem introduced to it by your instructor. You will locate the problem using a logical approach and instruments. THE INTENT IS NOT FOR YOU TO MAKE THE REPAIR BUT TO LOCATE THE PROBLEM AND ASK YOUR INSTRUCTOR FOR DIRECTION BEFORE COMPLETION.

Text References: Paragraphs 24.1, 24.2, 24.3, 24.4, 24.5, 24.6, 24.7, 24.8, 24.14, 24.15, and 24.16.

Tools and Materials: A VOM, ammeter, gage manifold, thermometer, goggles, gloves, straight-blade and Phillips screwdrivers, electrical tape, short piece of insulation, valve wrench, and $1/4$" and $5/16$" nut drivers.

Safety Precautions: Use care while touch testing the unit. Do not allow your hands close to the fan blades. DO NOT JUMP OUT ANY HIGH-PRESSURE CONTROL OR OVERLOAD. Wear goggles and gloves when connecting gage lines.

PROCEDURES

1. Put on your goggles and gloves. Connect a gage manifold and purge the gage lines.

2. Start the unit. If the unit will not run, use the VOM to determine why. It may be off because of temperature, low pressure, or high pressure. If the unit is off because of low pressure, you may jump the low-pressure control for a short moment only to see whether the unit will start. Gages should be watched at this time. Do not allow the low-side pressure to go into a vacuum.

3. With the unit running, touch test the suction line at the compressor. NOTE: The unit must have been running for at least 15 minutes. How does the suction line feel in relation to your hand temperature? Is it warmer or cooler?

4. Carefully touch the compressor discharge line. How does it feel? Is it hot or warm?

5. Touch test the suction line leaving the evaporator. Does it feel like the line entering the compressor?

6. Touch test the line leaving the TXV expansion valve. How does it feel?

7. What conclusions did you draw from the above touch test?

8. Is liquid refrigerant moving in the liquid-line sight glass?

9. Describe the problem and your recommended repair.

Maintenance of Workstation and Tools: Return all tools and equipment to their respective places.

Summary Statement: Describe how you decided whether you had an electrical or a mechanical problem.

QUESTIONS

1. Describe a starved evaporator.

2. What should the superheat be at the end of the evaporator in a typical system?

3. Why is superheat vital at the end of the evaporator?

4. What would be the symptoms of too much superheat?

5. What could the problem be when the liquid-line sight glass is full of liquid at the expansion valve, but only a small amount of liquid will feed through the valve?

6. How can a restriction occur in an expansion valve?

7. What is the usual symptom when the power element of a TXV loses its charge?

8. What would be the symptom if the TXV sensing element were not fastened correctly to the suction line?

9. Why should you allow the unit to run for 15 minutes before drawing conclusions?

10. What does it mean when the compressor discharge line is not hot?

Capillary Tube System Evaluation

Name	Date	Grade

Objectives: Upon completion of this exercise, you will be able to evaluate a system, using a capillary tube as a metering device.

Introduction: You will use a thermometer and gage manifold to check the evaporator and capillary tube performance under light-load and full-load conditions for the correct charge under these changing conditions.

Text References: Paragraphs 24.1, 24.29, and 24.30.

Tools and Materials: A gage manifold, thermometer, goggles, gloves, straight-blade and Phillips screwdrivers, electrical tape, short piece of insulation, valve wrench, $1/4$" and $5/16$" nut drivers, and a refrigeration system with service valves.

Safety Precautions: Wear goggles and gloves while working with refrigerant under pressure.

PROCEDURES

1. Put on your goggles and gloves. Connect the high- and low-side gage lines to the unit. The connections will probably be Schrader-type connectors for a capillary tube system.

2. Purge a small amount of refrigerant from each gage line to clean the gages of contaminants.

3. Fasten a thermometer lead to the suction line at the end of the evaporator but before any heat exchanger in the line.

4. Insulate the thermometer lead.

5. Start the system and let it run for 15 minutes and record the following information:

   - Suction pressure _____ psig
   - Suction-line temperature _____ °F
   - Refrigerant boiling temperature (converted from suction pressure) _____ °F
   - Superheat (subtract boiling temperature from line temperature) _____ °F
   - Discharge pressure _____ psig
   - Refrigerant condensing temperature (converted from discharge pressure) _____ °F

6. Block the condenser airflow until the condensing temperature is 125°F and the discharge pressure reads one of the following:

   - R-12   169 psig
   - R-22   278 psig
   - R-502   301 psig
   - R-134a   185 psig
   - R-404A   332 psig
   - R-410A   446 psig

7. Allow the system to run for at least 15 minutes before taking the next readings. You may have to turn the thermostat down to a lower setting to prevent the system from stopping. Take the following readings:

- Suction pressure _____ psig
- Refrigerant condensing temperature _____ *F
- Suction-line temperature _____ °F
- Discharge pressure _____ psig
- Refrigerant condensing temperature _____ °F
- Superheat (subtract boiling temperature from line temperature) _____ °F

8. Shut the unit off and remove the gages. Replace all panels with the correct fasteners.

Maintenance of Workstation and Tools: Return all tools to their respective places. Ensure that your work area is left clean.

Summary Statements: Describe what the superheat in the evaporator did when the head pressure was raised to the condensing temperature of 125°F. Explain why.

QUESTIONS

1. Why is a system using a capillary tube thought of as a dry-type or direct-expansion system?

2. Would an orifice-type expansion device perform much the same as a capillary tube metering device?

3. What causes refrigerant to flow through a capillary tube metering device?

4. What are the symptoms of a capillary tube metering device system with an overcharge of refrigerant?

5. What are the symptoms of a capillary tube metering device system with an undercharge of refrigerant?

6. Name two advantages of a capillary tube metering device.

Troubleshooting Exercise: Hermetic Compressor and Capillary Tube System

Name	Date	Grade

Objectives: Upon completion of this exercise, you will be able to evaluate a system for correct mechanical operation.

Introduction: This exercise should have a specified problem introduced to the unit by your instructor. You will locate the problem using a logical approach and instruments. The problem in this exercise may be discovered without the use of gages. Gages may then be used to verify the problem. THE INTENT IS NOT FOR YOU TO MAKE THE REPAIR BUT TO LOCATE THE PROBLEM AND ASK YOUR INSTRUCTOR FOR DIRECTION BEFORE COMPLETION.

Text References: Paragraphs 24.1, 24.29, and 24.30.

Tools and Materials: A VOM, ammeter, gage manifold, thermometer, goggles, gloves, straight-blade and Phillips screwdrivers, electrical tape, short piece of insulation, valve wrench, $1/4$" and $5/16$" nut drivers.

Safety Precautions: Wear goggles and gloves while transferring any refrigerant. DO NOT JUMP OUT THE HIGH-PRESSURE CONTROL.

PROCEDURES

1. Start the system and allow it to run for about 15 minutes to establish stable conditions.

2. Touch test the suction line at the compressor. HOLD THE LINE TIGHTLY, CLOSE TO THE PALM OF YOUR HAND. Describe how the line feels compared to your hand temperature.

3. Touch test the compressor housing. Is it the correct temperature at the correct places? Describe the touch test of the compressor.

4. Carefully touch test the compressor discharge line. How does it feel?

5. What conclusions have you drawn from the touch test?

6. Put on your goggles and gloves. Fasten gages to the system to verify the problem. SEE YOUR INSTRUCTOR ABOUT THE RECOMMENDED CORRECTIVE ACTION.

7. Describe your recommended repair, listing any parts or materials needed.

Maintenance of Workstation and Tools: Return the system to the condition your instructor recommends. Return all tools to their proper places.

Summary Statement: Describe what symptoms led you to the problem.

QUESTIONS

1. What causes a cool discharge line on a running compressor?

2. How should the crankcase of a running compressor feel compared to hand temperature?

3. What would the outcome be of prolonged liquid refrigerant slowly moving into the compressor?

4. How does an overcharge affect compressor efficiency?

5. Why does an overcharge of refrigerant affect compressor efficiency?

6. How does an overcharge of refrigerant affect suction pressure?

7. Would liquid refrigerant slowly moving through the compressor cause oil migration?

8. Does the discharge line of a compressor typically contain superheat?

9. Would the condenser have more or less subcooling if the unit had an overcharge of refrigerant?

10. Does a capillary tube show the same symptoms as a TXV system operating with an overcharge? Explain.

Troubleshooting Exercise: Hermetic Compressor and Capillary Tube System

Name	Date	Grade

Objectives: Upon completion of this exercise, you will be able to evaluate a system for correct mechanical operation.

Introduction: This exercise should have a specific problem introduced to the unit by your instructor. You will locate the problem using a logical approach and instruments. The problem in this exercise may be discovered without the use of gages. Gages may then be used to verify the problem. THE INTENT IS NOT FOR YOU TO MAKE THE REPAIR BUT TO LOCATE THE PROBLEM AND ASK YOUR INSTRUCTOR FOR DIRECTION BEFORE COMPLETION.

Text References: Paragraphs 24.1, 24.29, and 24.30.

Tools and Materials: A VOM, ammeter, gage manifold, thermometer, goggles, gloves, straight-blade and Phillips screwdrivers, electrical tape, short piece of insulation, valve wrench, and $1/4$" and $5/16$" nut drivers.

Safety Precautions: Wear goggles and gloves while transferring refrigerant.

PROCEDURES

1. Start the unit and allow it to run for about 15 minutes to establish stable conditions.

2. Touch test the suction line leaving the evaporator. Describe how it feels compared to your hand temperature.

3. Touch test the compressor housing. How does it feel compared to your hand temperature?

4. Place a thermometer lead on the compressor suction line 6" before the compressor, and one on the discharge line 6" from the compressor. Suction-line temperature _____ °F Discharge-line temperature _____ °F

5. Describe the problem with the evidence you have up to now.

6. Put on your goggles and gloves. Fasten gages to the system and record the pressures. Suction pressure _____ psig Discharge pressure _____ psig

7. Record what the pressures should be under normal conditions. Suction pressure _____ psig Discharge pressure _____ psig

8. What would you recommend as a repair for this system?

Maintenance of Workstation and Tools: Your instructor will direct you as to how to leave the unit. Return all tools to their proper places.

Summary Statement: Describe how the problem with this unit affected system performance.

QUESTIONS

1. Was the discharge line operating at too high a temperature on this system?

2. What would be the normal discharge-line temperature for this unit while operating normally? _____ °F

3. If the discharge line is overheating, what else is overheating?

4. What is the problem with overheating a discharge line?

5. What may cause a discharge line to overheat?

6. What is the maximum discharge-line temperature recommended by most manufacturers? _____ °F

7. What cools the compressor in the unit used for this exercise?

8. What should the suction pressure be for this system when operating correctly?

9. What determines the head pressure on an air-cooled system?

10. Will a unit evaporator freeze up due to a low charge?

Changing a Thermostatic Expansion Valve

Name	Date	Grade

Objectives: Upon completion of this exercise, you will be able to change a thermostatic expansion valve with a minimum loss of refrigerant.

Introduction: You will change the TXV on a working unit. Because there may not be a unit with a defective valve, you will go through the exact changeout procedure by removing the valve and then replacing it. The system should have service valves and a TXV that is installed with flare fittings.

Text References: Paragraphs 24.2, 24.3, 24.15, and 24.17.

Tools and Materials: A gage manifold, goggles, gloves, a refrigerant recovery system, a cylinder of the same type of refrigerant as is in the system, leak detector (soap or halide will do), service valve wrench, two 8" adjustable wrenches, assortment of flare nut wrenches, a vacuum pump, and flat-blade and Phillips screwdrivers.

Safety Precautions: Watch the high-pressure gage while pumping the system down to be sure the condenser will hold the complete charge. You may need to jump the low-pressure control out to pump the unit down. DO NOT JUMP OUT THE HIGH-PRESSURE CONTROL.

PROCEDURES

1. Put on your goggles and gloves. Fasten gages to the compressor service valves and purge the gage manifold.

2. Front seat the "king" valve at the receiver, or close the liquid-line service valve.

3. Jump the low-pressure control if there is one.

4. Start the unit and allow it to run until the low-side pressure drops to 0 psig. Turn the unit off and allow it to stand for a few minutes. If the pressure rises, pump it down again. You will normally need to pump the system down to 0 psig three times to remove all refrigerant from the low side. Remember, you are pumping the drier and the liquid line down also. You only need to evacuate the system to 0 psig because you are not removing a major component from the system.

5. Remove enough panels from the evaporator to comfortably reach the TXV valve.

6. Remove the TXV sensing element from the evaporator outlet.

7. Using the adjustable wrenches or the flare nut wrenches, loosen the inlet and outlet connections to the TXV. Then loosen the external equalizer line. Remove the valve from the system.

8. When the valve is completely clear of the evaporator compartment, you may then replace it as though the valve you removed were a new one. THIS SHOULD BE DONE QUICKLY SO THE SYSTEM IS NOT OPEN ANY LONGER THAN ABSOLUTELY NECESSARY, OR THE DRIER SHOULD BE CHANGED.

9. When the valve is replaced, pressurize the system and leak check. Pressure checking can be done with pure nitrogen or a mixture of nitrogen and R-22. The R-22 is used as a trace gas for electronic leak detection purposes. R-22 is the only refrigerant the EPA allows for use as a trace gas for leak detection purposes. The mixture of nitrogen and R-22 should never be recovered with a recovery unit; it can be blown to the atmosphere when the service technician finishes leak checking.

10. Connect the vacuum pump. Triple evacuate using the procedures in Unit 8.

11. After evacuation, open the "king" valve and start the system. The charge should be correct.

12. Return the unit to the condition your instructor advises.

Maintenance of Workstation and Tools: Return all tools to their correct places.

Summary Statement: Describe the triple evacuation process.

QUESTIONS

1. How did you change the TXV and get by without changing the filter-drier?

2. Why is triple evacuation better than one short vacuum?

3. Why should you NEVER jump out the high-pressure control?

4. What would the symptom of a TXV that had lost its charge in the sensing element be?

5. What is the advantage of a flare nut wrench over an adjustable wrench?

6. How was the sensing element fastened to the suction line on the unit you worked on?

7. Was the valve sensing bulb insulated?

8. What is the purpose of the insulation on the sensing bulb?

9. Is it necessary that all sensing bulbs be insulated?

10. Did this unit have a low-pressure control?

Special Refrigeration System Components

Unit Overview

There are many devices and components that can enhance the performance and reliability of the refrigeration system. Some of these protect components and some improve the reliability. Some of these devices and components are:

Two-temperature control
Evaporator pressure regulator (EPR)
Crankcase pressure regulator (CPR)
Spring-loaded relief valve
Oil safety controller
Defrost termination and fan delay control

Current sensing relay
Low ambient controls
Solenoid valves
Pressure switches
Low-ambient fan control
Defrost timer

Other accessories are: receiver, king valve on the receiver, filter-drier, sight glass, refrigerant distributor, heat exchanger, suction-line accumulator, service valves, oil separator, and pressure access ports.

Key Terms

Automatic Pumpdown
Ball Valve
Bi-Polar Step Motor
Condenser Flooding
Crankcase Pressure Regulator (CPR)
Current Sensing Relay
Defrost Termination and Fan Delay Control
Diaphragm Valve
Discharge Service Valve
Electronic Evaporator Pressure Regulator
Electronic Oil Safety Controller
Evaporator Pressure Regulator (EPR)

External Heat Defrost
Fan Cycling Head-Pressure Control
Fan Speed Control
Filter-Drier
Heat Exchanger
High-Pressure Control
Internal Heat Defrost
King Valve
Liquid Floodback
Liquid Refrigerant Distributor
Low Ambient Control
Low-Pressure Control
Migration
Multiple Evaporators
Net Oil Pressure

Oil Separator
One-Time Relief Valve
Planned Defrost
Pressure Access Ports
Random or Off-Cycle Defrost
Receiver
Refrigerant Sight Glass
Shutters or Dampers
Solenoid Valve
Spring-Loaded Relief Valve
Suction-Line Accumulator
Suction-Line Filter-Drier
Suction Service Valve
Two-Temperature Valve

REVIEW TEST

Name	Date	Grade

Circle the letter that indicates the correct answer.

1. **Two-temperature operation of a refrigeration system is necessary when:**
   A. the compressor must be protected from overloading.
   B. more than one evaporator, each at a different temperature, is operating from one compressor.
   C. the coil in the evaporator is excessively long.
   D. a multiple-circuit distributor is used.

2. **The EPR valve is installed in the suction line at the evaporator outlet and it:**
   A. modulates the suction gas to the compressor.
   B. will let the evaporator pressure go only as low as a predetermined point.
   C. will provide the same service as an automatic expansion valve when coupled with a TXV.
   D. all of the above.

3. **Normally there is a gage port known as a _____ gage port installed in the EPR valve body.**
   A. two-temperature
   B. Schrader
   C. low ambient
   D. one-time

4. **The crankcase pressure regulator (CPR) is installed in the suction line near the:**
   A. compressor.
   B. evaporator.
   C. EPR.
   D. multiple-circuit distributor.

5. **A hot pulldown is when:**
   A. there is excessive ice on the evaporator.
   B. the condenser has been flooded.
   C. the metering device is starving the evaporator.
   D. the temperature of the load is high and the compressor has to run for a long period of time to lower the temperature.

6. **A valve used to protect the compressor during a hot pulldown is the:**
   A. CPR valve.
   B. EPR.
   C. TXV.
   D. gage manifold.

7. **A one-time relief valve is constructed with:**
   A. a fusible plug.
   B. an adjustable spring.
   C. a Schrader port.
   D. a two-temperature mechanical valve.

8. **A low ambient control is used to _____ in cold weather.**
   A. maintain an acceptable head pressure
   B. lower the compressor crankcase temperature
   C. route refrigerant to the receiver
   D. reduce refrigerant flow to the evaporator

9. **_____ is a type of low-ambient temperature head pressure control:**
   A. CPR
   B. Fan cycling
   C. EPR
   D. Off-cycle defrost

10. **_____ is another type of low-ambient temperature head pressure control:**
   A. Air volume control using shutters and dampers
   B. Planned defrost
   C. Pumping down the system using the king valve
   D. Internal heat

11. **The solenoid valve is frequently used to control fluid flow. This valve:**
   A. has a slow reaction time.
   B. has a fast or snap action.
   C. is used to control fan speed.
   D. is a normally open valve.

12. **Condenser flooding may be used:**
   A. for compressor protection.
   B. to ensure sufficient oil in the compressor crankcase.
   C. to keep liquid refrigerant from the compressor.
   D. to control head pressure in low-ambient temperature conditions.

13. **One use for the low-pressure switch is:**
   A. for low charge protection.
   B. to keep liquid refrigerant from the compressor.
   C. to throttle refrigerant to the compressor.
   D. to keep the compressor from flooding.

14. **The high-pressure control:**
    A. shuts off the condenser fan when a high pressure exists.
    B. shuts down the compressor when the head pressure is excessive.
    C. ensures that enough refrigerant reaches the evaporator.
    D. controls the CPR at the compressor.

15. **Random defrost is accomplished:**
    A. with a timer.
    B. using a high-pressure control.
    C. using a low-pressure control.
    D. during the normal off cycle.

16. **Defrost using internal heat is accomplished with:**
    A. a sail switch.
    B. hot gas from the compressor.
    C. electric heating elements.
    D. mullion heaters.

17. **The receiver is located in the:**
    A. suction line.
    B. compressor discharge line.
    C. liquid line.
    D. evaporator outlets.

18. **The king valve is normally located:**
    A. between the receiver and the expansion valve.
    B. at the evaporator outlets.
    C. between the compressor and the condenser.
    D. between the condenser and the receiver.

19. **The suction line accumulator is located in the suction line and its primary purpose is to collect:**
    A. liquid refrigerant and oil.
    B. contaminants and moisture.
    C. excess vapor refrigerant.
    D. acid produced by the compressor.

20. **The oil separator is normally located between the:**
    A. condenser and the receiver.
    B. evaporator and the compressor.
    C. receiver and the expansion valve.
    D. compressor and the condenser.

21. **Electronic evaporator pressure regulating valves use a sensor in the refrigerated case's:**
    A. discharge air.
    B. return air.
    C. evaporator outlet.
    D. evaporator inlet.

22. **Electronic evaporator pressure regulating valves are powered by a(an):**
    A. magnetic valve.
    B. electric valve.
    C. pressure.
    D. bi-polar step motor.

23. **Low charge protection and control of refrigerated space temperature are the two major applications for a low-pressure switch in refrigeration.**
    A. true   B. false

24. **On an automatic pump-down system, the thermostat controlling the liquid-line solenoid is located:**
    A. on the evaporator.
    B. on the liquid line.
    C. in the outside ambient.
    D. in the refrigerated space.

25. **Oil pump discharge pressure minus crankcase pressure is:**
    A. gross oil pressure.
    B. differential oil pressure.
    C. net oil pressure.
    D. none of the above.

26. **A relay wired to the load side of a contactor or motor starter which acts like an inductive-type ammeter and is sensitive to current flow is:**
    A. the current sensing relay.
    B. the potential relay.
    C. the magnetic relay.
    D. none of the above.

27. **Electronic oil safety controllers use a _____ to sense net oil pressure.**
    A. thermistor
    B. transistor
    C. capillary tube
    D. pressure transducer

28. **The remote sensing bulb on a defrost termination and fan delay switch is located:**
    A. on the evaporator.
    B. on the fan housing.
    C. on the defrost heater.
    D. none of the above.

29. **The oil return line connecting the oil separator to the compressor's crankcase should be just above _____ temperature most of the time.**
    A. compressor crankcase
    B. discharge
    C. evaporating
    D. room

Pumping Refrigerant into the Receiver

Name	Date	Grade

Objectives: Upon completion of this exercise, you will be able to pump the refrigerant into the receiver so that the low-pressure side of the system may be serviced.

Introduction: You will work with a refrigeration or air-conditioning system with a receiver and pump the refrigerant into the receiver. The complete low-pressure side of the system may then be serviced as a separate system.

Text References: Paragraphs 25.19, 25.32, 25.33, and 25.43.

Tools and Materials: A gage manifold, goggles, gloves, adjustable wrench, refrigeration valve wrench, and a refrigeration system with a receiver and service valves.

Safety Precautions: Wear goggles and gloves while connecting gages and turning valve stems. DO NOT ALLOW THE COMPRESSOR TO OPERATE IN A VACUUM.

PROCEDURES

1. Put on your goggles and gloves. Connect and purge the gage manifold to the high and low sides of the system at the compressor service valves.

2. Slowly turn the valve stems away from the back seat until you have gage readings.

3. Start the system and observe the gage readings.

4. When the system has stabilized, slowly front seat the king valve. Watch the low-pressure gage and do not let the low-side pressure go below the atmosphere's into a vacuum.

5. If the system has a low-pressure control, the control will shut the compressor off before the pressure reaches 0 psig. You may use a short jumper wire to keep the compressor on. See the text, Figure 10-9.

6. When the system has been pumped down to 0 psig and the compressor is shut off, the pressure in the low-pressure side of the system will increase above 0. The compressor will have to be started again to pump it back to 0 psig. This is residual boiling. You may have to pump the low side down as many as three times to completely remove all refrigerant from the low-pressure side.

7. The system should now hold the 0 psig. This isolates the low side of the system for service and preserves the refrigerant in the receiver and condenser. If the pressure keeps increasing and will not stay at 0 psig, refrigerant is leaking into the low side either from the compressor or through the king valve.

8. Slowly open the king valve and allow the refrigerant to equalize back into the low side.

9. Start the system and verify that it is operating correctly.

10. Remove the gage lines from the compressor and replace all valve stem covers and gage port caps.

Maintenance of Workstation and Tools: Return all tools to their respective places and clean up the workstation. Leave the system as you found it or as your instructor directs you.

Summary Statement: Describe what components may be serviced when the low side of the system is at 0 psig. Hint: Look at the EPA rules.

QUESTIONS

1. What other substance circulates in the refrigeration system with the refrigerant?

2. What would happen if there were more refrigerant in the system than the condenser and receiver would hold?

3. If air enters the low side of the refrigeration system during servicing, how could it be removed?

4. What causes the pressure to rise after the system is pumped down, causing you to have to pump it down again?

5. Could you pump the refrigerant into the receiver if the compressor valves leaked?

6. If an expansion valve is changed using the pumpdown method, is it always necessary to evacuate the system before startup?

7. What is the name of the valve on the receiver that is used to stop refrigerant flow for pump-down purposes?

8. Can the gages be removed while the refrigerant is pumped into the condenser and receiver?

9. Is the crankcase of the compressor considered on the high- or the low-pressure side of the system?

Adjusting the Crankcase Pressure Regulator (CPR) Valve

Name	Date	Grade

Objectives: Upon completion of this exercise, you will be able to adjust a CPR valve in a typical refrigeration system.

Introduction: You will use gages and an ammeter to properly adjust a CPR valve to prevent compressor overloading on the startup of a hot refrigerated box (sometimes called a hot pulldown). This will more than likely be a low-temperature system.

Text References: Paragraphs 25.7 and 25.8.

Tools and Materials: Two gage manifolds, goggles, gloves, clamp-on ammeter, valve wrench, large straight-blade screwdriver, Allen head wrench set, Phillips and medium-size straight-blade screwdrivers, and a refrigeration system with a CPR (crankcase pressure regulator).

Safety Precautions: Wear goggles and gloves while working with refrigerant under pressure.

PROCEDURES

1. Put on your goggles and gloves. Connect and high- and low-side gage lines to the compressor and obtain a gage reading, either from Schrader connections or service valves.

2. Connect the other low-side gage line to the connection on the CPR valve. **NOTE:** This connection is on the evaporator side of the valve.

3. Purge a small amount of refrigerant from each gage line to clean any contaminants from the line.

4. Make sure the refrigerated box is under a load from being off, or shut it off long enough for the load to build up. The box temperature must be at least 10°F above its design temperature.

5. Clamp the ammeter on the common wire to the compressor. Make sure that no other current, such as the condenser fan, is flowing through the wire.

6. Record the compressor full-load amperage (from the nameplate) here: _____ A

7. Start the system and record the following information:
   - Suction pressure at the evaporator (from the gage attached to the CPR gage connection) _____ psig
   - Suction pressure at the compressor _____ psig
   - Difference in suction pressure _____ psig
   - Actual amperage of the compressor _____ A

8. Remove the protective cover from the top of the CPR valve.

9. If the compressor amperage is too high, slowly adjust the valve until the compressor amperage compares to the amperage on the nameplate. The valve is now set correctly.

10. If the valve is set correctly, turn the adjusting stem until the amperage is too high and then adjust it to the correct reading, as in step 9.

11. Remove the gages and replace all panels with their correct fasteners.

Maintenance of Workstation and Tools: Return all tools to their respective places. Leave the system in the condition you found it.

Summary Statement: Describe the action of the CPR valve and why this type of valve is necessary for a low-temperature system when it is started up after being off for some time.

QUESTIONS

1. The CPR valve acts as a restriction in the suction line until the evaporator is operating within its correct temperature range. (true or false)

2. How did service people start up low-temperature systems during a hot pulldown before the CPR valve?

3. Is a CPR valve normally found on a high-temperature system?

4. Does the oil circulating through the system pass through the CPR valve?

5. Is a CPR valve an electromechanical control?

6. Which of the following is used to correctly adjust a CPR valve, amperage or pressure?

7. What would be the symptoms if the CPR valve were to close and not open?

Application of Refrigeration Systems

Unit Overview

Retail stores use reach-in refrigeration units for merchandising their products. These display cases are available in high-, medium-, and low-temperature ranges, open or closed display, chest or upright type. These can be self-contained, with the condensing unit built inside the fixture, or they can have a remote condensing unit. Self-contained units reject the heat back into the store. They can be moved around to new locations without much difficulty. Equipment with remote condensing units may have a compressor for each unit or a compressor that serves several units. This procedure of using one compressor on several fixtures has the advantage of utilizing larger, more efficient compressors. It also more easily utilizes the heat from the equipment for heating the store or the hot water.

A cylinder unloading design may be used on the compressor by keeping certain cylinders from pumping when the load varies. Another method is to use more than one compressor manifolded with a common suction, discharge line, and receiver. The compressors can be cycled on and off as needed for capacity control.

If multiple evaporators are used, EPR valves may be located in each of the higher temperature evaporators. When the load varies, the compressors can be cycled on and off to maintain the lowest suction pressure.

Other types of refrigeration applications are specialized vending machines, water coolers, and refrigerated air driers.

Key Terms

Heat Reclaim
Individual Condensing Units
Line Set
Mullion Heaters

Refrigerated Air Driers
Remote Condensing Units
Self-Contained Reach-In Fixtures
Water Coolers

REVIEW TEST

Name	Date	Grade

Circle the letter that indicates the correct answer.

1. Remote condensing units for reach-in refrigeration units may consist of:
   A. the evaporator and condenser.
   B. the evaporator and metering device.
   C. the compressor and evaporator.
   D. a condenser and one or more compressors manifolded together.

2. An advantage in using one compressor for several refrigeration units is:
   A. that the compressor and evaporator are located near each other.
   B. the compressor is larger and more efficient.
   C. that one expansion valve can be used.
   D. that compressor cycling is easier.

3. If four compressors were manifolded together (parallel compressor system) to serve 12 refrigeration cases, they would have the responsibility of providing the suction pressure necessary for:
   A. the coldest temperature evaporator.
   B. the medium-temperature evaporator.
   C. the warmest temperature evaporator.
   D. a single thermostatic expansion valve.

4. When multiple evaporators are used with different temperature requirements, an evaporator pressure regulator must be used on:
   A. all evaporators.
   B. only the lowest temperature evaporator.
   C. all but the lowest temperature evaporators.
   D. only the highest temperature evaporators.

5. When hot gas is used for low-temperature refrigeration, defrost with remote condensing units, the hot gas is piped:
   A. to the evaporator from the receiver.
   B. from the compressor discharge line.
   C. from the evaporator to the receiver.
   D. to the EPR from the receiver.

6. The heat used in a heat reclaim system comes from the:
   A. compressor discharge line.
   B. condensing unit.
   C. liquid line.
   D. compressor motor housing.

7. Refrigerated shelves often use:
   A. plate-type evaporators.
   B. cooling from the liquid line.
   C. refrigerant piped from the CPR.
   D. remote EPR valves.

8. The sweat from surfaces around doors in display cases is normally:
   A. evaporated with hot discharge gas.
   B. removed by exposure to the condenser fan.
   C. kept from forming with mullion heaters.
   D. kept from forming with air currents from the evaporator fan.

9. Precharged tubing with quick-connect fittings is called:
   A. a ready charged set.   C. type L tubing.
   B. ABS tubing.   D. a line set.

10. The advantage of one large compressor for several refrigerated cases is:
    A. larger motors are more efficient.
    B. larger motors last longer.
    C. larger motors are more quiet.
    D. larger motors are less expensive to buy.

11. The advantage of multiple compressors or parallel compressors operating several refrigerated cases is:
    A. they are more efficient.
    B. if one compressor is defective, the others can carry the load until repairs are made.
    C. they are larger.
    D. all of the above.

12. The panels on a walk-in cooler are held together by:
    A. nails.
    B. sheet metal screws.
    C. masonry fasteners.
    D. cam-action locks.

Application and Evaluation of a Refrigeration System

Name	Date	Grade

Objectives: Upon completion of this exercise, you will be able to determine the application and the approximate temperature range of a refrigeration system by its features.

Introduction: You will be assigned a refrigerated box. You will record information about the box, from which you will be able to determine its application. You will look for evaporator location, fin spacing, type of defrost, and style of box.

Text References: Paragraphs 26.1, 26.2, 26.3, 26.4, 26.11, 26.12, and 26.14.

Tools and Materials: Phillips and straight-blade screwdrivers, $1/4$" and $5/16$" nut drivers, flashlight, adjustable wrench, and a refrigeration box.

Safety Precautions: Care must be taken to avoid injury to your hands when disassembling any piece of equipment.

PROCEDURES

1. Make sure the power on your refrigeration box is off and the panel is locked and tagged. Fill in the following information:
   - Type of box (open, chest, or upright) _____
   - If closed, are doors single, double, triple pan, or metal? _____
   - Package or split system? _____

2. Fill in the following information from the unit nameplate:
   - Manufacturer's name _____
   - Model number _____
   - Serial number _____
   - Type of refrigerant _____
   - Quantity of refrigerant (if known) _____

3. From the compressor nameplate:
   - Manufacturer _____
   - Model number _____
   - Full-load amperes (if available) _____
   - Locked rotor amperes _____

4. From the evaporator examination:
   - Evaporator location _____
   - Fin spacing _____
   - Forced draft (yes or no) _____
   - Number of fans _____

5. From the condenser examination:

- Condenser location   _____
- Forced draft (yes or no)   _____
- Number of fans   _____
- Type of head-pressure control (if any)   _____

6. Control data:

- Type of temperature control   _____
- Type of defrost (off-cycle, other)   _____
- Does unit have a defrost timer?   _____
- Does timer have an "X" terminal?   _____

Maintenance of Workstation and Tools: Replace all panels with the correct fasteners.

Summary Statement: From the above information and the information in the text, what is the application of this cooler and why?

QUESTIONS

1. How is ice melted with off-cycle defrost?

2. State some advantages of electric defrost.

3. Why is electric defrost not commonly used in medium-temperature refrigeration?

4. State an advantage of hot gas defrost.

5. What is the "X" terminal on the timer used for in some defrost systems?

6. What would the symptoms be for a medium-temperature cooler that did not get enough off-cycle for correct defrost?

7. What keeps frost and sweat from forming on the doors and panels of refrigerated boxes?

8. Where does the melted ice from defrost go in a package refrigerated box?

Commercial Ice Machines

Unit Overview

Ice-making equipment operates between low- and medium-temperature refrigeration. Ice is made with evaporator temperatures about 10°F, with the ice at or slightly below 32°F. Ice making is accomplished by accumulating the ice on some type of evaporator surface.

Ice flake machines are usually critically charged refrigeration systems that come either in a split or self-contained unit. Ice flake machines use an auger turning inside a flooded cylinder to scrape ice off the sides of the cylinder. The auger has cutting surfaces called flights that do the ice cutting. When water is frozen in the evaporator or freezing cylinder, minerals from the water are left behind and often become attached to the interior cylinder walls. An ice-machine cleaner will dissolve many minerals from the freezing section of an ice flake machine. A water flush cycle is a way to control unwanted mineral buildup in the water reservoir. A mechanical timer or a microprocessor can initiate these cycles.

Ice flake machines usually have a gear reduction drive that reduces the auger's drive motor speed. Mineral buildup on a freezing cylinder of an ice flake machine will cause the auger motor, gears, and auger bearings to be stressed. Worn bearings on the auger will cause the auger to wobble and rub against the freezing cylinder walls. This often causes scrape marks on the cylinder walls and metal shavings in the ice.

Ice bin controls are designed to terminate the operation of the ice machine when the bin is full of ice. Ice production depends on the ice machine's water inlet temperature and the surrounding temperature around the ice machine. As the water temperature or air surrounding the ice machine rises, ice production drops.

Making crescent-shaped ice involves running water over an evaporator plate in which copper evaporator tubing is soldered onto the plates at regular intervals. The coldest points on the evaporator are where the copper coil contacts the evaporator plate. Ice starts to form here first. During a hot gas harvest, the cubes slide down the evaporator plate by gravity. Ice then falls into a bin for storage.

Making cell-type ice cubes involves having a vertical evaporator plate with water flowing over the individual cells by means of a water pump. A serpentine evaporator coil is soldered to the back of the plate. Ice will gradually form in the cells and soon bridge with the neighboring cell. A hot gas harvest is then initiated, and a hot gas solenoid is energized in the discharge line. The compressor keeps running and pumps superheated hot gas through the evaporator. A hot gas defrost loosens the cubes from the evaporator plate. Either by gravity or a pushrod, the ice comes off of the evaporator plate and falls into a storage bin.

Cylindrical ice is made inside a tube-within-a-tube evaporator. As water flows through the center tube, it freezes on the outside walls of the tubes. Defrost lets the ice shoot out the end of the tube as a long cylinder.

Many different control schemes and controls from different ice-machine manufacturers are used for water sump levels, ice thickness, harvest initiation, defrost termination, and ice storage bin controls.

Microprocessors are often the main controller on modern ice machines. They have stored programs or algorithms within them. This is actually a stored sequence of events in their memories. Microprocessors have input and output terminals where they receive and transmit information, respectively. Inputs and outputs can be either digital or analog devices such as switches or thermistors, respectively. Knowing the microprocessor's sequence of events and how to input/output troubleshoot them is what a service technician should know.

Ice quality depends on the quality of the water entering the ice machine. Good water quality will produce a crystal-clear, hard cube with good cooling capacity. Poor water quality can produce soft, cloudy ice with less cooling capacity. Ice quality is measured in percent of hardness. Hardness is a measurement that represents the thermal cooling capacity of the ice. The higher the percent hardness, the more cooling capacity the ice possesses. The cleaner the ice machine is kept and the purer the water is entering the machine, the harder the ice will be.

Preventive maintenance is the action of regularly scheduled maintenance performed on an ice machine. It includes inspecting, cleaning, sanitizing, and servicing the ice machine and external water filtration/treatment system. Cleaning an ice machine with an acid-based cleaner simply removes scale from the mineral content in the water.

Sanitizing an ice machine with an approved commercial ice-machine sanitizing fluid kills viruses, bacteria, and protozoa.

Key Terms

Alkaline
Analog
Auger
Bacteria
Bin Switch
Cam Switch
Cell-Type Ice
Chloramines
Crescent-Shaped Ice
Cylindrical-Shaped Ice
Digital
Electronic Erasable
   Programmable Read-Only
   Memory (EEPROM)
Error Code

Flights
Flush Cycle
Freezing Cylinder
Hardness
Ice Flake Machine
Ice Quality
Infrared Electric Eye
Inputs
Internet
Light Emitting Diode (LED)
Liquid Floodback
Microprocessor
Outputs
pH
Preventive Maintenance

Protozoa
Purge Valve
Reverse Osmosis
Sanitizing
Scale Inhibitor
Self-Diagnostic
Sequence of Events
Siphon
Sonar
Total Dissolved Solids
Turbidity
Viruses
Weep Holes
Wet Compression

REVIEW TEST

Name	Date	Grade

Circle the letter that indicates the correct answer.

1. A type of material often used on evaporators because it is durable, very sanitary, and resists corrosion is:
   A. brass.
   B. stainless steel.
   C. copper.
   D. plastic.

2. Cutting surfaces on an auger for a flake ice machine are called:
   A. flights.
   B. edges.
   C. knives.
   D. blades.

3. An evaporator dirty from mineral deposits on a flake ice machine will cause:
   A. high suction pressure.
   B. low suction pressure.
   C. high head pressure.
   D. quiet operations.

4. The _____ of an ice flake machine has the highest failure rate.
   A. bin control
   B. dump solenoid
   C. microprocessor
   D. gear motor assembly

5. A periodic way of ridding unwanted mineral buildup in the water of an ice machine is:
   A. a flush cycle.
   B. bin cleaning.
   C. a chemical treatment.
   D. none of the above.

6. _____ is made using a gear motor and auger.
   A. Cube ice
   B. Cylinder ice
   C. Flake ice
   D. Block ice

7. Methods used for proper ice bin control are:
   A. sonar.
   B. thermostats.
   C. infrared eye.
   D. all of the above.

8. Ice production for a flake ice machine depends on:
   A. inlet water temperature.
   B. surrounding air temperature.
   C. inlet water purity.
   D. both A and B.

9. Which type of ice is made inside a tube within a tube evaporator?
   A. cell-type ice
   B. sheet ice
   C. cylindrical ice
   D. flake ice

10. Pre-chilling the evaporator in an ice machine before the water pump comes on helps prevent _____ in the water sump.
   A. air bubbles
   B. mineral deposits
   C. slush
   D. all of the above

11. Most ice cube machines are defrosted by:
   A. electric defrost.
   B. natural defrost.
   C. ambient defrost.
   D. hot gas defrost.

12. Liquid refrigerant returning to the compressor while the compressor is operating is called:
    A. liquid slugging.
    B. migration.
    C. liquid floodback.
    D. none of the above.

13. On some ice machines, air is introduced by way of _____ between the ice slab and the evaporator plate to release the suction force between the evaporator and the ice slab.
    A. weep holes
    B. air pump
    C. gravity
    D. water stream

14. As the low-side pressure drops on an ice machine, a reverse-acting low-pressure control will _____ its contacts at a set low pressure.
    A. open
    B. close

15. A small computer that has a sequence of events or an algorithm stored in its memory for controlling the functions of many ice machines is called a:
    A. diode.
    B. transducer.
    C. transistor.
    D. microprocessor.

16. The microprocessor will associate an ice maker malfunction with a(n) _____ and display it on a light emitting diode (LED) display.
    A. error code
    B. mistake code
    C. malfunction code
    D. none of the above

17. Ice production, cycle times, and system operating pressures for an air-cooled ice machine depend on what factors?
    A. air temperature entering the condenser
    B. water temperature entering the ice machine
    C. relative humidity of the ambient air
    D. both A and B

18. A buildup of minerals that forms a flaky coating on the surface of the evaporator and water system on many ice makers is:
    A. slime.
    B. sediment.
    C. scale.
    D. none of the above.

19. The thermal cooling capacity of ice is referred to as:
    A. hardness.
    B. compactness.
    C. coolness.
    D. density.

20. Sanitizing an ice machine gets rid of:
    A. bacteria.
    B. viruses.
    C. protozoa.
    D. all of the above.

21. Cleaning an ice machine with an approved ice-machine cleaner gets rid of:
    A. viruses.
    B. bacteria.
    C. scale.
    D. sediment.

22. Water conditions can be broken down into:
    A. suspended solids.
    B. dissolved minerals and metals.
    C. chemicals.
    D. all of the above.

23. Chemicals in water are removed by:
    A. sanitizing.
    B. cleaning.
    C. carbon filtration.
    D. reverse osmosis.

24. A dirty condenser on an ice machine will cause:
    A. high head pressure.
    B. low head pressure.
    C. low suction pressure.
    D. none of the above.

25. A restricted metering device on an ice machine will cause:
    A. high head pressure.
    B. high suction pressure.
    C. low suction pressure.
    D. none of the above.

Ice-Maker Data and Familiarization

Name	Date	Grade

Objectives: Upon completion of this exercise, you will become familiar with ice-machine controls, settings, and ice production.

Introduction: By filling out the ice maker information sheet, you will record important data.

Text References: Unit 27.

Tools and Materials: Pencil and eraser, safety glasses, compound gage set, two thermometers, flashlight, straight and Phillips screwdriver, adjustable wrench, $1/4$" and $5/6$" nut drivers, valve wrench, scale, temperature/pressure chart, and an ice maker of any type with instruction and specification manual.

Safety Precautions: Wear safety glasses and be careful of any electrical shock hazard especially with wet surfaces near the ice machine. Make sure the ice machine is properly grounded.

PROCEDURES

Complete the information sheet for ice makers below.

INFORMATION SHEET FOR ICE MAKERS

- Unit manufacturer _____
- Date of manufacturer _____
- Model number _____
- Serial number _____
- Refrigerant type _____
- Type of system _____
- Make of compressor _____
- Compressor model number _____
- Metering device control type _____
- Low-pressure control settings   Range ____ Differential _____ Cut-in/out ___/___
- Condenser Type   Water cooled ___   Air cooled ___
- Proper level of water in sump: _____ in.
- City water temperature _____°F
- Pound of ice per day (measured) _____
- Type of harvest _____
- Harvest time __:__
- Freeze time __:__

- Ice condition   Clear ____   Cloudy ____
- Recommended cleaning interval _____
- Evaporator superheat _____
- Ice bin capacity ____ lb
- Rated ice production per day _____

Maintenance of Workstation and Tools: Make sure all panels are placed back on the ice machine in their proper places. Clean up any water spills on the ice machine and floor area. Return all tools to their respective places.

Summary Statement: Describe why it is important to know the data you recorded in the INFORMATION SHEET FOR ICE MAKERS.

QUESTIONS

1. Explain why the freeze time of an ice machine would increase.

2. What would happen if the ice dropped off the evaporator and the ice machine became stuck in a hot gas defrost?

3. Why is most nameplate data on ice machines important to the service technician?

4. How will the entering water temperature affect the ice capacity of an ice machine?

5. What may cause cloudy ice?

Ice-Making Methods

Name	Date	Grade

Objectives: Upon completion of this exercise, you will be able to follow the ice-making cycle of a particular ice-making machine.

Introduction: You will use gages and a thermometer to follow the process by which ice is made. You will record the temperature at which ice is being made.

Text References: Paragraphs 27.2 and 27.3.

Tools and Materials: A gage manifold, goggles, gloves, straight-blade and Phillips screwdrivers, adjustable wrench, valve wrench, thermometer, and package ice maker of any type.

Safety Precautions: Turn off and lock out the power while installing gages. Wear goggles and gloves.

PROCEDURES

1. Turn off the power to the ice maker. Lock and tag the panel and keep the single key in your possession. Put on your goggles and gloves. Fasten the high- and low-side gage lines to the service valves or Schrader ports. Some ice makers do not have a high-pressure port. If yours does not, just ignore all references to it. Make sure that the pressure is indicated on your gages.

2. Remove enough panels from the unit so that you may fasten a thermometer lead to the suction line leaving the evaporator but before a heat exchanger, if there is one.

3. Place a thermometer lead in the ice storage bin.

4. Replace any panels that may be required for normal operation. Run the thermometer lead under the panel.

5. Start the ice maker and record the following information:
   - Type of ice made (flake or cube type) _____
   - Type of refrigerant used _____
   - Refrigerant charge required _____

6. When ice is being made at a steady rate, record the following information: (Cube makers make ice in batches about 10 to 20 minutes apart. Wait for at least three batches to be made before a steady rate is established.)
   - Suction pressure at harvest (for cube makers only) _____ psig
   - Suction pressure while making ice _____ psig
   - Suction-line temperature, making ice _____ °F
   - Superheat, making ice _____ °F
   - Discharge pressure _____ psig
   - Temperature in ice storage bin _____ °F

7. Remove gages and thermometer leads and replace panels with correct fasteners.

Maintenance of Workstation and Tools: Return all tools to their respective places and make sure that your work area is left orderly and clean.

Summary Statement: Describe the process by which ice is made in the equipment you worked with.

QUESTIONS

1. How is flake ice made?

2. How is cube ice made?

3. What is the difference between an ice-holding bin and an ice maker?

4. What is the purpose of a flush cycle in an ice machine?

5. What is the approximate evaporator temperature when ice harvests in a cube ice maker?

6. How is ice cut into cubes when it is made in sheets?

7. Explain why mineral deposits on an evaporator of an ice machine will cause low suction pressure.

8. How is ice kept refrigerated in the holding bin of a typical ice maker?

9. Explain the difference between cleaning and sanitizing an ice machine.

10. Why does an ice maker have a drain line?

Cleaning and Sanitizing an Ice Machine

Name	Date	Grade

Objectives: Upon completion of this exercise, you will become familiar with cleaning and sanitizing an ice machine.

Introduction: The student will choose an ice machine. With the help of the instructor, directions on the cleaning solution bottle, and the service manual, the student will clean and sanitize the machine.

Text Reference: Paragraph 27.6.

Tools and Materials: Pencil and eraser, goggles, flashlight, straight-blade and Phillips screwdrivers, adjustable wrench, $1/4$" and $5/6$" nut drivers, rubber gloves, soft scrubbing bottle brushes, clean dry rags, an ice maker of any type, ice-machine cleaner and sanitizer, and an instruction and specification manual.

Safety Precautions: Wear goggles and rubber gloves and be careful of any electrical shock hazard, especially with wet surfaces near the ice machine. Make sure the ice machine is properly grounded. Make sure the area is properly ventilated when using cleaning and sanitizing solutions. Always read the instructions on the cleaning and sanitizing solution bottles. Always refer to the service and maintenance manual for any specific cleaning and sanitizing instructions.

PROCEDURES

NOTE: Many ice-machine manufacturers recommend cleaning and sanitizing every 6 months to prevent harvest problems and unsafe consumption of the ice. However, conditions may exist where the ice machine will need cleaning and sanitizing more or less often. Ice machines located in bar environments may be exposed to a number of airborne contaminants and may have to be cleaned more frequently. However, never clean and sanitize an ice machine until it needs to be cleaned and/or sanitized.

Cleaning

1. Remove the ice machine's front panel. Make sure all ice is off the evaporator. You may have to initiate a harvest cycle or wait for the ice-making cycle to end. Turn the machine to the OFF position.

2. Remove any ice that may be in the ice machine's bin.

3. Following the manufacturer's instructions, add the proper amount of ice-machine cleaner to the water trough.

4. Start a WASH cycle at the three-position ICE/OFF/WASH switch by moving the switch to the WASH position. This should start the water pump but leave the compressor off. The cleaner must circulate for at least 15 to 20 minutes in order for it to be effective on any mineral deposits.

5. After 15 to 20 minutes, the diluted cleaning solution must be washed down the ice maker's drain and replaced by fresh incoming water. If the ice maker has a PURGE switch, press and hold this switch until the cleaner has been flushed down the drain and has been replaced by fresh incoming water.

6. End the WASH cycle by switching the three-position ICE/OFF/WASH switch to the OFF position. Remove the splash curtain and inspect the evaporator and associated spillways for any mineral deposits. Many times, mineral deposits will not show up until the evaporator and associated part have been air dried. A blow drier and/or a soft absorbent cloth will help the drying procedure.

7. Sometimes, the water distribution tubes, spillway, and header will have to be removed, disassembled, and cleaned with a bottle brush. They can be soaked in a solution of ice-machine cleaner and water mixed in a 3- to 5-gallon bucket. Always follow the manufacturer's directions or labeled instructions on the ice-machine cleaner container for the proper concentrations of cleaner.

8. The water trough and float assembly area may need special cleaning, especially if scale or slime buildup is visible. Disassembling, soaking, and scrubbing with a soft bottle brush is recommended.

9. Reassemble the ice maker.

Sanitizing

1. Mix the proper amount and concentration of approved sanitizing solution according to the manufacturer's recommendations or labeled instructions on the sanitizing solution's container.

2. Add enough sanitizing solution to the water trough for it to overflow the trough.

3. Place the three-position ON/OFF/WASH switch to the WASH position and circulate the sanitizing solution for 10 to 15 minutes or whatever time interval the manufacturer's instructions recommend.

4. During the sanitizing time period in step 3, use the remaining sanitizing solution to wipe down the ice machine's bin interior surface, ice deflector, door, and splash zones. Once finished, rinse all these areas with fresh water.

5. Inspect that all parts and assemblies are in their proper place and are leak free.

6. After the 10- to 15-minute sanitizing period is over, flush the sanitizing solution down the drain. If the ice maker has a PURGE switch, press and hold this switch until the sanitizer solution has been flushed down the drain and has been replaced by fresh incoming water. Flush the ice machine for another 3 to 4 minutes to assure all sanitizing solution has been removed from the ice maker.

7. Place the ICE/OFF/WASH switch to the ICE position and replace the front cover.

8. Discard the first four ice harvests.

Maintenance of Workstation and Tools: Return all tools to their proper places and make sure that your work area is left orderly and clean.

Summary Statement: Describe the difference between cleaning and sanitizing an ice machine.

QUESTIONS

1. Define "scale" as it applies to ice machines.

2. Define "preventive maintenance" as it applies to ice machines.

3. What is meant by the term "ice hardness?"

4. Why is it important to clean an ice machine?

5. What does sanitizing do to an ice machine that cleaning does not?

6. How can cleaning an ice machine help keep the harvest cycles operating smoothly and trouble free?

Special Refrigeration Applications

Unit Overview

Transport refrigeration is the process of refrigerating products while they are being transported from one place to another by truck, rail, air, or water.

Truck refrigeration systems include the use of ice for short periods of time, dry ice (solidified and compressed carbon dioxide), liquid nitrogen, liquid carbon dioxide (CO_2), eutectic solutions, and compression-type refrigeration systems.

Products shipped by rail are refrigerated with a self-contained unit located at one end of the refrigerated car. These are generally diesel-driven motor generator units similar to those units used on trucks.

Extra-low-temperature refrigeration, below $-10°F$, is generally used to fast freeze foods on a commercial basis.

These temperatures may be as low as $-50°F$. Often two stages of compression are used to achieve temperatures this low. For even lower temperatures, cascade refrigeration may be used, which might include two or three stages of refrigeration. When food is frozen, the faster the temperature of the product is lowered, the better the quality of the food.

Ships that must have refrigeration systems for their cargo may have large on-board refrigeration systems or may use self-contained units that are plugged into the ship's electrical system.

For air cargo hauling, ice or dry ice is generally used with specially designed containers. On-board refrigeration systems would be too heavy.

Key Terms

Ammonia
Cascade Refrigeration
Carbon Dioxide (CO_2)
Dry Ice
Eutectic
Extra-Low-Temperature

Refrigeration
Liquid Nitrogen
Marine Refrigeration
Marine Water Boxes
Nose-Mount
Piggyback Cars

Quick Freeze
Railway Refrigeration
Sublimation
Transport Refrigeration
Truck Refrigeration
Underbelly Mount

REVIEW TEST

Name	Date	Grade

Circle the letter that indicates the correct answer.

1. **Dry ice is composed of:**
   A. liquid nitrogen.
   B. liquid carbon dioxide.
   C. solidified carbon dioxide.
   D. solidified ammonia.

2. **Sublimation is to change from a:**
   A. liquid to a vapor.
   B. solid to a vapor.
   C. vapor to a solid.
   D. solid to a liquid.

3. **Air in the atmosphere is approximately:**
   A. 78% nitrogen.
   B. 21% nitrogen.
   C. 78% oxygen.
   D. 78% carbon dioxide.

4. **An eutectic solution is:**
   A. a phase change solution.
   B. a very warm solution.
   C. an acid solution.
   D. an explosive solution.

5. **Liquid nitrogen is used to:**
   A. clean the inside of refrigerated trucks.
   B. clean the tubes in marine refrigeration systems.
   C. refrigerate products in a truck.
   D. provide the cooling in a standard kitchen refrigerator.

6. **Extra-low-temperature refrigeration is used to:**
   A. ensure that ice cream remains hard.
   B. quick freeze meat and other foods.
   C. provide refrigeration in large ships.
   D. provide refrigeration for meat in large supermarkets.

7. **Cascade refrigeration:**
   A. utilizes flowing water to provide refrigeration.
   B. utilizes two or three stages for low-temperature refrigeration.
   C. provides medium-temperature refrigeration.
   D. provides high-temperature refrigeration.

8. **Condensers that have marine water boxes have:**
   A. removable water box covers.
   B. tubes made of cupronickel.
   C. tubes that can be cleaned.
   D. all of the above.

9. **Dry ice changes state at:**
   A. −44°F.   C. −320°F.
   B. 25°F.   D. −109°F.

10. **Liquid nitrogen changes state at:**
    A. −44°F.   C. −320°F.
    B. 25°F.   D. −109°F.

11. **Vapors given off from liquid nitrogen are:**
    A. odorless.
    B. oxygen depriving.
    C. colorless.
    D. all of the above.

12. **A compressor with capacity control will:**
    A. operate at reduced capacity.
    B. operate at a higher capacity than rated.
    C. pump excess oil.
    D. need more horsepower than rated.

13. **Most truck refrigeration systems:**
    A. will never shut off.
    B. must be started up manually.
    C. will not operate in cold climates.
    D. start and stop automatically.

14. **Most transport refrigeration is sized:**
    A. to hold the load temperature only.
    B. to pull the load temperature down.
    C. to use maximum fuel for refrigeration.
    D. with no capacity control.

15. **Quick freezing of foods:**
    A. is never done because it changes the flavor.
    B. preserves the flavor and quality of foods.
    C. is only done in other countries.
    D. none of the above.

16. **Chillers for ship refrigeration often circulate _____ for low-temperature refrigeration.**
    A. water
    B. nitrogen
    C. fish oil
    D. brine

Troubleshooting and Typical Operating Conditions for Commercial Refrigeration

Unit Overview

To service refrigeration systems, the technician should know the typical operating conditions of a particular system. This would include whether it is a high-, medium-, or low-temperature unit, the type of refrigerant, and the approximate desired box temperature.

Air-cooled condensers operate in many different environments. These may include inside a conditioned space such as in a store, outside in hot weather, or outside in cold weather. It must be assumed that the head pressure is controlled by some means so that typical operating conditions exist.

Water-cooled condensers operate with normal pressures and temperatures different from air-cooled equipment.

Some typical problems that can be found in refrigeration systems are:

1. low refrigerant charge.
2. excess refrigerant charge.
3. inefficient evaporator.
4. inefficient condenser.
5. restriction in the refrigerant circuit.
6. inefficient compressor.
7. low ambient conditions.

The preceding problems do not include problems with the electrical system. There are many techniques for troubleshooting, testing, and servicing refrigeration systems, many of which can be found in Unit 29 of the text.

Key Terms

Coil-to-Air Temperature Differences
Compressor Short Cycling
Defrost Termination Switch
Dehydration
Ice Conductivity
Inefficient Compressor
Inefficient Condenser

Inefficient Evaporator
Product Load Level
Refrigeration Recycle Unit
Split-System Refrigeration Equipment
Start Capacitor
Stress Crack

REVIEW TEST

Name	Date	Grade

Circle the letter that indicates the correct answer.

1. **The refrigerated box temperature of a high-temperature unit ranges from:**
   A. 75°F–95°F.   C. 45°F–60°F.
   B. 60°F–75°F.   D. 20°F–30°F.

2. **In a high-temperature refrigeration unit, the evaporator coil is normally _____ lower than the refrigerated box temperature.**
   A. 10°F–20°F   C. 35°F–45°F
   B. 25°F–35°F   D. 5°F–15°F

3. **Medium-temperature refrigerated box temperatures range from about:**
   A. 20°F–32°F.   C. 40°F–50°F.
   B. 30°F–45°F.   D. 50°F–60°F.

4. **Condensers located outside that operate year-round must have:**
   A. larger fan motors.
   B. CPR valves.
   C. head pressure control.
   D. larger diameter liquid-line piping.

5. When the temperature is above 70°F, the temperature difference between the ambient and the condensing temperature for most standard units is approximately:
   A. 15°F.
   B. 20°F.
   C. 30°F.
   D. 45°F.

6. In a wastewater water-cooled system, the temperature difference between the condensing refrigerant and the leaving water temperature:
   A. is approximately 10°F.
   B. is approximately 20°F.
   C. is approximately 30°F.
   D. varies.

7. In a recirculating water-cooled system, the temperature difference between the water entering and the water leaving the condenser should be approximately:
   A. 10°F
   B. 20°F
   C. 30°F.
   D. 40°F.

8. If there is a low refrigerant charge in a system using a capillary tube, the:
   A. suction and discharge gages will both indicate a high pressure.
   B. suction and discharge gages will both indicate a low pressure.
   C. suction gage will indicate a higher than normal reading and the discharge gage will indicate a lower than normal reading.
   D. suction gage will indicate a lower than normal reading and the discharge gage will indicate a higher than normal reading.

9. When the compressor in a system with a capillary tube is sweating, it is a sign of:
   A. refrigerant overcharge.
   B. a malfunctioning crankcase heater.
   C. a malfunctioning king valve.
   D. all of the above.

10. Three major functions of a condenser are to desuperheat the hot gas, condense the refrigerant, and:
    A. pump the refrigerant to the expansion valve.
    B. reduce the pressure of the refrigerant entering the evaporator.
    C. absorb heat from the ambient air.
    D. subcool the refrigerant before it leaves the coil.

11. The condenser should not be located where the fan discharge air is:
    A. subcooled.
    B. recirculated.
    C. partially vaporized.
    D. superheated.

12. If a sight glass shows bubbles during a normal operation, it indicates:
    A. an overcharge of refrigerant.
    B. a low charge of refrigerant.

13. When an evaporator has a water coil to transfer heat from, the temperature difference between the evaporating temperature and the leaving water temperature is called the:
    A. arrival.
    B. delta.
    C. approach.
    D. none of the above.

Situational Service Ticket

Name	Date	Grade

NOTE: Your instructor must have placed the correct service problem in the system for you to successfully complete this exercise.

Technician's Name:_____ Date: _____

CUSTOMER COMPLAINT: No cooling.

CUSTOMER COMMENTS: The temperature started rising this morning.

TYPE OF SYSTEM:

OUTDOOR UNIT:
MANUFACTURER_____   MODEL NUMBER_____   SERIAL NUMBER_____

INDOOR UNIT:
MANUFACTURER_____   MODEL NUMBER_____   SERIAL NUMBER_____

COMPRESSOR (where applicable):
MANUFACTURER_____   MODEL NUMBER_____   SERIAL NUMBER_____

TECHNICIAN'S REPORT

1. SYMPTOMS: _____

2. DIAGNOSIS: _____

3. ESTIMATED MATERIALS FOR REPAIR: _____

4. ESTIMATED TIME TO COMPLETE THE REPAIR: _____

SERVICE TIPS FOR THIS CALL

1. This is an electrical problem. Gages are not needed.

2. Touch test the suction line and the compressor housing before drawing any conclusions.

Situational Service Ticket

Name	Date	Grade

NOTE: Your instructor must have placed the correct service problem in the system for you to successfully complete this exercise.

Technician's Name:_____ Date: _____

CUSTOMER COMPLAINT: No cooling.

CUSTOMER COMMENTS: When we came in this morning, the cooler temperature was 65°F instead of 45°F.

TYPE OF SYSTEM:

OUTDOOR UNIT:
MANUFACTURER_____   MODEL NUMBER_____   SERIAL NUMBER_____

INDOOR UNIT:
MANUFACTURER_____   MODEL NUMBER_____   SERIAL NUMBER_____

COMPRESSOR (where applicable):
MANUFACTURER_____   MODEL NUMBER_____   SERIAL NUMBER_____

TECHNICIAN'S REPORT

1. SYMPTOMS: _____

2. DIAGNOSIS: _____

3. ESTIMATED MATERIALS FOR REPAIR: _____

4. ESTIMATED TIME TO COMPLETE THE REPAIR: _____

SERVICE TIP FOR THIS CALL

1. This unit has an electrical problem. Use your VOM and clamp-on ammeter to draw conclusions.

Situational Service Ticket

Name	Date	Grade

NOTE: Your instructor must have placed the correct service problem in the system for you to successfully complete this exercise.

Technician's Name:_____ Date: _____

CUSTOMER COMPLAINT: No cooling.

CUSTOMER COMMENTS: The cooler temperature started rising earlier in the day. The compressor is not running.

TYPE OF SYSTEM:

OUTDOOR UNIT:
MANUFACTURER_____   MODEL NUMBER_____   SERIAL NUMBER_____

INDOOR UNIT:
MANUFACTURER_____   MODEL NUMBER_____   SERIAL NUMBER_____

COMPRESSOR (where applicable):
MANUFACTURER_____   MODEL NUMBER_____   SERIAL NUMBER_____

TECHNICIAN'S REPORT

1. SYMPTOMS: _____

2. DIAGNOSIS: _____

3. ESTIMATED MATERIALS FOR REPAIR: _____

4. ESTIMATED TIME TO COMPLETE THE REPAIR: _____

SERVICE TIP FOR THIS CALL

1. This unit has an electrical problem. Use your VOM to track the problem.

Electric Heat

Unit Overview

Electric heat is produced by placing a special wire called nichrome, made of nickel chromium, in an electrical circuit. This wire causes a resistance to electron flow and produces heat at the point of resistance.

Some electric heat is produced by small portable space heaters that transfer much of the heat produced by radiation to the solid objects in front of the heater. These objects in turn warm the space around them. Other space heaters use fans to move the air over the heating elements and out into the space. There are radiant heating panels, electric baseboard heaters, and unit heaters suspended from the ceiling. All of these produce electric resistance heating. There are electrically heated hot water boilers (hydronic) that pump hot water out to baseboard or radiator-type terminal units.

Central forced-air furnaces are used with ductwork to distribute the heated air to rooms or spaces away from the furnace. These heaters are usually controlled by a single thermostat, resulting in no individual room or zone control. This type of forced-air heating system can usually have cooling air conditioning and humidification added because of the duct-type air-handling feature.

Key Terms

Central Forced-Air Electric Furnace
Contactor
Electric Baseboard Heating
Electric Hydronic Boiler
Electric Radiant Heating Panel

Infrared Rays
Nichrome
Sequencer
Unit Heater

REVIEW TEST

Name	Date	Grade

Circle the letter that indicates the correct answer.

1. **The resistance wire often used in an electric heating element is made of:**
   A. copper.
   B. stainless steel.
   C. nickel chromium.
   D. magnesium.

2. **Radiant heat is transferred by:**
   A. ultraviolet rays.   C. ionization.
   B. infrared rays.   D. diffraction.

3. **Most electric baseboard units are _____ heaters.**
   A. natural-draft convection
   B. forced-draft convection
   C. forced-draft conduction
   D. radiant

4. **Central forced-air electric furnaces frequently use _____ to start and stop multiple heating circuits.**
   A. capacitors   C. limit controls
   B. sequencers   D. cool anticipators

5. **A sequencer utilizes a _____ to initiate its operation.**
   A. bimetal heat motor
   B. bimetal snap disc
   C. capacitor
   D. nichrome resistor

6. **The component located in the thermostat that shuts the electric furnace down prematurely so that the space will not be overheated is the:**
   A. sequencer.
   B. cool anticipator.
   C. heat anticipator.
   D. mercury bulb.

7. **The disadvantage in using a contactor to open and close the circuit to the electric heating elements is the:**
   A. noise it makes.
   B. low-control voltage.
   C. heating elements would all be started at the same time unless separate time delay relays were used.
   D. both A and C.

8. **A limit switch or control is used to:**
   A. start the furnace if the thermostat malfunctions.
   B. ensure that the fan stops when the furnace cools down.
   C. start the furnace prematurely on a cold day.
   D. open the circuit to the heating elements if the furnace overheats.

9. **Electric unit heaters are usually installed:**
   A. hanging from the ceiling.
   B. as baseboard heat.
   C. as forced-air furnaces.
   D. for hot water systems.

10. **Electric ceiling panel heat:**
    A. easily adapts to central air conditioning.
    B. is radiant-type heat.
    C. uses circulating hot water.
    D. can easily have humidity equipment added to the system.

11. **The _____ thermostat is commonly used for forced-air electric furnace installations.**
    A. low-voltage
    B. all-voltage
    C. no-voltage
    D. high-voltage

12. **The _____ thermostat is commonly used for baseboard electric heat installations.**
    A. low-voltage
    B. all-voltage
    C. no-voltage
    D. high-voltage

13. **The control voltage for the controls on a sequencer are typically:**
    A. 24 V.
    B. 208 V.
    C. 460 V.
    D. 12 V.

14. **An electric furnace is operating on 225 V and is drawing 83 amps. What heat capacity is this furnace putting out?**
    A. 55,787 Btu/h
    B. 33,467 Btu/h
    C. 98,953 Btu/h
    D. 63,738 Btu/h

Electric Furnace Familiarization

Name	Date	Grade

Objectives: Upon completion of this exercise, you will be familiar with the components in an electric heating system and will be able to list the specifications for these components.

Introduction: You will inspect all the components in an electric furnace and read specifications off the component nameplates or from available literature.

Text References: Paragraphs 30.7, 30.8, 30.9, 30.10, 30.11, 30.12, 30.13, 30.14, and 30.15.

Tools and Materials: Straight-blade and Phillips screwdrivers, $1/4$" and $5/16$" nut drivers, VOM, thermometer, calculator, wire size chart, and a complete electric furnace.

Safety Precautions: Turn the power off and lock out the power to the heating unit before beginning your inspection. You should keep the only key to the locked panel in your possession. After turning the power off, check the line voltage with your VOM to make sure that it is turned off.

PROCEDURES

1. Locate each of the following components. Place a checkmark after the name when you have located it.

   - Thermostat _____
   - Heat anticipator _____
   - Control device(s) for energizing heating elements (contactor or sequencer) _____
   - Heating elements _____
   - Fan motor and fan _____

2. Provide the following information from available literature or component nameplates:

   - Fan motor horsepower _____
   - Fan motor shaft size _____
   - Capacitor rating of fan motor (if any) _____
   - Fan motor FLA _____
   - Fan motor LRA _____
   - Fan motor operating voltage _____
   - Fan diameter _____
   - Number of heating elements _____
   - Amperage draw of each heating element _____
   - KW rating of each heating element _____
   - Resistance of each heating element _____
   - Current draw in the control circuit _____

- Setting of the heat anticipator　　　　　_____
- Minimum wire size of each heating element　　_____
- Btu capacity of each heating element　　_____
- Minimum wire size to furnace　　　　_____
- Minimum wire size to fan motor　　　_____

Maintenance of Workstation and Tools: Replace all panels on the furnace or leave them as you are instructed. Return all tools, the VOM, and other materials to their proper locations.

Summary Statement: Make a sketch of an electric furnace indicating the approximate location of each component.

QUESTIONS

1. Electric heating elements are rated in which of the following terms, kW, Btu/h, or gal/h?

2. The individual heater wire size is dependent on which of the following: voltage, current, kW, or horsepower?

3. Electric heat is supposed to glow red hot in the air stream. True or false?

4. When a fuse blows several times, a larger fuse should be installed. True or false?

5. What control component starts the fan on a typical electric heat furnace?

6. When the furnace air filter is stopped up, which of the following controls would cut off the electric heat first, the limit switch or the fuse link?

7. Which instrument is used to check the current draw in the electric heat circuit: ohmmeter, voltmeter, or ammeter?

8. Is electric heat normally less expensive or more expensive to operate than gas heat?

Determining Airflow (cfm) by Using the Air Temperature Rise

Name	Date	Grade

Objectives: Upon completion of this exercise, you will be able to calculate the quantity of airflow in cubic feet per minute (cfm) in an electric furnace by measuring the air temperature rise, amperage, and voltage.

Introduction: You will measure the air temperature at the electric furnace supply and return ducts and the voltage and amperage at the main electrical supply. By using the formulae furnished, you will calculate the cubic feet per minute supplied by the furnace. The power supply must be single phase.

Text References: Paragraphs 30.7 and 30.15.

Tools and Materials: A VOM, ammeter, wattmeter (if available), electronic thermometer, calculator, Phillips and straight-blade screwdrivers, $1/4$" and $5/16$" nut drivers, and an operating electric furnace.

Safety Precautions: Use proper procedures in making electrical measurements. Your instructor should check all measurement setups before you turn the power on. Use VOM leads with alligator clips. Keep your hands away from all moving or rotating parts. Lock and tag the panel where you turn the power off. Keep the single key in your possession.

PROCEDURES

1. Turn the power to the furnace off and lock out the panel. Under the supervision of your instructor and with the power off, connect the ammeter and voltmeter where they will measure the voltage and total amperage supplied to the electric furnace. See Figure 30-1 as an example problem.

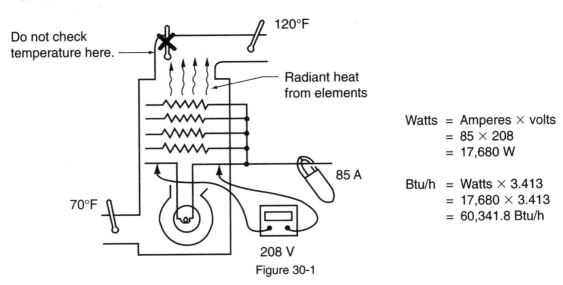

Figure 30-1

Watts = Amperes × volts
= 85 × 208
= 17,680 W

Btu/h = Watts × 3.413
= 17,680 × 3.413
= 60,341.8 Btu/h

2. Turn the unit on. Wait for all elements to be energized. Record the current and voltage below:

Current _____ A (Amperes × Volts = Watts)

Voltage _____ V (_____ × _____ = _____ Watts)

3. Convert watts to Btu per hour, Btu/h. This is the input. Watts × 3.413 = Btu/h

_____ × 3.413 = _____ Btu/h

4. Find the temperature difference (TD) in the supply and return duct. Place a temperature sensor in the supply duct (around the first bend in the duct to prevent radiant heat from the elements from hitting the probe) and a temperature sensor in the return duct. Record the temperatures below:

Supply _____°F   (Supply − return = Temperature difference)

Return duct _____°F   (_____ − _____ = _____°F TD)

5. To calculate the cubic feet per minute (CFM) of airflow, use the following formulae and procedures:

$$\text{Cfm airflow} = \frac{\text{Total heat input (step 3)}}{1.1 \times \text{TD (step 4)}} = \frac{\_\_\_\_ \text{ Btu/h}}{1.1 \times \_\_\_\_°F} = \_\_\_\_\_$$

6. If a wattmeter is available, you could measure the wattage of the circuit to check your voltage and amperage measurements. There will be a slight difference because the motor is an inductive load and does not calculate perfectly using an ammeter and voltmeter to measure power. Do not measure the wattage until your instructor has checked your setup.

7. Replace all panels on the unit with the correct fasteners.

Maintenance of Workstation and Tools: Return all tools and instruments to their proper places. Leave the furnace as you are instructed.

Summary Statement: Describe why it is necessary for a furnace to provide the correct cfm.

QUESTIONS

1. The kW rating of an electric heater is the _____. (temperature range, power rating, voltage rating, or name of the power company)

2. The Btu rating is a rating of the _____. (time, wire size, basis for power company charges, or heating capacity)

3. The factor for converting watts to Btu is _____.

4. A kW is _____. (1000 watts, 1200 watts, 500 watts, or 10,000 watts)

5. The typical air temperature rise across an electric furnace is normally considered (high or low) compared to a gas furnace.

6. The air temperature at the outlet grilles of an electric heating system is considered (hot or warm) compared to a gas furnace.

7. The circulating fan motor normally runs on (high or low) speed in the heating mode.

8. What is the current rating of a 7.5 kW heater operating at 230 volts?

9. A 12 kW heater will put out _____ Btu per hour.

Setting the Heat Anticipator for an Electric Furnace

Name	Date	Grade

Objectives: Upon completion of this exercise, you will be able to take an amperage reading with a 10-wrap loop and properly adjust a heat anticipator in an electric furnace low-voltage control circuit.

Introduction: You will be working with a single-stage and a multistage electric heating furnace, taking amperage readings in the low-voltage circuit, and checking or adjusting the heat anticipator.

Text References: Paragraphs 15.5, 30.8, 30.9, and 30.10.

Tools and Materials: Phillips and straight-blade screwdrivers, $1/4$" and $5/16$" nut drivers, clamp-on ammeter, single strand of thermostat wire, a single-stage electric furnace, and a multistage electric furnace.

Safety Precautions: Make sure that the power is turned off, the panel is locked, and you have the only key before removing panels on the electric furnace. Turn the power on only after your instructor approves your setup.

PROCEDURES

1. Turn the power off, lock and tag the panel. With the power off on a single-stage electric heating unit, remove the panels. Remove the red wire and install a 10-wrap loop around one jaw of a clamp-on ammeter. This is a loop of thermostat wire wrapped ten times around the ammeter jaw, which will amplify the amperage reading ten times. Connect back into the circuit, Figure 30-2.

2. Turn the power on. Switch the thermostat to heat and adjust it to call for heat. Check the amperage at the loop and record: _____ A

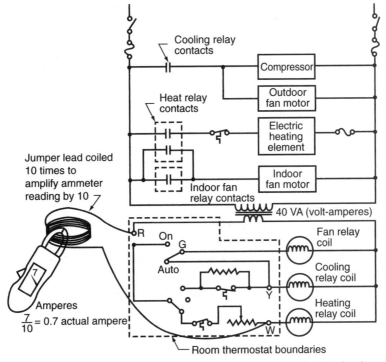

Figure 30-2 Measuring the amperage in the heat anticipator circuit.

3. This should be approximately 5 amperes. Divide the reading by 10 to get the actual amperage. Record it here: _____ A

4. Adjust the heat anticipator under the thermostat cover to reflect the actual amperage if it is not set correctly.

5. With the power on a multistage electric heating unit off, and the panels locked and tagged, remove the panels. Remove the red wire and install a 10-wrap loop around the jaw of an ammeter as in step 1. Connect back into the circuit, Figure 30-2.

6. Turn the power on. Switch the thermostat to heat and adjust to call for heat. Check the amperage and record: _____ A

7. Divide this reading by 10 to get the actual amperage and record: _____ A

8. Adjust the heat anticipator to reflect the actual amperage if it is not set correctly.

Maintenance of Workstation and Tools: Replace all panels. Return all tools and instruments, leaving your work area neat and orderly.

Summary Statements: Write a description of how the heat anticipator works, telling what its function is. Describe why the amperage is greater when more than one sequencer is in the circuit.

QUESTIONS

1. Most heat anticipators are _____. (fixed resistance or variable resistance)

2. A heat anticipator that is set wrong may cause _____. (no heat, excessive temperature swings, or noise in the system)

3. If the service technician momentarily shorted across the sequencer heater coil, it would cause which of the following, blown fuse, burned heat anticipator, the heat turned on, or the cooling turned on?

4. All heating thermostats have heat anticipators. True or false?

5. When a system has several sequencers that would cause too much current draw through the heat anticipator, which of the following could be done to prevent the anticipator from burning up?
   A. installing a higher amperage rated thermostat
   B. wiring the sequencer so that all of the coils did not go through the anticipator
   C. changing to a package sequencer.
   D. B or C.

6. The heat anticipator differs from the cooling anticipator. True or false?

7. When the furnace is two-stage, there are how many heat anticipators?

8. The heat anticipator is located in which component, the thermostat or the subbase?

9. If the heat anticipator was burned so that it was in two parts, the symptoms would be which of the following?
   A. no heat
   B. too much heat
   C. the cooling coming on
   D. the residence overheating

Low-Voltage Control Circuits Used in Electric Heat

Name	Date	Grade

Objectives: Upon completion of this exercise, you will be able to describe and draw diagrams illustrating the low-voltage control circuits in single- and multistage electric furnaces.

Introduction: You will study the low-voltage wiring in a single- and a multistage electric furnace. You will then draw a pictorial and ladder diagram for each of these circuits.

Text References: Paragraphs 30.7, 30.8, 30.9, 30.10, 30.11, 30.12, 30.13, and 30.14.

Tools and Materials: Phillips and straight-blade screwdrivers, $\frac{1}{4}$" and $\frac{5}{16}$" nut drivers, and single- and multistage electric furnaces.

Safety Precautions: Ensure that the power is turned off, that the panel is locked and tagged, and that you have the only key before removing panels on the electric furnaces. Turn the power on only after your instructor approves your setup.

PROCEDURES

1. Turn the power off, lock and tag the electrical panel, and remove the panels from an electric heater that has one stage of heat. The electric circuits should be exposed.

2. Study the wiring in the low-voltage circuit until you understand how it controls the heat strip.

3. Draw a pictorial diagram, including the thermostat of the low-voltage circuit, in the space below. Draw a ladder diagram of the circuit.

    PICTORIAL   LADDER

4. Make sure that the power is off and the electrical panel is locked out, and remove the panels from an electric furnace that has more than one stage of heat. The electric circuits should be exposed.

5. Study the wiring in the low-voltage circuit until you understand how it controls the strip heaters.

6. Draw a pictorial diagram including the thermostat of the low-voltage circuit in the space below. Draw a ladder diagram of the same circuit.

PICTORIAL   LADDER

Maintenance of Workstation and Tools: Replace all panels. Return tools and leave the workstation as instructed.

Summary Statement: Describe the sequence of events in the complete low-voltage circuit for each system studied.

QUESTIONS

1. Low-voltage controls are used for which of the following reasons?
   A. less expensive
   B. easier to troubleshoot
   C. more reliable
   D. safety

2. Low-voltage wire has an insulation value for which of the following?
   A. 50 V   C. 500 V
   B. 100 V   D. 24 V

3. Wire size is determined by which of the following?
   A. voltage   C. ohms
   B. amperage   D. watts

4. The gage of wire that is normally used in residential low-voltage control circuits is which of the following?
   A. No. 12   C. No. 14
   B. No. 18   D. No. 22

5. Low-voltage wire always has to be run in conduit. True or false?

6. The nominal low voltage for control circuits is which of the following?
   A. 50 V   C. 115 V
   B. 24 V   D. 230 V

7. All units have only one low-voltage power supply. True or false?

Checking a Package Sequencer

Name	Date	Grade

Objectives: Upon completion of this exercise, you will be able to check a package sequencer by energizing the operating coil and checking the contacts of the sequencer with an ohmmeter.

Introduction: You will energize the operating coil of a package sequencer for electric heat with a low-voltage power source to cause the contacts to make. You will then check the contacts with an ohmmeter to make sure that they close as they are designed to do, similarly to those in Figure 30-3.

Text References: Paragraphs 30.7, 30.8, 30.9, 30.10, and 30.12.

Tools and Materials: Low-voltage power supply, a VOM, and a package sequencer used in an electric furnace.

Safety Precautions: You will be working with low voltage (24 V). Do not allow the power supply leads to arc together.

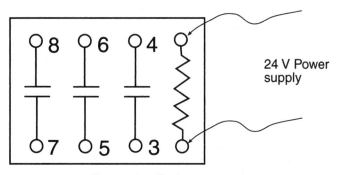

Figure 30-3 Package sequencer.

PROCEDURES

1. Turn the VOM function switch to AC voltage and the range selector switch to the 50 V or higher scale.

2. Turn the power off and lock out the panel. Fasten one VOM lead to one of the 24 V power supply leads and the other VOM lead to the other supply lead.

3. Turn the power on and record the voltage: _____ V It should be very close to 24 V.

4. Turn the power off.

5. Fasten the two leads from the 24 V power supply to the terminals on the sequencer that energize the contacts. They are probably labeled heater terminals.

6. Turn the ohms selector range switch to R × 1. Fasten the VOM leads on the terminals for one of the electric heater contacts. See the text, Figures 30-13, 30-14, 30-15, 30-16, 30-17, and 30-18, for examples of how a sequencer is wired.

7. Turn the 24 V power supply on. You should hear a faint audible click each time a set of contacts makes or breaks.

8. After about 3 minutes, check each set of contacts with the VOM to see that all sets are made. You may want to disconnect the power supply and allow all contacts to open, and then repeat the test to determine the sequence in which the contacts close.

9. Turn the power off and disconnect the sequencer.

Maintenance of Workstation and Tools: Return the VOM to its proper storage place. You may need the sequencer while preparing the Summary. Put it away in its proper place when you are through with it.

Summary Statements: Draw a wiring diagram of the package sequencer, showing all components. See the text, Figure 30-17, for an example.

QUESTIONS

1. What type of device is inside the sequencer to cause the contacts to close when voltage is applied to the coil?

2. What type of coil is inside the sequencer, magnetic or resistance?

3. What is the advantage of a sequencer over a contactor?

4. How many heaters can a typical package sequencer energize?

5. How is the fan motor started with a typical package sequencer?

6. What is the current-carrying capacity of each contact on the sequencer you worked with?

7. Do all electric furnaces use package sequencers?

8. If the number 2 contact failed to close on a package sequencer, would the number 3 contact close?

9. How should you determine the heat anticipator setting on the room thermostat when using a package sequencer?

10. How should you determine the heat anticipator setting on the room thermostat when using individual sequencers?

Checking Electric Heating Elements Using an Ammeter

Name	Date	Grade

Objectives: Upon completion of this exercise, you will be able to check an electric heating system by using a clamp-on ammeter to make sure that all heating elements are drawing power.

Introduction: You will identify all electric heating elements, then start an electric heating system. You will then use an ammeter to prove that each element is drawing power.

Text References: Paragraphs 30.7, 30.8, 30.10, 30.11, 30.12, 30.13, and 30.14.

Tools and Materials: A VOM, clamp-on ammeter, goggles, straight-blade and Phillips screwdrivers, flashlight, $1/4$" and $5/16$" nut drivers, and electric heating system.

Safety Precautions: You will be placing a clamp-on ammeter around the conductors going to the electric heat elements. Wear goggles. Do not pull on the wires or they may pull loose at the terminals.

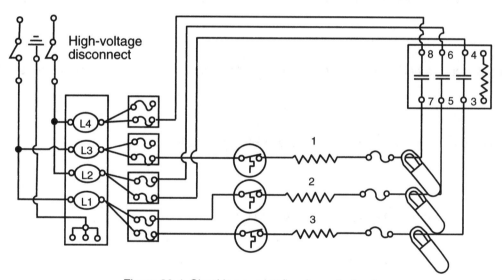

Figure 30-4 Checking current flow in each circuit.

PROCEDURES

1. Turn the power to an electric furnace off and remove the door to the control compartment. Lock and tag the panel when the power is off and keep the key in your possession.

2. Locate the conductor going to each electric heating element, Figure 30-4. Consult the diagram for the furnace you are working with. Make sure that you can clamp an ammeter around each conductor.

3. Turn the power on and listen for the sequencers to start each element. Wait for about 3 minutes; the fan should have started.

4. Clamp the ammeter on one conductor to each heating element and record the readings below. When the element is drawing current, it is heating.

- Element number 1 _____ A   Element number 2 _____ A
- Element number 3 _____ A   Element number 4 _____ A
- Element number 5 _____ A   Element number 6 _____ A

NOTE: Your furnace may only have two or three elements.

5. Turn the room thermostat off and using the clamp-on ammeter, observe which elements are deenergized first.

6. Replace all panels with the correct fasteners.

Maintenance of Workstation and Tools: Return all tools to their respective places. Replace all panels, if appropriate, on the electric furnace.

Summary Statement: Describe the function of an electric heating element, telling how the current relates to the resistance in the actual heating element wire.

QUESTIONS

1. What are the actual heating element wires made of?

2. If the resistance in a heater wire is reduced, what will the current do if the voltage is a constant?

3. How much heat would a 5 kW heater produce in Btu per hour?

4. What started the fan on the unit you were working on?

5. What is the difference between a contactor and a sequencer?

6. Why are sequencers preferred over contactors in ductwork applications?

7. How much current would a 5 kW heater draw with 230 V as the applied voltage?

8. What is the advantage of a package sequencer over a system with individual sequencers?

9. If the number 1 sequencer coil were to fail, would all the heat be off?

Troubleshooting Exercise: Electric Furnace

Name	Date	Grade

Objectives: Upon completion of this exercise, you will be able to troubleshoot electrical problems in an electric furnace.

Introduction: This unit should have a specified problem introduced to it by your instructor. YOU WILL LOCATE THE PROBLEM USING A LOGICAL APPROACH AND INSTRUMENTS. THE INTENT IS NOT FOR YOU TO MAKE THE REPAIR BUT TO LOCATE THE PROBLEM AND ASK YOUR INSTRUCTOR FOR DIRECTION BEFORE COMPLETION.

Text References: Paragraphs 12.11, 30.8, 30.9, 30.10, 30.11, 30.12, 30.13, and 30.14.

Tools and Materials: Straight-blade and Phillips screwdrivers, $1/4$" and $5/16$" nut drivers, VOM, and an ammeter.

Safety Precautions: You will be working with live voltages. Use only insulated meter leads and be very careful not to touch any energized terminals.

PROCEDURES

1. With the power off and the electrical panel locked out, remove the panel to the electric heat controls. WHILE THE POWER IS OFF, IDENTIFY HOW MANY STAGES OF STRIP HEAT ARE IN THIS UNIT. IDENTIFY BOTH SIDES OF LINE VOLTAGE FOR EACH STAGE OF STRIP HEAT. YOU ARE GOING TO ATTACH AN AMMETER AND VOLTMETER TO EACH LATER. THE TIME TO IDENTIFY THEM IS WHILE THEY ARE NOT ENERGIZED.

2. Set the room thermostat to call for heat and turn the unit on.

3. Give the unit a few minutes to get up to temperature. If the unit is controlled by sequencers and has several stages of heat, it takes from 20 to 45 seconds per stage of heat to be fully energized.

4. Check the current to each stage of strip heat and record here:

   1. _____ A   2. _____ A   3. _____ A

   If there are more than three stages, add them to the end.

5. Check and record the voltage for each heater here:

   1. _____ V   2. _____ V   3. _____ V

   If there are more than three stages, add them to the end.

6. Were all stages drawing current?

7. Were all stages energized with line power?

8. Describe the problem and your recommended repair.

Maintenance of Workstation and Tools: Return the unit and tools as per the instructor's directions.

Summary Statement: Describe Ohm's law and how it applies to electric heat.

QUESTIONS

1. What material is an electric heating element made of?

2. Why doesn't the wire leading up to the electric furnace get hot?

3. How is the heat concentrated at the electric heating element?

4. What would happen to the electric heating element if the heat were not dissipated by the fan?

5. What is the advantage of a sequencer over a contactor?

6. Why is electric heat not used exclusively instead of gas and oil?

7. What is the efficiency of electric heat?

8. Describe how an ohmmeter may have been used in the above exercise.

9. What voltage does the typical room thermostat operate on?

10. How would reduced airflow affect the electric furnace?

Troubleshooting Exercise: Electric Furnace

Name	Date	Grade

Objectives: Upon completion of this exercise, you will be able to check the performance of an electric furnace.

Introduction: This unit should have a specified problem introduced to it by your instructor. You will locate the problem using a logical approach and instruments. THE INTENT IS NOT FOR YOU TO MAKE THE REPAIR BUT TO LOCATE THE PROBLEM AND ASK YOUR INSTRUCTOR FOR DIRECTION BEFORE COMPLETION.

Text Reference: Unit 30.

Tools and Materials: Straight-blade and Phillips screwdrivers, $1/4$" and $5/16$" nut drivers, ammeter, VOM, and a thermometer.

Safety Precautions: Be careful while taking any voltage readings across live electrical terminals. Lock and tag the electrical panel, and keep the key in your possession when the power is turned off and service is performed.

PROCEDURES

1. With the power off and the electrical panel locked, remove the panels to the electric heat control box.

2. Identify all electric heating elements. You will be taking voltage and ammeter readings in a few moments, and now is the time to locate where these readings should be taken.

3. Start the unit and allow time for all stages to become energized.

4. Place a temperature lead in the return air and one in the supply air. Be sure the radiant heat from the unit does not affect the lead in the supply air. See text, Figure 30-23, for correctly placing the probes.

5. Measure the voltage:

   1. _____ V 2. _____ V 3. _____ V

   If there are more than three stages, add them to the end.

6. Describe any problems you have found and the recommended repair.

7. Measure the amperage to each heater and record:

   1. _____ A 2. _____ A 3. _____ A

   If there are more than three stages, add them to the end.

8. Record the temperature difference between the return and supply air here: _____ °F

9. Describe the probable problem.

Maintenance of Workstation and Tools: Return all tools to their places and leave the unit as your instructor directs you.

Summary Statement: Describe the difference in a package and individual sequencer control circuit.

QUESTIONS

1. Why is electric heat used even where the energy cost is more than gas or oil?

2. Name three types of electric heat.

3. What would happen to the heating elements if the fan were to stop on a forced-air system?

4. Is there any protection from overheating?

5. Describe the type of overheat protection the unit in this exercise has.

6. What is the purpose of the heat anticipator in the room thermostat?

7. Is all control voltage 24 V for electric heat?

8. What causes a sequencer to function when low voltage is applied?

9. How does this differ from a contactor?

10. Describe a fusible link.

Troubleshooting Exercise: Electric Furnace

Name	Date	Grade

Objectives: Upon completion of this exercise, you will be able to troubleshoot the control circuit for a forced-air electric furnace.

Introduction: This unit should have a specified problem introduced to it by your instructor. You will locate the problem using a logical approach and instruments. THE INTENT IS NOT FOR YOU TO MAKE THE REPAIR BUT TO LOCATE THE PROBLEM AND ASK YOUR INSTRUCTOR FOR DIRECTION BEFORE COMPLETION.

Text Reference: Unit 30.

Tools and Materials: Straight-blade and Phillip screwdrivers, $1/4$" and $5/16$" nut drivers, VOM, and an ammeter.

Safety Precautions: You will be troubleshooting live voltage. Make sure you do not come in contact with any electrical terminals while the unit is on. When the power is off, the electrical panel should be locked and tagged, and you should keep the only key in your possession.

PROCEDURES

1. With the power off and the electrical panel locked, remove the door to the electrical controls for the heat strips and identify all heat strip terminals. You will have to take voltage readings; now is the time to find them.

2. Turn the power on and set the thermostat to call for heat.

3. Using the ammeter, check the amperage at all heaters and record the amperage here.

   1. _____ A   2. _____ A   3. _____ A

   If there are more than three stages, record them at the end.

4. Are all heaters pulling current? If not, check the line voltage to the heater that is not pulling current. Is voltage present at both terminals? Follow the voltage path back toward the source until you discover where the interruption is.

5. Describe the problem with the unit and any parts that may need replacing to repair the unit.

Maintenance of Workstation and Tools: Return all tools to their places and leave the unit as the instructor directs you.

Summary Statement: Describe the current flow from the power supply to the power-consuming device and back to the power supply.

QUESTIONS

1. What is the typical line voltage for a residence using electric heat?

2. What is the typical control voltage used for forced-air electric furnaces?

3. Why is the lower control voltage used?

4. What would the symptoms be for a system when the heat anticipator is set at the wrong value?

5. What is the disadvantage of ceiling panel electric heat?

6. What is the disadvantage of baseboard electric heat?

7. What control voltage does panel and baseboard electric heat use?

8. Describe how a unit heater works.

9. How does the low-voltage thermostat prevent arcing when the contacts break circuit?

10. What is the purpose of the fusible link used in some electric heating elements?

Changing a Sequencer

Name	Date	Grade

Objectives: Upon completion of this exercise, you will be able to change a sequencer in an electric furnace.

Introduction: You will remove a sequencer in an electric furnace and replace and rewire it just as though you have a new one.

Text References: Paragraphs 12.8 and 12.9 and Unit 30.

Tools and Materials: Straight-blade and Phillips screwdrivers, $1/4$" and $5/16$" nut drivers, VOM, ammeter, and needle-nose pliers.

Safety Precautions: Make sure the power is off and the panel is locked and tagged before you start this project. Use the voltmeter to verify that the power is off.

PROCEDURES

1. Turn the power off and lock out the electrical panel.

2. Check to make sure the power is off using the VOM.

3. Draw a wiring diagram showing all wires on the sequencer to be changed. If some of the wires are the same color, make a tag and mark them for easy replacement.

4. Remove the wires from the sequencer one at a time. Use care when removing wires from spade or push on terminals and do not pull the wire out of the connector. Check for properly crimped terminals.

5. Remove the sequencer to the outside of the unit and reuse as if it were a new one.

6. Now, remount the sequencer.

7. Replace the wiring one at a time. Be very careful that each connection is tight. Electric heat pulls a great deal of current and any loose connection will cause a problem.

8. When the sequencer is installed to your satisfaction, ask the instructor to approve the installation before you turn the power on.

9. Turn the power on and set the thermostat to call for heat.

10. After the unit has been on long enough to allow the sequencers to energize the heaters, use the ammeter to verify that the heaters are operating.

11. Turn the unit off when you are satisfied that it is working properly.

Maintenance of Workstations and Tools: Return all tools to their places and leave the unit as your instructor directs you.

Summary Statement: Describe the internal operation of a sequencer.

QUESTIONS

1. What would low voltage do to the capacity of an electric heater?

2. What would high voltage do to the capacity of an electric heater?

3. An electric heater has a resistance of of 11.5 ohms and is operating at 230 volts. What would the amperage reading be for this heater? Show your work.

4. What would the power be for the above heater? Show your work.

5. What would the Btu/h rating be on the above heater? Show your work.

6. If the voltage for the above heater were to be increased to 245 volts, what would the wattage and Btu/h be? Show your work.

7. If the voltage were reduced to 208 volts, what would the wattage and Btu/h be for the above unit? Show your work.

8. Using the above voltages, would the heater be operating within the ±10% recommended voltage?

9. What is the unit of measure the power company charges for electricity?

10. What starts the fan motor in the unit you worked on?

Situational Service Ticket

Name	Date	Grade

NOTE: Your instructor must have placed the correct service problem in the system for you to successfully complete this exercise.

Technician's Name:_____ Date: _____

CUSTOMER COMPLAINT: No heat.

CUSTOMER COMMENTS: It just quit last night.

TYPE OF SYSTEM:

OUTDOOR UNIT:
MANUFACTURER_____   MODEL NUMBER_____   SERIAL NUMBER_____

INDOOR UNIT:
MANUFACTURER_____   MODEL NUMBER_____   SERIAL NUMBER_____

COMPRESSOR (where applicable):
MANUFACTURER_____   MODEL NUMBER_____   SERIAL NUMBER_____

TECHNICIAN'S REPORT

1. SYMPTOMS: _____

2. DIAGNOSIS: _____

3. ESTIMATED MATERIALS FOR REPAIR: _____

4. ESTIMATED TIME TO COMPLETE THE REPAIR: _____

SERVICE TIP FOR THIS CALL

1. Check all power supplies.

Situational Service Ticket

Name	Date	Grade

NOTE: Your instructor must have placed the correct service problem in the system for you to successfully complete this exercise.

Technician's Name:_____ Date: _____

CUSTOMER COMPLAINT: No heat.

CUSTOMER COMMENTS: The heat went off this morning.

TYPE OF SYSTEM:

OUTDOOR UNIT:
MANUFACTURER_____   MODEL NUMBER_____   SERIAL NUMBER_____

INDOOR UNIT:
MANUFACTURER_____   MODEL NUMBER_____   SERIAL NUMBER_____

COMPRESSOR (where applicable):
MANUFACTURER_____   MODEL NUMBER_____   SERIAL NUMBER_____

TECHNICIAN'S REPORT

1. SYMPTOMS: _____

2. DIAGNOSIS: _____

3. ESTIMATED MATERIALS FOR REPAIR: _____

4. ESTIMATED TIME TO COMPLETE THE REPAIR: _____

SERVICE TIPS FOR THIS CALL

1. Check the complete low-voltage circuit.

2. Read Paragraphs 15.4, 15.5, 15.6, 30.9, 30.12, and 30.13 in the text.

Gas Heat

Unit Overview

The gas furnace manifold and controls meter the gas to the burners where the gas is burned. This creates flue gases in the heat exchanger and heats the air in and surrounding the heat exchanger. The venting system vents the flue gases to the atmosphere. The blower moves the heated air from around the heat exchanger through the ductwork to the areas where the heat is wanted.

Natural gas and liquified petroleum (LP) are the fuels most commonly used in gas furnaces. Liquified petroleum is liquified propane, butane, or a combination of propane and butane.

For combustion to take place, there must be fuel, oxygen, and heat. It is the reaction between these three that produces rapid oxidation or burning. The fuel or gas is fed through a gas regulator to a gas valve, through a manifold to an orifice and a gas burner, where it is ignited. It may be ignited by a standing pilot, electric spark, or hot-surface igniter.

Conventional gas furnaces generally use natural convection to vent the flue gases. High-efficiency furnaces recirculate the flue gases through an extra heat exchanger and use a fan to push them out the flue.

Modern gas furnaces use either induced or forced draft systems. Induced draft systems have a combustion blower motor located on the outlet of the heat exchanger. They pull or suck combustion gases through the heat exchanger, usually causing a slight negative pressure in the heat exchanger itself. Forced draft systems have a combustion blower motor on the inlet of the heat exchanger. They push or blow combustion gases through the heat exchanger and cause a positive pressure in the heat exchanger.

Multipoise or multiple-position furnaces can be installed in any position, which adds versatility to the furnace. Multiple safety controls like bimetallic flame rollout or over-temperature limits must be designed into the multipoise furnace for safety.

Modern heat exchangers come in many sizes, shapes, and materials. Many modern heat exchangers are L- and S-shaped and are made of aluminized steel. There is actually a serpentine path for combustion gases to travel. Serpentine heat exchangers use inshot burners instead of ribbon or slotted port burners.

Modern condensing furnaces may have two or even three heat exchangers. The last heat exchanger is usually made of stainless steel or aluminum to resist corrosion. It is designed to intentionally handle condensate and eventually drain it away from the furnace.

The dew point temperature is the temperature at which the condensation process begins in a condensing furnace. The dew point temperature varies depending on the composition of the flue gas and the amount of excess air. As excess air decreases, the dew point temperature increases.

Many modern gas furnaces use electronic modules or integrated furnace controllers (IFC) to control the operation of the furnace. These controllers often have dual-in-line pair (DIP) switches, which can be switched to reprogram the furnace to make it more versatile.

Flame rectification is a popular way to prove the presence of a flame in a modern furnace. In a flame rectification system, the pilot flame, or main flame, can conduct electricity because it contains ionized combustion gases which are made up of positively and negatively charged particles. The flame is located between two electrodes of different sizes. The electrodes are fed with an alternating current from the furnace's electronic module or controller. Current will flow greater in one direction than the other because of the different sized electrodes. The flame actually acts as a switch. When the flame is present, the switch is closed and will conduct electricity. Flame rectification systems can be classified as either single rod (local sensing) or dual rod (remote sensing). Single-rod systems consist of an igniter and a sensor all in one rod. Dual-rod systems use a separate igniter and flame-sensing rod.

A direct-vented high-efficiency gas furnace has a sealed combustion chamber. This means that the combustion air is brought in from the outside through an air pipe, which is usually made of PVC plastic. A nondirect-vented high efficiency gas furnace uses indoor air for combustion.

A positive pressure system can be vertically or sidewall vented. It is a system whose flue pipe pressure is positive the entire distance to its terminal end.

Excess air consists of combustion air and dilution air. Combustion air is primary air and/or secondary air. Primary air enters the burner before combustion takes place. Secondary air happens after combustion and it supports combustion. Dilution air is excess air after combustion and usually enters at the end of the heat exchanger.

Furnace efficiency ratings are determined by the amount of heat that is transferred to the heated medium. Factors that determine the efficiency of a furnace are the type of draft, the amount of excess air, the temperature difference of the air or water entering versus leaving the heating medium side of the heat exchanger, and the flue stack temperature.

Electronic ignition modules come in 100% shutoff, non-100% shutoff, continuous retry with 100% shutoff, and many other custom control schemes, which are manufacturer dependent. The combustion blower can pre-purge, inter-purge, and/or post-purge the heat exchanger depending on the module's control scheme.

Twinning of a furnace involves the operation of two furnaces connected side by side by a common ducting system. Twinning is usually used when a heating capacity is needed greater than the largest capacity furnace manufactured. It is also used when the airflow of one furnace, for air-conditioning purposes, is greater than a single furnace can provide.

Modern two-stage gas furnaces use a two-stage gas valve and a two-speed combustion blower motor with two pressure switches to prove the draft.

Modern modulating gas furnaces follow the heat loss of a structure. They use a modulating gas valve instead of a staged valve. They also come with variable speed warm air blowers to vary the speed and amount of warm air. Variable output thermostats that send a proportional signal instead of an on-off signal to the furnace controller often control modulating furnaces.

Key Terms

Automatic Combination Gas Valve
Bimetallic Safety Device
Burners
Carbon Dioxide
Carbon Monoxide
Combustion
Combustion Air
Condensing-Gas Furnace
Continuous Retry
Dew Point
Diaphragm Gas Valve
Dilution Air
Direct-Spark Ignition (DSI)
Direct Vented
Downflow
Dual-In-Line Pair (DIP) Switch
Dual-Rod (Remote) Sensing
Excess Air
Fan Switch
Flame Impingement
Flame-Proving Devices
Flame Rectification
Forced Draft
Gas Furnace Burner
Gas Regulator

Gas-Type Solenoid Valve
Glow Coil
Hard Lockout
Heat Exchanger
Heat-Motor-Controlled Valve
High-Efficiency Gas Furnace
Hot Surface Ignition
Horizontal Furnace
Ignition or Flame Velocity
Induced Draft
Inshot Burners
Integrated Furnace Controller (IFC)
Inter-Purge
Limit Switch
Liquid-Filled Remote Bulb
Liquified Petroleum (LP)
Lockout
Low-Boy
Manifold
Manifold Pressure
Methane
Modulating Gas Furnace
Multi-Poise (Multipositional) Furnace
Natural Gas

Non-100% Shutoff
Non-Direct Vented
100% Shut-Off
Orifice
Post-Purge
Pre-Purge
Servo-Pressure Regulator
Single-Rod (Local) Sensing
Smart Valve
Soft Lockout
Spark-to-Pilot (Intermittant Pilot) Ignition
Standing Pilot
Thermocouple
Thermopile
Troubleshooting Flowchart
Upflow
Upshot Burners
Variable Output Thermostat
Venting Products of Combustion
Venturi
Water Column (W.C.)
Water Manometer

REVIEW TEST

Name	Date	Grade

Circle the letter that indicates the correct answer.

1. **Natural gas is composed of 90 to 95%:**
   A. oxygen.
   B. ethanol.
   C. methane.
   D. propane.

2. **Liquid petroleum (LP) is composed of liquified:**
   A. natural gas.
   B. propane or butane.
   C. methane.
   D. sulfur compounds.

3. **Natural gas:**
   A. is lighter than air.
   B. has approximately the same specific gravity as air.
   C. is heavier than air.

4. **Combustion is sometimes expressed as:**
   A. a venturi effect.
   B. hydrocarbonization.
   C. rapid oxidation.
   D. the mixing of propane and butane.

5. **A water manometer is used to:**
   A. check combustion.
   B. measure gas pressure.
   C. determine the gas burning temperature.
   D. determine the gas burning color.

6. **The natural gas flame should be:**
   A. yellow.
   B. yellow with some orange tips.
   C. orange with some yellow tips.
   D. blue with some orange tips.

7. It takes approximately _____ cubic feet of air to produce 2 cubic feet of oxygen to mix with 1 cubic foot of natural gas to produce 1050 Btu when burned.
   A. 10
   B. 20
   C. 30
   D. 40

8. A standing pilot in a conventional gas furnace:
   A. is ignited by an electric spark when the thermostat calls for heat.
   B. is used as a flame-proving device.
   C. burns continuously.
   D. burns only when the main burner is lit.

9. A thermocouple is used in a conventional gas furnace as a:
   A. flame-proving device.
   B. standing pilot.
   C. limit switch.
   D. spark igniter.

10. The orifice is a precisely sized hole in the:
    A. manifold.
    B. spud.
    C. burner tube.
    D. combustion chamber.

11. The limit switch is a safety device that:
    A. shuts the gas off at the pressure regulator when a leak is detected.
    B. shuts the gas off if there is a power failure.
    C. closes the gas valve if the furnace overheats.
    D. closes the gas valve if the pilot is blown out.

12. In a conventional gas furnace, the hot flue gases are:
    A. forced out a PVC vent with a blower.
    B. recirculated through a second heat exchanger and then vented through the draft diverter.
    C. forced through the type B vent by a blower.
    D. vented by natural convection.

13. A glow coil is used to:
    A. ignite the main gas burner.
    B. ignite the pilot when the thermostat calls for heat.
    C. prove the pilot flame.
    D. reignite a pilot light if it goes out.

14. A direct-spark ignition (DSI) system is designed to:
    A. reignite a pilot light if it is blown out.
    B. ignite the pilot when the thermostat calls for heat.
    C. provide ignition to the main burners.
    D. ignite the pilot following a furnace shutdown over the summer.

15. A condensing gas furnace uses _____ from a condenser-type heat exchanger to improve the furnace efficiency.
    A. sensible heat
    B. latent heat
    C. superheat
    D. subcooling

16. A furnace that can be positioned in more that one position is a:
    A. variable furnace.
    B. multipoise furnace.
    C. flexible furnace.
    D. diverse furnace.

17. Where are auxiliary limit switches always found in heating applications?
    A. Up-flow furnaces
    B. Down-flow furnaces
    C. Horizontal furnaces
    D. None of the above

18. Two gas valves physically in series but wired in parallel are:
    A. series valves.
    B. redundant valves.
    C. parallel valves.
    D. none of the above.

19. A small microprocessor with integrated circuits mounted on a board to control the logic or sequence of events of the furnace is a(n):
    A. microboard.
    B. integrated furnace controller (IFC).
    C. transducer.
    D. transformer.

20. When a draft blower is located at the outlet of the heat exchanger and causes a negative pressure in the heat exchanger, it is called
    A. induced draft
    B. forced draft
    C. inlet draft
    D. outlet draft

21. Flame current in a flame rectification system is measured in _____ units.
    A. milli-amp
    B. milli-volt
    C. micro-volt
    D. micro-amp

22. A furnace that has a sealed combustion chamber and its combustion air comes from outside the conditioned space is a:
    A. direct-vented furnace.
    B. sealed furnace.
    C. unitary furnace.
    D. dual-function furnace.

23. **A combination gas valve that has a built-in electronic module in its body is:**
   A. a smart valve.
   B. an intelligent valve.
   C. a control valve.
   D. none of the above.

24. **Another name for remote sensing in a flame rectification system is a:**
   A. single-rod system.
   B. dual-rod system.
   C. triple-rod system.
   D. none of the above.

25. **A process where the combustion blower motor runs for a specified time period before the furnace lights to ensure safe ignition is called:**
   A. post-purge.
   B. power-purge.
   C. pre-purge.
   D. none of the above.

26. **Air that mixes with combustion byproducts after combustion at the beginning of the flue pipe is:**
   A. excess air.
   B. primary air.
   C. secondary air.
   D. dilution air.

27. **A(n) _____ is a furnace's electronic module timed safety shutdown period, which may reoccur many times.**
   A. soft lockout
   B. dead band
   C. intermission
   D. hard lockout

28. **When there is a pilot-proving device failure and both the pilot and main gas valves are shut down, there is:**
   A. less than 100% shutoff.
   B. 100% shutoff.
   C. intermittent shutoff.
   D. none of the above.

29. **An ignition system where the pilot light is only lit during the on-cycle of the furnace is called:**
   A. a direct burner ignition.
   B. a hot surface ignition.
   C. an intermittent pilot ignition.
   D. none of the above.

Gas Furnace Familiarization

Name	Date	Grade

Objectives: Upon completion of this exercise, you will be able to recognize and state the various components of a typical gas furnace.

Introduction: You will remove the front panels and possibly the blower and motor for the purposes of identifying the characteristics of all parts of the furnace.

Text References: Paragraphs 31.1, 31.2, 31.4, 31.5, 31.6, 31.13, and 31.14.

Tools and Materials: Straight-blade and Phillips screwdrivers, $1/4$" and $5/16$" nut drivers, a 6" adjustable wrench, Allen wrenches, a flashlight, and a gas furnace.

Safety Precautions: Turn the power to the furnace off before beginning this exercise. Lock and tag the power at the disconnect panel. There should be only one key. Keep this key on your person while performing this exercise.

PROCEDURES

1. With the power off and the electrical panel locked, remove the front burner and blower compartment panels.

2. Fan information:
   - Motor full-load current _____ A
   - Type of motor _____
   - Diameter of motor _____ in.
   - Shaft diameter _____ in.
   - Motor rotation (looking at motor shaft) _____
   - Fan wheel diameter _____ in.   Width _____ in.
   - Number of motor speeds _____ high rpm _____ low rpm _____

3. Burner information:
   - Type of burner _____
   - Number of burners _____
   - Type of pilot safety _____
   - Gas valve voltage _____ V
   - Gas valve current _____ A

4. Unit nameplate information:
   - Manufacturer _____
   - Model number _____
   - Serial number _____
   - Type of gas _____
   - Input capacity _____ Btu/h/
   - Output capacity _____ Btu/h
   - Voltage _____ V
   - Control voltage _____ V

5. Heat exchanger information:

   What is the heat exchanger made of? _____   Type of metal _____
   Number of burner passages _____   Flue size _____ in.
   Type of heat furnace? (upflow, downflow, or horizontal)

6. Replace all panels with the correct fasteners.

Maintenance of Workstation and Tools: Return all tools to their respective places and be sure the work area is clean.

Summary Statement: Describe the combustion process for the furnace that you worked on.

QUESTIONS

1. What component transfers the heat from the products of combustion to the room air?

2. What is the typical gas manifold pressure for a natural gas furnace?

3. Which of the following gases requires a 100% gas shutoff, natural or propane?

4. Why does this gas require a 100% shutoff?

5. What is the typical control voltage for a gas furnace?

6. Name two advantages for this particular control voltage.

7. What is the purpose of the drip leg in the gas piping located just before a gas appliance?

8. What is the typical line voltage for gas furnaces?

9. What is the purpose of the vent on a gas furnace?

10. What is the purpose of the draft diverter on a gas furnace?

Identification of the
Pilot Safety Feature of a Gas Furnace

Name	Date	Grade

Objectives: Upon completion of this exercise, you will be able to look at the controls of a typical gas furnace and determine the type of pilot safety features used.

Introduction: You will remove the panels from a typical gas-burning furnace, examine the control arrangement, and describe the type of control that prevents the burner from igniting if the pilot light is out.

Text References: Paragraphs 31.17, 31.18, 31.19, 31.20, and 31.22.

Tools and Materials: A VOM, straight-blade and Phillips screwdrivers, a flashlight, $1/4$" and $5/16$" nut drivers, a box of matches, and a typical gas-burning furnace.

Safety Precautions: Turn the power off. Lock and tag the electrical power panel and keep the key in your possession. Use caution while working with any gas-burning appliance. If there is any question, ask your instructor.

PROCEDURES

1. Select a gas-burning furnace. Shut off and lock out the power.

2. Remove the panel to the burner compartment.

3. Remove the cover from the burner section if there is one.

4. Using the flashlight, examine the sensing element that is in the vicinity of the pilot light and compare it to the text, Figures 31-53 through 31-61. Follow the sensing tube to its termination point to help identify it.

5. What type of pilot safety feature does this furnace have?

6. If the pilot light is lit, blow it out for the purpose of learning how to relight it.

7. Turn the power on and light the pilot light. You may find directions for the specific procedure in the furnace. If not, ask your instructor.

8. Turn the room thermostat to call for heat. Make sure that the burner ignites.

9. Allow the furnace to run until the fan starts, then turn the room thermostat to off and make sure the burner goes out.

10. Stand by until the fan stops running.

11. Replace all panels with the correct fasteners.

Maintenance of Workstation and Tools. Return all tools to their respective places. Leave the area around the furnace clean.

Summary Statement: Describe the exact sequence used for the pilot safety on this furnace.

QUESTIONS

1. Why is it necessary to have a pilot safety shutoff on a gas furnace?

2. How long could gas enter a heat exchanger if a pilot light were to go out during burner operation of a gas furnace with a thermocouple pilot safety?

3. Describe how a thermocouple works.

4. Describe a mercury sensor application.

5. What is the typical heat content of 1 cubic foot of natural gas?

6. Where does natural gas come from?

7. What is the typical pressure for a natural gas furnace manifold?

8. What component reduces the main pressure for a typical gas furnace?

9. How does the gas company make the consumer aware of a gas leak?

10. How does the gas company charge the consumer for gas consumption?

Gas Furnace Heat-Sensing Elements

Name	Date	Grade

Objectives: Upon completion of this exercise, you will be able to state the use of the various sensing elements on a gas furnace with a thermocouple pilot safety control.

Introduction: On a gas furnace with a thermocouple pilot safety device, you will locate the fan-limit and pilot safety controls, cause each sensing element to function, and record the function.

Text References: Paragraphs 31.15, 31.16, 31.17, 31.18, 31.28, and 31.29.

Tools and Materials: A gas furnace (upflow is preferred) with a temperature-operated fan-limit control, straight-blade and Phillips screwdrivers, 4" and 6" adjustable wrenches, a millivolt meter, a thermocouple adapter, and a thermometer.

Safety Precautions: Turn the power to the furnace off before installing the millivolt adapter and placing the thermometer probe. Lock and tag the panel where the power is turned off. There should be only one key. Keep this key on your person while performing this exercise.

PROCEDURES

1. With the power off and the electrical panel locked, remove the burner compartment panel.

2. Locate the fan-limit control, Figures 31-44, 31-46, 31-47, 31-48.

3. Locate the gas valve and the thermocouple connection to the gas valve. Install a thermocouple adapter for taking millivolt readings, Figure 31-2.

4. Follow the instructions on the furnace and light the pilot light.

5. When the pilot light is burning correctly, fasten the millivolt meter to the adapter and record the millivolt reading: _____ mV

6. With only the pilot light burning, blow out the pilot light and record the voltage of the thermocouple at the time the pilot safety valve drops out: _____ mV

7. With the power off, insert a temperature lead in the furnace outlet plenum close to the heat exchanger.

8. Turn the power on and start the furnace. Record the temperature at 2-minute intervals. Circle the time the fan motor starts:

   2 minutes _____ °F   4 minutes _____ °F
   6 minutes _____ °F   8 minutes _____ °F
   10 minutes _____ °F   12 minutes _____ °F

9. When the fan motor starts, shut the power off, remove one fan motor lead and tape it for safety. Turn the power back on.

10. Let the burner operate and watch for the limit switch to shut the burner off because there is no air circulation. Record the temperature here: _____ °F

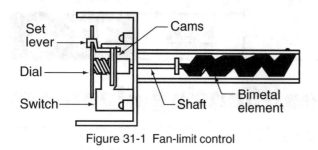

Figure 31-1 Fan-limit control

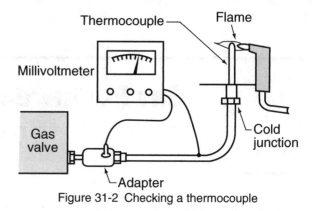

Figure 31-2 Checking a thermocouple

11. Turn the power off and replace the fan wire. Remove the temperature lead. Be careful not to burn your hands.

Maintenance of Workstation and Tools. Replace all panels and make sure the furnace is left as instructed. Return all tools to their places and make sure the work area is clean.

Summary Statement: Describe the action and principle of a thermocouple.

QUESTIONS

1. What are the circumstances that would cause the limit control to shut the main burner off?

2. Approximately how long does it take the thermocouple to shut the gas supply off when the pilot light goes out?

3. What materials are some thermocouples made of?

4. Is the thermocouple in contact with the pilot flame or the main burner flame?

5. Does the thermocouple open a solenoid valve or hold a solenoid valve open?

6. Name two pilot safety devices other than the thermocouple.

7. What is the sensing element in a typical fan-limit switch made of?

8. Are all fan-starting elements activated by the furnace temperature?

Gas Valve, Blower Fan Motor, Blower, and Draft Analysis

Name	Date	Grade

Objectives: Upon completion of this exercise, you will be able to understand and analyze a gas furnace's gas valve, blower fan motor, blower, and draft, and perform calculations on furnace capacity.

Introduction: You will remove the panels of the furnace for the purpose of identification and analysis of many components within the furnace.

Text References: Paragraphs 31.5, 31.6, 31.7, 31.8, 31.9, 31.10, 31.13, and 31.14.

Tools and Materials: A gas furnace, straight-blade and Phillips screwdrivers, 1/4" and 5/16" nut drivers, a 6" adjustable wrench, velometer and/or anemometer, flashlight, calculator, mirror, Volt-Ohm-Amp meter, manometer, two thermometers, and a pencil with eraser.

Safety Precautions: Make certain the furnace is properly grounded and that you have safety glasses and the proper protective clothing for making these observations and tests.

PROCEDURES

Gas Valve Information

1. What is the amp draw of the gas valve? _____ A

2. Does the gas valve have one or two electric operators?

3. Is the gas valve a redundant gas valve?

4. Is there any time delay built into the valve?

5. Is the gas valve one or two stage?

6. Does the main gas valve body contain the pressure regulator?

Fan Motor and Blower Information

1. What is the rated amperage of the fan motor? _____ A

2. What is the actual amperage draw of the fan motor with the access panels installed? _____ A

3. With the fan running and the access panel removed, does the fan motor overload? Record the amperage reading: _____ A

4. With the fan motor running, block the return air inlet. What happened to the ampere draw? Why did this happen?

5. With the fan motor running, block the discharge air. What happened to the amp draw of the motor? Why did this happen?

6. With the information you obtained from questions 4 and 5, explain how you would determine that an air blockage was in the return or discharge duct.

7. Is the fan motor single or dual voltage?

8. How many speeds can the fan motor be wired for? If the fan motor has more than one speed, which speed is used for heating purposes? Which speed is being used for the fan relay circuit?

9. What type of fan blades are in the blower?

10. Is the fan motor reversible?

Draft Information

1. List the type of draft used on the unit. (atmospheric, forced, induced)

2. If the unit is induced or forced draft, what type of component is used to prove that the draft has been established? Also explain how this type of component proves that the draft has been established.

General Checks and Test Data

1. Level the thermostat if it is a mercury bulb type.

2. Check the thermostat for proper calibration. If out of calibration, calibrate it.

3. Make sure the anticipator in the thermostat is set properly.

4. Adjust primary air for proper flame if there is an adjustment.

5. Hook the manometer up to the unit and check the gas pressure before and after the regulator. Adjust if necessary.

6. Properly adjust the fan control and high limit setting if the controls are adjustable.

7. Check to make sure electrical components are tight. Make sure the electrical circuit operates the unit properly.

8. Check all types of proving devices on the unit to make sure they will protect the unit if a malfunction exists.

9. Check the temperature rise across the heat exchanger and record it: _____ °F

10. Take a velometer and/or anemometer and check the velocity of the heating air across the unit. Determine the CFM of air across the unit. Show your work in the space below. Show all work and units.

11. Cover up ⅓ of the discharge air opening and repeat step 10. Show all work and units.

12. Now, cover up ½ of the discharge air opening and repeat step 10. Show all work and units.

13. Now, cover up ¾ of the discharge air opening and repeat step 10. Show all work and units.

14. Compare the velocity and the CFM in steps 11, 12, and 13. If the CFM decreased at a fast rate, explain why.

15. Take the data you collected in step 10 along with the rated output of the unit and the delta-T of the unit and determine whether the heating unit is operating within the specifications of the nameplate. Show all work and units.

16. Be prepared to explain the operation of the unit and the function of each component both electrically and mechanically.

Wiring for a Gas Furnace

Name	Date	Grade

Objectives: Upon completion of this exercise, you will be able to follow the control sequence and wiring, and state the application of the various controls.

Introduction: You will trace the wiring on a gas furnace and develop a pictorial and a ladder-type diagram for a gas furnace.

Text Reference: Paragraph 31.27.

Tools and Materials: A typical gas furnace with a thermocouple safety pilot, a flashlight, colored pencils, $1/4$" and $5/16$" nut drivers, and straight-blade and Phillips screwdrivers.

Safety Precautions: Before starting this exercise, turn the power to the furnace off, lock and tag the electrical panel, and keep the key on your person.

PROCEDURES

1. Make sure the power is off and the electrical panel is locked. Remove the front cover to the furnace.

2. Remove the cover to the junction box where the power enters the furnace.

3. Study the wiring of the furnace. After you feel you understand the wiring, in the space below draw a pictorial and a ladder wiring diagram of the complete high- and low-voltage circuit of the gas furnace. Hint: Draw each component showing all terminals in the approximate location first, and then draw the wire connections. DO NOT try to draw the thermocouple circuit.

PICTORIAL DIAGRAM   LADDER DIAGRAM

4. Replace all covers and panels.

Maintenance of Workstation and Tools. Return all tools and clean the work area.

Summary Statements: Using the ladder diagram, describe the complete sequence of events required to operate the gas valve and the fan, starting with the thermostat. Explain what happens with each control.

QUESTIONS

1. What is the job of the thermocouple?

2. What is a thermopile? Why are they used?

3. Why should the thermocouple not be tightened down too tightly at the gas valve?

4. Does the limit control shut off power to the transformer or to the gas valve in the furnace you used?

5. Why is the typical line voltage for a gas furnace 120 volts?

6. Did the furnace in the diagram above have a door switch?

7. If a furnace had a door switch, how would the technician troubleshoot the furnace electrical circuit if the door had to be removed?

8. At how many speeds will the fan motor in the above diagram operate?

9. Why is the typical low voltage for a gas furnace 24 volts?

10. Does the furnace in the diagram above have one or two limit controls?

Maintenance of Gas Burners

Name	Date	Grade

Objectives: Upon completion of this exercise, you will be able to perform routine maintenance on an atmospheric gas burner.

Introduction: You will remove the burners from a gas furnace, perform routine maintenance on the burners, place them back in the furnace, and put the furnace back into operation.

Text References: Paragraphs 31.4, 31.13, and 31.30.

Tools and Materials: A gas furnace, straight-blade and Phillips screwdrivers, a flashlight, two 6" adjustable wrenches, two 10" pipe wrenches, $1/4$" and $5/16$" nut drivers, soap bubbles for leak testing, compressed air, goggles, and a vacuum cleaner.

Safety Precautions: Turn the power and gas to the furnace off before beginning. Lock and tag the electrical panel and keep the key with you. Wear safety glasses when using compressed air.

PROCEDURES

1. With the power and gas off and the electrical panel locked, remove the front panel of the furnace.

2. Disconnect the burner manifold from the gas line.

3. Remove any fasteners that hold the burner or burners in place and remove the burner or burners.

4. Examine the burner ports for rust and clean as needed. Put on safety goggles and use compressed air to blow the burners out.

5. Vacuum the burner compartment and the manifold area.

6. Remove the draft diverter for access to the top or end of the heat exchanger. Vacuum this area.

7. Using the flashlight, examine the heat exchanger in every area that you can see, looking for cracks, soot, and rust.

8. Shake any rust or soot down to the burner compartment and vacuum again. A folded coat hanger or brushes may be used to break loose any debris. You can see how much debris came from the heat exchanger area by vacuuming it first and then again after cleaning the heat exchanger.

9. Replace the burners in their proper seats and all panels.

10. Turn the gas on and use soap bubbles for leak testing any connection you loosened or are concerned about.

11. Turn the power on and start the furnace. Wait for the burner to settle down and burn correctly. The burner may burn orange for a few minutes because of the dust and particles that were broken loose during the cleaning.

12. When you are satisfied that the burner is operating correctly, shut the furnace off and replace all panels.

Maintenance of Workstation and Tools. Return all tools to their places. Make sure the work area is clean and that the furnace is left as instructed.

Summary Statement: Describe the complete combustion process, including the characteristics of a properly burning flame.

QUESTIONS

1. How many cubic feet of air must be supplied to burn 1 cubic foot of gas?

2. How many cubic feet of air are normally furnished to support the combustion of 1 cubic foot of gas?

3. Why is there a difference in the amount of air in questions 1 and 2?

4. Name the three typical types of atmospheric gas burners.

5. What is the difference in an atmospheric burner and a forced-air burner?

6. What is the part of the burner that increases the air velocity to induce the primary air?

7. Where is secondary air induced into the burner?

8. What is the purpose of the draft diverter?

9. What type of air adjustment did the furnace you worked with have?

10. What are the symptoms of a shortage of primary air?

Checking a Gas Furnace for Correct Venting

Name	Date	Grade

Objectives: Upon completion of this exercise, you will be able to check a typical gas furnace to make sure that it is venting correctly.

Introduction: You will start up a typical gas furnace and use a match or smoke to determine that the products of combustion are rising up the flue.

Text Reference: Paragraph 31.25.

Tools and Materials: A small amount of fiberglass insulation, straight-blade and Phillips screwdrivers, $1/4$" and $5/16$" nut drivers, a flashlight, matches, and a typical operating gas furnace.

Safety Precautions: Following the directions with the furnace, light the pilot light. Take care while working around the hot burner and flue pipe. The vent pipe becomes very hot. It may be 350°F and will burn you.

PROCEDURES

1. Select any gas-burning appliance that is naturally vented through a flue pipe.

2. Remove the cover to the burner and make sure the pilot light is lit. If it is not, follow the appliance instructions and light it.

3. Turn the thermostat up to the point that the main burner lights.

4. While the furnace is heating up to temperature, you may carefully touch the flue pipe to see whether it is becoming hot.

5. Locate the draft hood. See the text, paragraph 31.25.

6. Place a lit match or candle at the entrance to the draft hood where the room dilution air enters the draft hood. The flame will follow the airflow and should be drawn toward the furnace and up the flue pipe, Figure 31-3.

7. Shut the furnace off with the thermostat, and allow the fan to run and cool the furnace.

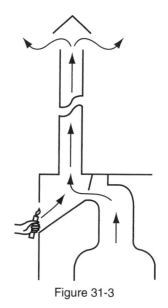

Figure 31-3

8. When the furnace has cooled, place the small amount of fiberglass insulation either at the top or the bottom of the flue pipe to restrict the flow of flue gases.

9. Start the furnace again and place the candle at the entrance to the draft regulator. The flame should sway out toward the room, showing that the appliance flue products are going into the room and not up the flue.

10. Stop the furnace and REMOVE THE FLUE BLOCKAGE.

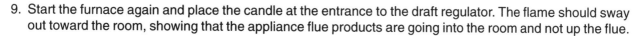

11. START THE FURNACE AND ONCE AGAIN PLACE THE MATCH FLAME AT THE ENTRANCE TO THE DRAFT DIVERTER TO PROVE THE FURNACE IS VENTING.

12. Replace all panels with the correct fasteners and MAKE SURE THE FLUE IS NOT BLOCKED.

Maintenance of Workstation and Tools. Return all tools to their respective places.

Summary Statement: Describe the function of the draft diverter.

QUESTIONS

1. What is the air that is drawn into the draft diverter called?

2. What gases are contained in the products of complete combustion?

3. What gases are contained in the products of incomplete combustion?

4. Name the component that transfers heat from the products of combustion to the circulated air.

5. Do all gas furnaces have products of combustion?

6. Describe a gas flame that is starved for oxygen.

7. When a gas flame is blowing and lifting off the burner head, what is the likely problem?

8. What does the draft diverter do in the case of a downdraft in the flue pipe?

9. What is the minimum size flue pipe that should be attached to a gas furnace?

10. Is it proper to vent a conventional gas furnace out the side of a structure without a vertical rise?

Combustion Analysis of a Gas Furnace

Name	Date	Grade

Objectives: Upon completion of this exercise, you will be able to perform a combustion analysis on a gas furnace.

Introduction: You will use a combustion analyzer to analyze a gas furnace for correct combustion.

Text References: Paragraphs 31.4 and 31.10.

Tools and Materials: Straight-blade and Phillips screwdrivers, $1/4$" and $5/16$" nut drivers, a flue gas analysis kit including thermometer, and an operating gas furnace.

Safety Precautions: Be very careful working around live electricity. You will be working near a flame and a hot flue gas pipe. Do not burn your hands.

PROCEDURES

1. Remove the panel to the burner section of the furnace.

2. Light the pilot light if needed.

3. Start the furnace burner and watch for ignition.

4. Let the furnace burner burn for 10 minutes and then examine the flame. Compare the flame to the explanation in the text, Figure 30-10. If the flame needs adjusting, make the adjustments.

5. Following the instructions in the particular flue gas kit that you have, perform a flue gas analysis. Record the results here:

   • Percent carbon dioxide content _____ %
   • Flue gas temperature _____ °F
   • Efficiency of the burner _____ %

6. Adjust the air shutter for minimum air to the burner and wait 5 minutes.

7. Perform another combustion analysis and record the results here:

   • Percent carbon dioxide content _____ %
   • Flue gas temperature _____ °F
   • Efficiency of the burner _____ %

8. Adjust the burner air back to the original setting and wait 5 minutes. Perform one last test and record the results here:

   • Percent carbon dioxide content _____ %
   • Flue gas temperature _____ °F
   • Efficiency of the burner _____ %

9. Turn the furnace off and replace any panels with the correct fasteners.

Maintenance of Workstation and Tools: Return all tools to their respective places.

Summary Statement: Describe the difference between perfect combustion and ideal or typical accepted combustion.

QUESTIONS

1. What is CO_2?

2. What is CO?

3. Name the products of complete combustion.

4. What is excess air?

5. Why is perfect combustion never obtained?

6. What is the ideal CO_2 content of flue gas for a gas furnace?

7. How much air is consumed to burn 1 cubic foot of gas in a typical gas burner?

8. Why must flue gas temperatures in conventional furnaces be kept high?

9. What do yellow tips mean on a gas flame?

10. How is the air adjusted on a typical gas burner?

Determining Airflow (cfm) Using the Air Temperature Rise Method

Name	Date	Grade

Objectives: Upon completion of this exercise, you will be able to determine the amount of air in CFM moving through a gas furnace, using the air temperature rise method.

Introduction: You will use a thermometer and the gas input for a standard efficiency gas furnace to determine the cfm of airflow in a system. A water manometer will be used to verify the correct gas pressure.

Text Reference: Paragraph 31.14.

Tools and Materials: A standard efficiency gas furnace, a thermometer, and a water manometer.

Safety Precautions: Shut the gas off before checking the manifold pressure of the gas furnace.

PROCEDURES

1. Make sure the gas is off and install the water manometer on the gas furnace.

2. Turn the gas on, start the furnace, and record the gas pressure at the manifold (between the gas valve and the burners): _____ inches of water column. This should be 3.5 inches of water column for natural gas. If you have a gas other than natural gas, your instructor will have to help you with the correct pressure and Btu content of the gas. If the pressure is not correct, adjust the regulator until it is correct.

3. When you have established the correct gas pressure at the manifold, stop the furnace, shut the gas off, and remove the water manometer.

4. Turn the gas on, start the furnace, and while it is getting up to temperature, install a thermometer lead in the supply and return ducts. You will achieve the best reading by placing the supply duct lead several feet from the furnace, where the air leaving the furnace will have some distance to mix, Figure 31-4.

5. Record the temperature difference after the furnace has operated for 15 minutes. _____ °F TD.

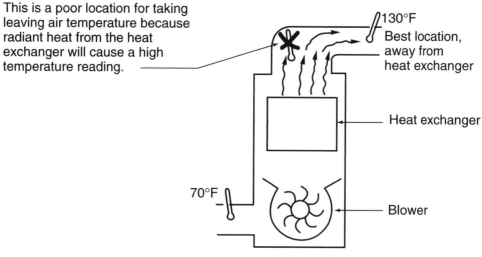

This is a poor location for taking leaving air temperature because radiant heat from the heat exchanger will cause a high temperature reading.

130°F
Best location, away from heat exchanger

Heat exchanger

70°F

Blower

Figure 31-4

6. Calculate the furnace gas output based on the nameplate rating in Btu/h × 0.80 (0.80 is the decimal equivalent of 80%, the efficiency of a standard gas furnace). For example, a 100,000 Btu/h furnace has a typical input of 100,000 × 0.80 = 80,000 Btu/h. The remaining heat goes up the flue of the furnace as products of combustion. Your furnace output is _____ nameplate rating × 0.80 = Btu/h.

7. Use the following formula to determine the airflow.

$$\text{Cfm} = \frac{\text{Total heat output}}{1.1 \times \text{TD (temperature difference)}} = \underline{\hspace{1cm}} \text{Cfm}$$

8. Stop the furnace and replace all panels.

Maintenance of Workstation and Tools: Return all tools to their places. Make sure the work area is clean and that the furnace is left as you are instructed.

Summary Statement: Describe how heat is transferred from the burning gas to the air leaving the furnace.

QUESTIONS

1. What would the temperature of the leaving air do if the air filter were to become stopped up?

2. What would the flue gas symptoms be if there were too much airflow?

3. What is the Btu heat content of 1 cubic foot of natural gas?

4. What is the typical high and low temperature difference across a gas furnace?
   _____ high − _____ low = _____ °F

5. What would cause too much temperature difference?

6. What are the effects on a furnace and duct system when many of the outlet air registers are shut off to prevent heat in a room?

7. What is the efficiency of a standard gas furnace?

8. What is the efficiency of a high-efficiency gas furnace?

9. How is high efficiency accomplished?

10. What is the manifold pressure of a typical natural gas furnace?

Main Burner, Pilot Burner, Fan and Limit Control, Transformer, and Thermostat Analysis

Name	Date	Grade

Objectives: Upon completion of this exercise, you will be able to understand and analyze the gas furnace's main burner, pilot burner, fan and limit control, and transformer.

Introduction: You will remove the panels of the furnace for the purpose of identification and analysis of many components within the furnace.

Text References: Paragraphs 31.3, 31.4, 31.13, 31.14, 31.15, 31.16, 31.17, and 31.18.

Tools and Materials: A pilot-type gas furnace, straight-blade and Phillips screwdrivers, ¼" and ⁵⁄₁₆" nut drivers, a 6" adjustable wrench, flashlight, mirror, and a pencil with eraser.

Safety Precautions: Make certain the furnace is properly grounded and that you have goggles and the proper protective clothing for making these observations and tests.

PROCEDURES

1. What is the Btu input of the unit?

2. What is the recommended temperature rise across the heat exchanger?

3. What are the recommended static pressure requirements for the unit?

4. Is the unit 100% shut down or less than 100% shut down in case of pilot failure?

5. Is the unit an up-flow, counterflow, low-boy, or horizontal?

Burner Information

1. How many burners are in the unit?

2. What sizes are the orifices in the unit?

3. How many Btus of fuel will be consumed with the present orifices?

4. Is the burner a power burner?

Pilot Information

1. Is the pilot manual freestanding, automatic freestanding, or automatic recycling?

2. If the pilot is automatic, does it use a glow coil, high-voltage arc, or hot-surface ignition process?

3. What type of pilot or flame-proving device is used on the unit? Is it millivolt, bimetallic, liquid-filled tube, or flame rectification?

4. Explain the principle of operation behind the proving device you chose in question 3 above.

5. Is the pilot gas flow controlled by a millivolt pilot stat or a 24-volt operator?

Fan Control and High Limit Information

1. How many limits are used in the unit? If more than one limit is used in the unit, explain the purpose each one is used for.

2. What types or high limits are used on the heating unit: helix, thermal disc, or fuse type?

3. If the limits are thermal disc or fuse type, at what temperature will they open the circuit?

4. What type of fan control is used on the unit: helix, thermal disc, or time delay?

5. If the fan control is a thermal disc, at what temperature will it close the circuit and energize the fan?

6. If the fan control is a time delay, list how long it takes to close the fan circuit after it has been energized.

Transformer Information

1. What is the VA rating of the control transformer?

2. Determine the maximum amperage used by the control circuit when the maximum loads are energized at any given time.

3. What is the maximum amount of current draw that could be added to the control circuit without overloading the control transformer?

Thermostat Information

1. Is the thermostat a heating thermostat or combination heating/cooling thermostat?

2. Does the thermostat have an anticipator?

3. If the thermostat has an anticipator, is it fixed or adjustable?

4. If the anticipator is adjustable, take an amp draw reading of the control circuit and set the anticipator to the correct setting. What is this correct setting?

5. What components in the unit determine the anticipator setting?

6. Explain what may happen with the operation of the unit if the anticipator is set at a higher amperage setting than it is supposed to be. Also explain why.

7. Explain what may happen with the operation of the unit if the anticipator were set at a lower amperage setting than it is supposed to be. Also explain why.

Troubleshooting Exercise:
Gas Furnace with Standing Pilot

Name	Date	Grade

Objectives: Upon completion of this exercise, you will be able to troubleshoot a gas furnace with a standing pilot.

Introduction: This exercise has a specific problem introduced to the furnace by your instructor. You will locate the problem using a logical approach and instruments. THE INTENT IS NOT FOR YOU TO MAKE THE REPAIR BUT TO LOCATE THE PROBLEM AND ASK YOUR INSTRUCTOR FOR DIRECTION BEFORE COMPLETION.

Text Reference: Unit 31.

Tools and Materials: Straight-blade and Phillips screwdrivers, $1/4$" and $5/16$" nut drivers, VOM, millivoltmeter, and a thermocouple adapter.

Safety Precautions: Be careful around energized electrical terminals. Use all safety precautions. Keep hands well back on meter leads when taking electrical measurements. Keep hands away from all electrical terminals.

PROCEDURES

1. Turn the power to the furnace on and set the thermostat to call for heat.

2. Describe what happens.

3. Turn the power off and remove the cover to the burner compartment. Is the pilot light burning?

4. If the pilot light is off, go through the pilot lighting procedures.

5. Describe what happened.

6. Using the thermocouple adapter, check the thermocouple. Remember, the pilot light will have to be lit for the thermocouple to generate power. You will need to hold the button down on some furnaces to allow pilot light gas to the pilot light for this test.

7. Determine the voltage without a load on the thermocouple _____ millivolts.

8. Screw the thermocouple back in the valve and with the pilot lit, check the voltage with a load on the thermocouple of _____ millivolts.

9. Describe the problem here.

10. Describe the recommended repair here.

11. Make the recommended repair if your instructor directs you to, and start the furnace.

Maintenance of Workstation and Tools: Leave the furnace as your instructor tells you and place all tools in their correct places.

Summary Statement: Describe how a thermocouple generates voltage.

QUESTIONS

1. What is the advantage of a standing pilot for a furnace located under a house?

2. Approximately how much heat does a standing pilot give off?

3. What is the difference in a 100% shutoff and non-100% shutoff pilot system?

4. Describe how a thermocouple works.

5. Can a pilot light flame be adjusted?

6. What is the recommended manifold pressure for a natural gas furnace?

7. What is the recommended manifold pressure for an LP gas furnace?

8. Which gas is heavier than air, natural or LP?

9. Describe a thermopile.

10. Name two different types of pilot lights.

Troubleshooting Exercise:
Gas Furnace with Spark Ignition

Name	Date	Grade

Objectives: Upon completion of this exercise, you will be able to troubleshoot a simple problem in a gas furnace with spark ignition.

Introduction: This unit has a specified problem introduced by your instructor. You will locate the problem using a logical approach and instruments. The manufacturer's troubleshooting procedures would help. THE INTENT IS NOT FOR YOU TO MAKE THE REPAIR BUT TO LOCATE THE PROBLEM AND ASK YOUR INSTRUCTOR FOR DIRECTION BEFORE COMPLETION.

Text Reference: Unit 31.

Tools and Materials: Straight-blade and Phillips screwdrivers, $1/4$" and $5/16$" nut drivers, and a VOM.

Safety Precautions: Use care while working with electricity.

PROCEDURES

1. Turn the power on and set the thermostat to call for heat.

2. Give the furnace a few minutes to start. It should not start if the correct problem is in the furnace. Describe the sequence of events while the furnace is trying to start.

3. Did the pilot light ignite?

4. Did the burner ignite?

5. Did the fan start?

6. Use a VOM to make sure the thermostat is calling for heat. Check for low voltage at the gas valve.

7. Is there a spark at the spark igniter for the pilot light?

8. If there is no spark, check the manufacturer's literature for the procedure for checking the circuit.

9. If you do not have the manufacturer's instruction, check the gap at the spark igniter. Look for a grounded lead or electrode.

10. Describe your findings.

11. If your instructor agrees, make the repair and start the furnace.

Maintenance of Workstation and Tools: Return the workstation and tools to the correct order.

Summary Statement: Describe the events of a correctly working spark ignition system.

QUESTIONS

1. What is the difference between spark ignition and direct spark ignition?

2. Why is spark ignition important for rooftop or outdoor equipment?

3. How does spark ignition save money over a standing pilot system?

4. What is the approximate voltage for the spark to pilot ignition system?

5. Did the spark ignition system on this furnace have a circuit board?

6. Were you able to follow the wiring diagram furnished with the furnace?

7. What would you suggest to make the wiring diagram more clear?

8. Was there a troubleshooting procedure furnished with the furnace?

9. Was it clear and did it direct you to the problem?

10. How does low voltage affect the spark ignition system?

Changing a Gas Valve on a Gas Furnace

Name	Date	Grade

Objectives: Upon completion of this exercise, you will be able to change a gas valve on a typical gas furnace using the correct tools and procedures.

Introduction: You will use the correct wrenches and procedures and change the gas valve on a gas furnace.

Text References: Paragraphs 31.6, 31.7, 31.8, 31.9, and 31.10.

Tools and Materials: Straight-blade and Phillips screwdrivers, needle-nose pliers, two 10" adjustable wrenches, $3/8$" and $7/16$" open-end wrenches, two 14" pipe wrenches, soap bubbles for leak checking a VOM, and thread seal compatible with natural gas.

Safety Precautions: Make sure the power and the gas are off before starting this exercise. The electrical panel should be locked and you should have the key.

PROCEDURES

1. Turn the power off and check it with the VOM. Lock and tag the electrical panel.

2. Turn the gas off at the main valve before the unit is serviced.

3. Remove the door from the furnace. Watch how it removes so you will know how to replace it.

4. Make a wiring diagram of the wiring to be removed. The wiring may need to be tagged for proper identification. Remove the wires from the gas valve. Use needle-nose pliers on the connectors if they are spade type. Do not pull the wires out of their connectors.

5. Look at the gas piping and decide where to take it apart. There may be a pipe union or a flare nut connection to the inlet of the system. Either one may be worked with.

6. Remove the pilot line connections from the gas valve. USE THE CORRECT SIZE END WRENCH AND DO NOT DAMAGE THE FITTINGS. THE PILOT LINE IS DELICATE, ALUMINUM.

7. Use the adjustable wrenches for flare fitting connections and pipe wrenches for gas piping connections and disassemble the piping to the gas valve. LOOK FOR SQUARE SHOULDERS ON THE GAS VALVE AND BE SURE TO KEEP THE WRENCHES ON THE SAME SIDE OF THE GAS VALVE. TOO MUCH STRESS CAN EASILY BE APPLIED TO A GAS VALVE BY HAVING ONE WRENCH ON ONE SIDE OF THE VALVE AND THE OTHER ON THE PIPING ON THE OTHER SIDE OF THE VALVE.

8. Remove the valve from the gas manifold. USE CARE NOT TO STRESS THE VALVE OR THE MANIFOLD.

9. While you have the valve out of the system, examine the valve for some of its features, such as where a thermocouple or spark igniter connects. Look for damage.

10. After completely removing the gas valve, treat it like a new one and fasten it back to the gas manifold. BE SURE TO USE THREAD SEAL ON ALL EXTERNAL PIPE THREADS. DO NOT OVERUSE AND DO NOT USE ON FLARE FITTINGS.

11. Fasten the inlet piping back to the gas valve.

12. Fasten the pilot line back using the correct wrench.

13. Fasten all wiring back using the wiring diagram.

14. Ask your instructor to look at the job and approve, then turn the gas on. IMMEDIATELY USE SOAP BUBBLES AND LEAK CHECK THE GAS LINE UP TO THE GAS VALVE.

15. Turn the thermostat to call for heat. WHEN THE BURNER LIGHTS, YOU HAVE GAS TO THE VALVE OUTLET AND THE PILOT LIGHT. LEAK CHECK IMMEDIATELY.

16. When you are satisfied there are no leaks, turn the furnace off.

17. Turn the power off, and lock and tag the electrical panel.

18. Use a wet cloth and clean any soap bubble residue off the fittings.

Maintenance of Workstation and Tools: Return the workstation to proper condition as the instructor directs you. Return the tools to their respective places.

Summary Statement: Describe what may happen if too much thread seal is used on the inlet fittings to the gas valve and the outlet to the gas valve.

QUESTIONS

1. Did the gas valve on this system have a square shoulder for holding with a wrench?

2. Did you have any gas leaks after the repair?

3. Why is thread seal used on threaded pipe fittings?

4. Why is thread seal not used on flared fittings?

5. Is thread seal used on the pipe union on the center connection?

Changing a Fan Motor on a Gas Furnace

Name	Date	Grade

Objectives: Upon completion of this exercise, you will be able to change a fan motor on a gas furnace and then restart the furnace.

Introduction: You will use the correct working practices and tools to change a fan motor on a gas furnace.

Text References: Paragraphs 17.8, 17.9, 17.14, and 17.15.

Tools and Materials: Straight-blade and Phillips screwdrivers, $\frac{1}{4}$" and $\frac{5}{16}$" nut drivers, Allen wrenches, small wheel puller, needle-nose pliers, a VOM, an ammeter, 6" adjustable wrench, and slip joint pliers.

Safety Precautions: Make sure the power is off, and the electrical panel is locked and tagged before starting this exercise. Be sure to properly discharge any fan capacitors.

PROCEDURES

1. Turn the power off and check it with a VOM. Lock and tag the electrical panel.

2. Remove the fan compartment door.

3. Make a wiring diagram for the motor wiring so replacement will be easy.

4. Discharge any fan capacitors, and disconnect the motor wires.

5. If the motor is direct drive, remove the blower from the compartment. If the motor is a belt drive, loosen the tension on the belt.

6. If the motor is direct drive, remove the motor from the blower wheel. If the motor is a belt drive, remove it from the mounting bracket and remove the pulley with a puller if needed.

7. Treat this motor as a new motor and replace it in the reverse order that you removed it. Be sure to select the correct motor speed.

8. Double-check the wiring.

9. When all is back together, ask your instructor to check the installation.

10. When the installation has been inspected, start the motor and check the amperage. REMEMBER TO CHECK THE AMPERAGE WITH THE FAN DOOR IN PLACE. YOU MAY NEED TO CLAMP THE AMMETER AROUND THE WIRE ENTERING THE FURNACE.

11. Turn the furnace off.

Maintenance of Workstation and Tools: Leave the furnace as specified by your instructor. Return all tools to their places.

Summary Statement: Describe the airflow through a gas furnace.

QUESTIONS

1. What type of fan motor did this furnace have, shaded pole or permanent split capacitor?

2. Which type of fan motor is the most efficient, shaded pole or permanent split capacitor?

3. Why would a manufacturer use an inefficient motor when an efficient one is available?

4. What type of motor mount did the motor in this exercise have, rigid or rubber?

5. What type of bearings did the motor have, sleeve or ball bearing?

6. Was the motor single or multiple speed?

7. What was the color of the capacitor leads on the unit (if applicable)?

8. What happens to the fan amperage if the leaving airflow is restricted?

9. What happens to the fan amperage if the entering airflow is restricted?

10. Did the motor have oil ports?

Situational Service Ticket

Name	Date	Grade

NOTE: Your instructor must have placed the correct service problem in the system for you to successfully complete this exercise.

Technician's Name:_____ Date: _____

CUSTOMER COMPLAINT: No heat.

CUSTOMER COMMENTS: Furnace will not start.

TYPE OF SYSTEM:

OUTDOOR UNIT:
MANUFACTURER_____   MODEL NUMBER_____   SERIAL NUMBER_____

INDOOR UNIT:
MANUFACTURER_____   MODEL NUMBER_____   SERIAL NUMBER_____

COMPRESSOR (where applicable):
MANUFACTURER_____   MODEL NUMBER_____   SERIAL NUMBER_____

TECHNICIAN'S REPORT

1. SYMPTOMS: _____

2. DIAGNOSIS: _____

3. ESTIMATED MATERIALS FOR REPAIR: _____

4. ESTIMATED TIME TO COMPLETE THE REPAIR: _____

SERVICE TIP FOR THIS CALL

1. Use your voltmeter.

Situational Service Ticket

Name	Date	Grade

NOTE: Your instructor must have placed the correct service problem in the system for you to successfully complete this exercise.

Technician's Name:_____ Date: _____

CUSTOMER COMPLAINT: No heat.

CUSTOMER COMMENTS: Furnace stopped in the night.

TYPE OF SYSTEM:

OUTDOOR UNIT:
MANUFACTURER_____ MODEL NUMBER_____ SERIAL NUMBER_____

INDOOR UNIT:
MANUFACTURER_____ MODEL NUMBER_____ SERIAL NUMBER_____

COMPRESSOR (where applicable):
MANUFACTURER_____ MODEL NUMBER_____ SERIAL NUMBER_____

TECHNICIAN'S REPORT

1. SYMPTOMS: _____

2. DIAGNOSIS: _____

3. ESTIMATED MATERIALS FOR REPAIR: _____

4. ESTIMATED TIME TO COMPLETE THE REPAIR: _____

SERVICE TIPS FOR THIS CALL

1. Use your voltmeter.

Oil Heat

Unit Overview

When the thermostat calls for heat on an oil furnace, the oil burner starts. The oil-air mixture is ignited. After the mixture has heated, the combustion chamber and the heat exchanger reach a predetermined temperature. The fan comes on and distributes the heated air from around the heat exchanger through the duct to the space to be heated. When this space has reached a specific temperature, the thermostat causes the burner to shut down. The fan continues to run until the heat exchanger cools to a set temperature.

There are six grades of fuel oil used as heating oils. The No. 2 grade is the one most commonly used in residential and light commercial oil heating. One gallon of No. 2 fuel oil produces about 140,000 Btu of heat energy when burned. Oil is a fossil fuel pumped from the earth. The lighter fuel oils are products of distillation at an oil refinery, during which the petroleum is vaporized, condensed, and the different grades are separated.

Proper oil storage is important to ensure that the quality of the oil is maintained prior to use. Oil can be stored either above ground or underground.

The condensing oil furnace is one of the latest product designs producing more efficiency. This type of furnace has one or more typical heat exchangers plus another coil-type exchanger. The flue gases are passed through the traditional heat exchangers where much of the heat is removed. Then they pass through the coil-type heat exchanger where nearly all of the remaining heat is removed. In this coil-type heat exchanger, the temperature is reduced below the dew point and moisture is condensed from the flue gases. This change of state is a latent heat high-efficiency exchange. The remaining flue gases are forced with a blower through a vent usually constructed of PVC.

Key Terms

Air Tube
Atomization
Bleeding a Fuel Line
Booster Pump
Burner Fan or Blower
Burner Motor
Cad Cell
Carbon Dioxide Test
Combustion Chamber

Condensing Oil
   Furnace
Downflow
Draft Test
Electrodes
Flame Impingement
Fuel-Line Filter
Fuel Oil
Fuel Oil Pump

Gun-Type Oil Burner
Heat Exchanger
Horizontal Furnace
Ignition Transformer
Intermittent Ignition
Interrupted Ignition
Low-Boy
Net Stack Temperature
Nozzle

One-Pipe Supply
   System
Primary Control
Smoke Test
Stack Switch
Two-Pipe Supply
   System
Upflow

REVIEW TEST

Name	Date	Grade

Circle the letter that indicates the correct answer.

1. **An upflow furnace:**
   A. is generally used where there is little head-room.
   B. has the return air at the bottom.
   C. has the return air at the top.
   D. has the air discharge at the bottom.

2. **Of the six regular grades of fuel oil, the grade most commonly used in residential and light commercial heating systems is No.:**
   A. 1.
   B. 2.
   C. 4.
   D. 6.

3. How much heat is produced when 2 gallons of No. 2 fuel oil are burned?
   A. 10,000 Btu
   B. 50,000 Btu
   C. 140,000 Btu
   D. 280,000 Btu

4. When hydrocarbons unite rapidly with oxygen, it is called:
   A. vaporization.
   B. condensation.
   C. atomization.
   D. combustion.

5. Assuming perfect combustion, how many pounds of oxygen are required to burn 1 pound of fuel oil?
   A. 1.44 lb
   B. 3.00 lb
   C. 14.4 lb
   D. 30.00 lb

6. Assuming 50% excess air is being provided, how many cubic feet of air will be supplied to an oil burner to burn 1 pound of fuel oil?
   A. 3.0 cu ft
   B. 14.4 cu ft
   C. 144 cu ft
   D. 288 cu ft

7. When fuel oil is broken up to form tiny droplets, it is called:
   A. heat of fusion.
   B. condensation.
   C. atomization.
   D. combustion.

8. The motor on a residential gun-type oil burner is usually a _____ motor.
   A. split-phase multihorsepower
   B. split-phase fractional horsepower
   C. three-phase
   D. shaded pole

9. The pump on a residential gun-type oil burner is a _____ pump.
   A. centrifugal
   B. single-stage or dual-stage rotary
   C. scroll-type
   D. air

10. The fuel oil enters the orifice of the nozzle through the:
    A. hollow cone.
    B. squirrel cage fan.
    C. swirl chamber.
    D. flame retention head.

11. The circular direction of the air in the air tube is in _____ direction as the fuel oil motion.
    A. the same
    B. the opposite
    C. either the same or opposite

12. The ignition transformer steps up the 120 V residential voltage to approximately _____ V.
    A. 208   C. 1000
    B. 440   D. 10,000

13. The stack switch:
    A. starts the burner when the thermostat calls for heat.
    B. passes power to the ignition transformer.
    C. provides for intermittent ignition.
    D. shuts the burner off if it does not sense heat in the flue.

14. The cad cell is a safety device that:
    A. senses heat in the flue.
    B. senses heat at the burner.
    C. senses light in the combustion chamber.
    D. senses heat in the heat exchanger.

15. The cad cell is often coupled with a:
    A. diode.   C. transistor.
    B. triac.   D. rectifier.

16. A booster pump is needed when the supply tank is _____ below the oil burner.
    A. any distance
    B. 5 feet or more
    C. 10 feet or more
    D. 15 feet or more

17. An oil deaerator:
    A. increases the pressure of the oil.
    B. allows a one-pipe oil delivery system to operate as a two-pipe oil delivery system.
    C. causes the temperature of the oil entering the fuel pump to drop.
    D. eliminates the need for an oil filter.

18. A fuel line filter is located between the:
    A. supply tank and the pump.
    B. pump and the nozzle.
    C. swirl chamber and the orifice.

19. **The heat exchanger in a forced-air oil furnace:**
   A. combines the flue gases and the air to be heated and circulated.
   B. transfers heat from the flue gases to the air to be heated and circulated and keeps them separate.
   C. is a heat sensor and shuts down the furnace if there is not enough heat in the flue.
   D. is a light sensor and shuts down the furnace if there is no flame burning in the combustion chamber.

20. **A condensing oil furnace achieves part of its higher efficiency because of a _____ exchange.**
   A. sensible heat   C. superheat
   B. latent heat   D. radiant heat

21. **Excess air is supplied at the burner to ensure that there will be:**
   A. a low net stack temperature.
   B. enough oxygen for complete combustion.
   C. enough hydrocarbons for complete combustion.
   D. sufficient atomization.

22. **How much excess air is typically supplied to the system to ensure that complete combustion occurs?**
   A. 10%
   B. 20%
   C. 50%
   D. 100%

23. **To make combustion efficiency tests, a hole must be made:**
   A. in the flue on the chimney side of the draft regulator.
   B. in the flue on the furnace side of the draft regulator.
   C. at the fuel manifold.
   D. under the draft hood.

24. **The net stack temperature is:**
   A. the difference between the flue temperature before the draft regulator and the temperature in the flue after it.
   B. the difference between the flue temperature and the temperature of the air surrounding the furnace.
   C. whether or not the airflow is correct across the heat exchanger.
   D. the amount of excess air at the burner.

25. **An air-temperature rise check is performed to determine:**
   A. the draft at the flue.
   B. the combustion efficiency.
   C. whether or not the airflow is correct across the heat exchanger.
   D. the amount of excess air at the burner.

Familiarization with Oil Furnace Components

Name	Date	Grade

Objectives: Upon completion of this exercise, you will be able to identify the various components of an oil furnace and describe their function.

Introduction: You will remove, if necessary, and study the components, and record their characteristics to help you become familiar with an oil furnace.

Text References: Paragraphs 32.2, 32.3, 32.7, 32.8, and 32.9.

Tools and Materials: Straight-blade and Phillips screwdrivers, two 6" adjustable wrenches, an Allen wrench set, an open end wrench set, a nozzle wrench, rags, an oil burner (Figure 32-1), and a flashlight.

Safety Precautions: Turn the power and oil supply line off before starting this exercise. Lock and tag the electrical panel and keep the key with you.

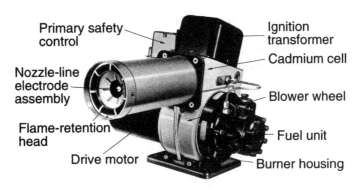

Figure 32-1 An oil burner. *Courtesy R.W. Beckett Corporation*

PROCEDURES

After turning the power and oil supply off, and locking the electrical panel, remove the front burner and blower compartment panels. Record the following:

1. Fan information (you may have to remove the fan):

   - Motor full-load amperage _____ A
   - Type of motor _____
   - Diameter of motor _____ in.
   - Shaft diameter _____ in.
   - Motor rotation (looking at motor shaft) _____
   - Fan wheel diameter _____ in.
   - Width _____ in.
   - Number of motor speeds _____ High rpm _____ Low rpm _____

2. Burner information (burner may have to be removed):
   - Nozzle size _____ gpm
   - Nozzle angle _____ degrees
   - Pump speed _____ rpm
   - Pump motor amperage _____ A
   - Number of pump stages _____
   - One- or two-pipe system _____
   - Type of safety control, cad cell or stack switch _____

3. Nameplate information:
   - Manufacturer _____
   - Model number _____
   - Serial number _____
   - Capacity _____ Btu/h
   - Furnace voltage _____ V
   - Control voltage _____ V
   - Recommended temperature rise, low _____, high _____

4. Is the oil tank above or below the furnace? _____

5. Is there an oil filter in the line leading to the furnace? _____ What is the oil line size? _____

6. Reinstall all components. Have your instructor inspect to ensure that the furnace is in working order.

Maintenance of Workstation and Tools: Replace all furnace panels and leave oil burner as you are instructed. Return all tools and materials to their respective places. Clean your work area.

Summary Statement: Describe the complete heating process from the oil supply to the heat transfer method, to the air in the conditioned space.

QUESTIONS

1. What is the heat content of a gallon of No. 2 fuel oil?

2. What unit of heat is an oil furnace capacity expressed in?

3. Are oil furnace capacities expressed in output or input?

4. What is the purpose of a two-pipe oil supply system?

5. How is the fan motor started in a typical oil furnace?

6. What is the typical voltage for an oil furnace?

7. What is the typical control voltage for an oil furnace?

8. What is the function of the limit control for an oil furnace?

9. What is the function of the stack switch on an oil furnace?

10. What is the function of a cad cell on an oil furnace?

Oil Burner Maintenance

Name	Date	Grade

Objectives: Upon completion of this exercise, you will be able to perform routine maintenance on a typical gun-type oil burner. You will also be able to describe the difference in resistance in a cad cell when it "sees" or does not "see" light.

Introduction: You will remove a gun-type oil burner from a furnace, and remove and inspect the oil nozzle. You will also check and set the electrodes.

Text References: Paragraphs 32.1, 32.2, 32.3, 32.7, 32.8, 32.9, 32.12, and 32.16.

Tools and Materials: Straight-blade and Phillips screwdrivers, a set of open end wrenches ($^3/_8$" through $^3/_4$"), two 6" adjustable wrenches, rags, a small shallow pan, a VOM, goggles, a flashlight, and a nozzle wrench.

Safety Precautions: Shut the power and oil supply off, and lock and tag the electrical panel before starting this exercise.

PROCEDURES

1. With the electrical power and oil supply off and the electrical panel locked, disconnect the oil supply line (and return if any). Let any oil droplets drop in the pan.

2. Disconnect the electrical connections to the burner. There should be both high- (120 V) and low-voltage connections.

3. Remove the burner from the furnace and place the burner assembly on a workbench. See Figure 32-31 and Figure 32-36 in the text for an example of a burner assembly.

4. Loosen the fastener that holds the transformer in place. The transformer will usually raise backward and stop. This exposes the electrode and nozzle assembly.

5. Loosen the oil-line connection on the back of the burner. This will allow the nozzle assembly to be removed from the burner. **NOTE:** This can be removed with the burner in place on the furnace, but we are doing it on a workbench for better visibility.

6. Examine the electrodes and their insulators. The insulators may be carefully removed and cleaned in solvent if needed. They are ceramic and very fragile.

7. Remove the oil nozzle using the nozzle wrench. Inspect and replace or reinstall the old nozzle. Follow instructions.

8. Set the electrodes in the correct position using an electrode gage, Figure 32-2, as a guide.

9. Replace the nozzle assembly back into the burner assembly.

10. Remove the cad cell (if this unit has a cad cell).

11. Connect your ohmmeter to the cad cell. Turn the cell eye toward a light and record the ohm reading here: _____ ohms

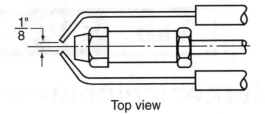

Top view

Height adjustment

$\frac{1"}{2}$ Residential installation

$\frac{3"}{8}$ Commercial installation

Side view

Position of electrodes in front of nozzle is determined by spray angle of nozzle.

Electrodes cannot be closer than $1/4$ in. to any metal part.

Figure 32-2

12. Cover the cell eye and record the ohm reading here: _____ ohms

13. Replace the cad cell in the burner assembly. Make sure the burner is reassembled correctly before replacing it on the furnace. Check to see that all connections are tight.

14. Return the burner assembly to the furnace and fasten the oil line or lines. Make all electrical connections.

Maintenance of Workstation and Tools: Return all tools to their places. Make sure there is no oil left on the floor or furnace.

Summary Statement: Describe the complete burner assembly, including the static disc in the burner tube and the cad cell.

QUESTIONS

1. Why should you be sure that no oil is left on any part, in addition to it being a possible fire hazard?

2. What is the color of No. 2 fuel oil?

3. What must be done to oil to prepare it to burn?

4. Why is the typical oil burner called a "gun-type" oil burner?

5. What is the typical pressure at which an oil burner operates?

6. Is there a strainer in an oil burner nozzle?

7. What is the purpose of the tangential slots in an oil burner nozzle?

Electronic Controls for Oil Burners

Name	Date	Grade

Objectives: Upon completion of this exercise, you will be able to follow the electrical circuit in an oil furnace and check a cad cell for correct continuity.

Introduction: You will use a VOM (volt-ohm-milliammeter) to follow the line voltage circuit through a primary control and a cad cell on a typical oil furnace.

Text References: Paragraphs 32.9 and 32.10.

Tools and Materials: A VOM with one lead with an alligator clip and the other lead with a probe, a screwdriver, electrical tape, and an oil furnace that has a cad cell and primary control.

Safety Precautions: Be sure to turn the power off while removing the primary control from its junction box and performing the cad cell check. Lock and tag the electrical panel and keep the key with you. Be very careful while making electrical measurements. Make these only under the supervision of your instructor.

PROCEDURES

1. Turn the power off, lock the electrical panel, and remove the primary control from its junction box. Suspend it to the side where voltage readings may be taken.

2. Set the voltmeter at the 250 V scale.

3. Use a lead with an alligator clip and place it on the white wire (neutral).

4. Start the oil furnace and make sure that ignition takes place, that the furnace is fired.

5. Prepare to take voltage readings on the black and orange wires. Remember the black wire is the hot lead feeding the primary and the orange wire is the hot lead leaving the primary to the burner motor and transformer.

6. Using the pointed end of the second probe, check for power at the black terminal and record here: _____ V

7. Check the voltage at the orange terminal and record here: _____ V

8. Turn the furnace power supply off, and lock the panel.

9. Remove the cad cell. This is usually accomplished by removing one screw that holds the transformer and raising the transformer to one side.

10. Using the ohm feature of the meter, check the resistance of the cad cell with the eye pointed at the room light source and record it here: _____ ohms. With the eye covered with electrical tape: _____ ohms

11. With the electrical tape still covering the eye, replace the cad cell in its mounting in the furnace and prepare the furnace for firing.

12. Place the alligator clip lead on the white wire. Start the furnace while measuring the voltage at the orange wire. The furnace should run for about 90 seconds before shutting off.

13. Record the running time here: _____ sec. If you miss the timing the first time, allow 5 minutes for the control to cool and reset the primary. **CAUTION:** DO NOT RESET THE PRIMARY MORE THAN THREE TIMES OR EXCESS OIL MAY ACCUMULATE IN THE COMBUSTION CHAMBER.

14. Turn the power off, remove the tape from the cad cell, start the furnace, and run it for 10 minutes to clear any excess oil from the combustion chamber.

15. Remove one lead from the cad cell and tape it to keep it from touching another circuit. This produces an open circuit in the cad cell.

16. Start the furnace while measuring the voltage from the white to the orange wire. Record the voltage here: _____ V

17. Remove the meter leads and place all wiring panels and screws back in their correct order.

Maintenance of Workstation and Tools: Return all tools to their respective places.

Summary Statement: Describe how the cad cell and the primary control work together to protect the oil furnace from accumulating too much oil in the combustion chamber.

QUESTIONS

1. What material is in a cad cell that allows it to respond to light?

2. What is the response that a cad cell makes to change in light?

3. What would the symptoms be of a dirty cad cell?

4. What is the sensing element in a stack switch?

5. What would the result of excess oil in the combustion chamber be if it were ignited?

6. What would the symptom of an open cad cell be?

7. What is the color of an oil burner flame?

8. What would cause a cad cell to become dirty?

Checking the Fan and Limit Control

Name	Date	Grade

Objectives: Upon completion of this exercise, you will be able to check a fan and limit switch on an oil furnace for correct action.

Introduction: You will use a thermometer probe in the area of the fan and limit control to determine the temperature and then observe the control action.

Text References: Paragraph 32.10.

Tools and Materials: Straight-blade and Phillips screwdrivers, electronic thermometer, $1/4$" and $5/16$" nut drivers, and a flashlight.

Safety Precautions: Turn the power off and lock out before installing the thermometer sensor. Be careful not to get burned on hot surfaces.

PROCEDURES

1. With the power off, locked, and tagged, remove the front cover of the furnace and locate the fan-limit switch.

2. Remove the cover from the fan-limit control to see whether it has a circular dial that turns as it is heated. If so, it can be observed during the test.

3. Slide the thermometer sensor in above the heat exchanger, close to the fan-limit sensor location.

4. Start the furnace and watch the temperature climb. Record the temperature every 2 minutes.

2 minutes _____°F	4 minutes _____°F
6 minutes _____°F	8 minutes _____°F
10 minutes _____°F	12 minutes _____°F

5. Record the temperature when the fan starts: _____°F

6. When the fan starts, shut the furnace off using the system switch. The furnace will not have time to cool.

7. Disconnect a wire from the fan motor so the fan will not start.

8. Start the furnace again and record the temperature when the limit switch stops the burner pump and fan: _____°F

9. Shut the power off and reconnect the fan wire.

10. Turn the power on and the fan should start and cool the furnace.

11. Wait until the fan cools the furnace to the point that the fan stops, then remove the temperature lead from the area of the fan-limit control.

12. Start the furnace again to make sure it will start for the next exercise.

Maintenance of Workstation and Tools: Return all tools to their proper places and make sure your work area is clean.

Summary Statement: Describe what happens within the fan-limit control from the time the stack starts to heat until it cools to the point where the fan shuts off.

QUESTIONS

1. State a common cause for a limit switch to stop a forced-air furnace.

2. Describe the sensor on the fan-limit control that senses heat.

3. What would happen if a technician wired around a limit switch and the fan became defective and would not run?

4. What would the symptoms be on an oil furnace if the fan setting on the fan-limit control were set too low?

5. Describe the location of the fan-limit control on the furnace you worked with.

6. How many wires are attached to the fan-limit control on the furnace you worked with?

7. What circuit does the limit control break when there is an overheat condition?

8. Is the limit switch normally in the high- or low-voltage circuit?

9. What is the typical supply voltage for an oil furnace?

10. What would be the symptom for a defective limit switch where the contacts remained open?

Adjusting an Oil Burner Pump Pressure

Name	Date	Grade

Objectives: Upon completion of this exercise, you will be able to install the gages on an oil pump and adjust it to the correct pressure.

Introduction: You will install a pressure gage on the oil pump discharge and a compound gage (that will read into a vacuum) on the suction side of the oil pump. You will then start the pump and adjust the discharge pressure to the correct pressure.

Text References: Paragraphs 32.9 and 32.16.

Tools and Materials: Straight-blade and Phillips screwdrivers, vacuum or compound gage, pressure gage (150 psig), any adapters to adapt gages to an oil pump, shallow pan to catch oil drippings, goggles, two adjustable wrenches, thread seal, set of Allen wrenches, and an operating oil-burning furnace.

Safety Precautions: Do not spill any fuel oil on the floor; use a pan under the pump while connecting and disconnecting gages. Turn the power to the furnace off when you are not calling for it to run. Lock and tag the electrical panel and keep the only key in your possession.

PROCEDURES

1. Shut off and lock out the power to the furnace. Wear safety goggles.

2. Shut the fuel supply valve off if the supply tank is above the oil pump.

3. Remove the gage plug for the discharge pressure. Usually this is the one closest to the small oil line going to the nozzle. Connect the high-pressure gage here. See the text, Figure 32-90, for an example.

4. Remove the gage plug for the suction side of the pump. This is usually located next to the inlet piping on the pump.

5. Remove the pressure-regulating screw protective cap and insert the correct Allen wrench in the slot.

6. Open the fuel valve, start the furnace, and quickly make sure that there are no oil leaks. If there are, shut the furnace off and repair them. IF THERE ARE NO LEAKS, MAKE SURE THAT THE OIL HAS IGNITED AND IS BURNING.

7. Record the pressures on the gages here:
   • Inlet pressure _____ psig or in. Hg
   • Outlet pressure _____ psig

8. Adjust the outlet pressure to 90 psig (this is 10 psig below the recommended pressure), then back to 100 psig.

9. Adjust the outlet pressure to 110 psig (this is 10 psig above the recommended pressure), then back to 100 psig. Count the turns per 10 psig.

10. Shut the furnace off and remove the gages.

11. Wipe any oil away from the fittings and dispose of any oil in the pan as your instructor advises.

12. Replace all panels with the correct fasteners.

Maintenance of Workstation and Tools: Make sure the workstation is clean and return all tools to their respective places.

Summary Statements: Describe how the oil burner pump responded to adjustment. Approximately how many turns per 10 psig change?

QUESTIONS

1. What is the typical recommended oil burner nozzle pressure?

2. Why should fuel oil not be spilled?

3. What was the size of the pressure tap in the oil pump for the high-pressure gage?

4. What was the size of the pressure tap in the oil pump for the low-pressure gage?

5. How many Btus are there in a gallon of fuel oil?

6. What is the typical efficiency for an oil furnace?

7. When is a two-pipe system necessary for an oil burner?

8. How far vertically can an oil burner pump without an auxiliary pump?

9. Name the two places that oil is filtered before it is burned.

10. What is the name of the slots in an oil burner nozzle?

Combustion Analysis of an Oil Burner

Name	Date	Grade

Objectives: Upon completion of this exercise, you will be able to perform a draft check and a smoke test, and analyze an oil burner for combustion efficiency.

Introduction: You will use a draft gage to measure the draft over the fire, a smoke tester to measure the smoke in the flue, and use a combustion analyzer to perform a combustion analysis on an oil-burning system.

Text Reference: Paragraph 32.17.

Tools and Materials: A portable electric drill, $1/4$" drill bit, $1/4$" and $5/16$" nut drivers, straight-blade and Phillips screwdrivers, draft gage, smoke test kit, goggles, combustion analysis kit, and an operating oil-burning furnace.

Safety Precautions: The flue pipe of an oil furnace is very hot while running; do not touch it.

PROCEDURES

1. Wear safety goggles. If holes are not already available, drill a $1/4$" hole in the flue pipe at least 6" before the draft regulator and one in the burner inspection door. Get advice from your instructor.

2. Insert the thermometer from the analysis kit in the hole in the flue pipe.

3. Start the furnace and determine that the oil ignites.

4. Wait for 10 minutes or until the thermometer in the flue stops rising; then use the draft gage and measure the draft above the fire, at the inspection door, Figure 32-3. Record the reading here: _____ in. wc.

5. Remove the thermometer from the flue pipe and perform a smoke test following the instructions in the kit, Figure 32-4. Record the smoke test number here: _____

6. Perform a combustion analysis at the $1/4$" hole in the flue pipe. Record the CO_2 reading here: _____

7. Record the combustion efficiency here: _____ % efficient

8. Turn the furnace off and remove the instruments.

9. Replace all panels with the correct fasteners.

10. Fill the hole with a $1/4$" cap if available.

Maintenance of Workstation and Tools: Return all tools to their respective places.

Summary Statements: Describe the efficiency of this oil furnace. If it was running below 70% efficiency, explain the probable causes.

Figure 32-3 Checking draft over fire. *Photo by Bill Johnson*

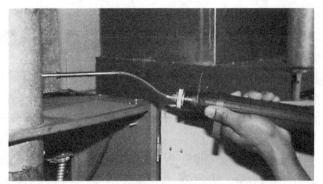

Figure 32-4 Making smoke test. *Photo by Bill Johnson*

QUESTIONS

1. What could be the cause of a high-stack temperature?

2. How would you prove a furnace was stopped up with soot in the heat exchanger?

3. What is the function of the heat exchanger?

4. What is the function of the combustion chamber?

5. What is the purpose of the restrictor in the blast tube?

6. What is the difference between a hollow core and a solid core oil nozzle?

7. What is the purpose of the tangential slots in the nozzle?

8. What must be done to oil to prepare it to burn?

9. What would be the result of a situation where a customer reset the primary control many times before the service person arrived?

10. What are the symptoms of an oil forced-air furnace with a hole in the heat exchanger?

Troubleshooting Exercise: Oil Furnace

Name	Date	Grade

Objectives: Upon completion of this exercise, you will be able to troubleshoot a problem in an oil furnace.

Introduction: This furnace should have a specific problem introduced by your instructor. You will locate the problem using a logical approach and instruments. THE INTENT IS NOT FOR YOU TO MAKE THE REPAIR BUT TO LOCATE THE PROBLEM AND ASK YOUR INSTRUCTOR FOR DIRECTION BEFORE COMPLETION.

Text Reference: Unit 32.

Tools and Materials: Straight-blade and Phillips screwdrivers, Allen wrenches, two 6" adjustable wrenches, set of socket wrenches, a flashlight, and an oil gage.

Safety Precautions: Make sure all power to the furnace is off. Lock and tag the panel before removing any parts.

PROCEDURES

1. Set the thermostat to call for heat and start the furnace.

2. Did the burner ignite?

3. If no, did the fan start?

4. Did the burner motor start?

5. Turn the power off and lock out. Install a gage on the burner outlet. See text, Figure 32-90.

6. With the gage in place, restart the furnace. Record the pressure reading here: _____ psig

7. If the burner is running and there is no pressure, what could the problem be?

8. Use the flashlight to view the pump coupling. Is it turning and turning the oil pump?

9. Describe what you think the problem could be.

10. What is your recommended next step?

Maintenance of Workstation and Tools: Return the furnace to the condition recommended by your instructor. Return all tools to their places.

Summary Statements: Describe the difference between a single-stage and a two-stage oil pump. Describe the application for both.

QUESTIONS

1. Does fuel oil burn as a liquid or a vapor?

2. What is the purpose of the oil pump and nozzle?

3. What is the typical oil pressure for a gun-type oil burner?

4. What is the purpose of the cutoff inside the oil pump?

5. At what pressure does the cutoff function?

6. What is the purpose of the coupling between the pump and motor?

7. How long can an oil furnace run after startup without firing?

8. What happens if the burner does not fire in the prescribed time?

9. If the burner does not fire in the prescribed time, how do you restart it?

10. Describe the cad cell.

Troubleshooting Exercise: Oil Furnace

Name	Date	Grade

Objectives: Upon completion of this exercise, you will be able to troubleshoot a control problem on an oil furnace.

Introduction: This furnace should have a specified problem introduced by your instructor. You will locate the problem using a logical approach and instruments. THE INTENT IS NOT FOR YOU TO MAKE THE REPAIR BUT TO LOCATE THE PROBLEM AND ASK YOUR INSTRUCTOR FOR DIRECTION BEFORE COMPLETION.

Text Reference: Unit 32.

Tools and Materials: Straight-blade and Phillips screwdrivers, a VOM, and two 6" adjustable wrenches.

Safety Precautions: Always turn the power off. Lock and tag the panel while working on electrical components if you do not need the power for testing purposes.

PROCEDURES

1. Set the thermostat to heat and start the furnace.

2. Did the burner ignite?

3. Describe the sequence of events.

4. If the furnace ignited but did not continue to run and shut off at the reset, allow it to cool for 5 minutes and start it again.

5. Shut the power off and raise the transformer for access to the cad cell.

6. Remove the leads to the cad cell and measure the resistance of the cad cell. Does the cad cell change resistance with changes in light? Refer to the "Cad Cell Relay" section on page 808 in the text.

7. Describe the problem.

8. Describe the repair.

9. Consult your instructor about making the repair and do so if directed to.

10. Return the furnace to the condition your instructor directs you.

Maintenance of Workstation and Tools: Return all tools to their respective places.

Summary Statement: Describe the cad cell and primary control and how they are used to verify ignition.

QUESTIONS

1. What would happen if too much oil were allowed to enter the combustion chamber and ignition occurred?

2. What should you do in the event you ignited a furnace with too much oil in the combustion chamber?

3. How long is a furnace allowed to run without ignition?

4. What does a cad cell see that allows it to verify ignition?

5. Can a cad cell become covered with soot and simulate a malfunction?

6. What should be done in the event the above happened?

7. What can cause excess soot in an oil furnace?

8. Do all furnaces use cad cells?

9. How does the efficiency affect soot in an oil furnace?

10. Would it be wise to run a combustion test every year?

Situational Service Ticket

Name	Date	Grade

NOTE: Your instructor must have placed the correct service problem in the system for you to successfully complete this exercise.

Technician's Name:_____ Date: _____

CUSTOMER COMPLAINT: Unit smoking.

CUSTOMER COMMENTS: Customer wants furnace tuneup and combustion analysis because smoke is coming out of the flue.

TYPE OF SYSTEM:

FURNACE UNIT:
MANUFACTURER_____   MODEL NUMBER_____   SERIAL NUMBER_____

TECHNICIAN'S REPORT

1. SYMPTOMS: _____

2. DIAGNOSIS: _____

3. ESTIMATED MATERIALS FOR REPAIR: _____

4. ESTIMATED TIME TO COMPLETE THE REPAIR: _____

SERVICE TIP FOR THIS CALL

1. Adjust the air to correct the smoke test before flue gas analysis.

Situational Service Ticket

Name	Date	Grade

NOTE: Your instructor must have placed the correct service problem in the system for you to successfully complete this exercise.

Technician's Name:_____ Date: _____

CUSTOMER COMPLAINT: No heat.

CUSTOMER COMMENTS: Unit quit today, has been reset three times, but still will not start.

TYPE OF SYSTEM:

FURNACE UNIT:
MANUFACTURER_____   MODEL NUMBER_____   SERIAL NUMBER_____

TECHNICIAN'S REPORT

1. SYMPTOMS: _____

2. DIAGNOSIS: _____

3. ESTIMATED MATERIALS FOR REPAIR: _____

4. ESTIMATED TIME TO COMPLETE THE REPAIR: _____

SERVICE TIP FOR THIS CALL

1. Use your voltmeter.

Hydronic Heat

Unit Overview

Hydronic heating systems are systems where water is heated and pumped to terminal units such as a radiator or a finned-tube baseboard unit. In some systems, the water is heated to form steam, which is then piped to the terminal units.

Hot-water hydronic systems are piped in with a number of system components including pressure-reducing valves, air separators, air vents, expansion tanks, circulator pumps, pressure relief valves, and low-water cutoffs.

These systems are designed to include more than one zone when necessary. If the system is heating a small home or area, there may be only one zone. If the house is large or multilevel, there may be several zones. Often there are separate zones for the bedrooms so that these temperatures can be kept lower than those in the rest of the house.

The water is heated in the boiler using an oil, gas, or electric heat source. These burners or heating elements are very similar to those discussed in other units. Sensing elements in the boiler start and stop the heat source according to the boiler temperature. The water is circulated with a centrifugal pump. A thermostatically controlled zone valve allows the heated water into the zone needing heat. Most residential installations use finned-tube baseboard units to transfer heat from the water to the air.

Many residential systems use a series loop system for each zone. All of the hot water flows through all of the units. The water temperature in the last unit will be lower than in the first because heat is given off as the water moves through the system. A one-pipe system uses

diverter tees to direct water to terminal units in a particular zone that requires heat. In the two-pipe reverse-return system, two pipes run parallel with each other and the water flows in the same direction in each pipe. The water entering the first heating unit is the last to be returned to the boiler. This equalizes the distance of the water flow. In the two-pipe direct-return system, the water flowing through the nearest heating unit to the boiler is the first back to the boiler. This is a difficult system to balance and is seldom used in residential applications. Some systems are installed as loops or coils in floors or ceilings. These are called panel systems and are normally designed with each coil as an individual zone, each balanced separately.

Radiant heating systems are intended to heat the shell of the structure as opposed to the air in the structure. Radiant heating systems circulate water that is cooler than in high-temperature applications and can be installed as a slab-on-grade, thin slab or concrete-free system. Primary-secondary pumping allows the boiler to provide different heating zones or loops with water of different temperatures for different applications. One popular application for primary-secondary pumping is to provide high-temperature water for high temperature terminal units and lower temperature water for a zone that is being heated with radiant heating.

Most oil- or gas-fired hydronic boilers can be furnished with a domestic hot-water heater consisting of a coil inserted into the boiler. The domestic hot water is contained within the coil and heated by the boiler quickly, which eliminates the need for a storage tank.

Key Terms

Balancing Values
Boiler
Centrifugal Pump (Circulator)
Diverter Tee
Expansion Joint
Expansion Tank (Air Cushion Tank)
Finned-Tube Baseboard Unit
Flow-Control Valves
Hydronic Heating
Impeller

Limit Control
Low-Water Cutoff
Mixing Valve
One-Pipe System
Outdoor Reset
Panel System
Point of No Pressure Change
Pressure-Relief Valve
Primary-Secondary Pumping
Radiant Heating
Series Loop

Tankless Domestic Hot-Water Heater
Terminal Unit
Two-Pipe Direct-Return System
Two-Pipe Reserve-Return System
Water-Regulating Valve
Zone
Zone-Control Valve

REVIEW TEST

Name	Date	Grade

Circle the letter that indicates the correct answer.

1. **A terminal unit in a hydronic heating system may be a _____ unit.**
   A. gas or oil-fired
   B. finned-tube baseboard
   C. one-pipe fitting
   D. any of the above

2. **Boilers that have efficiencies over 90% are commonly referred to as:**
   A. noncondensing boilers.
   B. conventional boilers.
   C. condensing boilers.
   D. inefficient boilers.

3. **In a wet-base boiler, the water being heating is located:**
   A. only above the combustion area.
   B. only below the combustion area.
   C. both above and below the combustion area.
   D. nowhere near the combustion area.

4. **All of the following are characteristics of a cast iron boiler except:**
   A. they heat up very quickly.
   B. they are classified as high-mass boilers.
   C. they have long run times and long off cycles.
   D. they typically contain between 15 and 30 gallons of water.

5. **Copper water-tube boilers:**
   A. contain more water than cast iron boilers.
   B. are low-mass boilers.
   C. heat water very slowly.
   D. all of the above are correct.

6. **A limit control:**
   A. regulates the amount of water to each unit.
   B. regulates the water pressure from the supply line.
   C. is used to balance the flow rate.
   D. shuts the heat source down if the temperature becomes too high.

7. **The pressure relief valve:**
   A. regulates the pressure of the water in the supply line.
   B. is set to relieve at or below the maximum allowable working pressure of the boiler.
   C. shuts the boiler down if the temperature becomes too high.
   D. relieves the pressure at the expansion tank.

8. **The expansion tank provides a place for air initially trapped in the system and:**
   A. space for excess water from the supply line.
   B. water for the domestic hot-water coil.
   C. space for the expanded water in the system.
   D. all of the above.

9. **If the height of a hot-water hydronic piping arrangement extends 23 feet above the inlet of the expansion tank, the pressure in the expansion tank should be about:**
   A. 9 psig.
   B. 12 psig.
   C. 15 psig.
   D. 20 psig.

10. **The part of the centrifugal pump that spins and forces water through the system is the:**
    A. heat motor.
    B. impeller.
    C. flow control.
    D. propeller fan.

11. **Centrifugal pumps are positive displacement pumps.**
    A. true
    B. false

12. **An expansion joint is often used in:**
    A. an air-eliminating device.
    B. the installation of a manual air vent.
    C. a finned-tube baseboard unit.
    D. a flow-control system.

13. **It is desirable to have the inlet of the circulator connected to the:**
    A. return side of the boiler.
    B. point of no pressure change.
    C. inlet of the water pressure regulating valve.
    D. inlet of the terminal units.

14. **The air separator is the hot-water hydronic system component that:**
    A. removes air from the hydronic system.
    B. separates air from the water in the hydronic system.
    C. both A and B.
    D. neither A nor B.

15. **The air vent is the hot-water hydronic system component that:**
    A. removes air from the hydronic system.
    B. separates air from the water in the hydronic system.
    C. both A and B.
    D. neither A nor B.

16. **Where is the water pressure reducing valve located in a hot-water hydronic system?**
    A. between the terminal unit and the return line on the boiler
    B. between the main water supply and the point of no pressure change
    C. between the boiler and the pressure relief valve
    D. at the outlet of the circulator

17. **Zone-control valves are controlled by the:**
    A. limit control.
    B. thermostat.
    C. air eliminator.
    D. water-regulating valve.

18. **A flow-control valve is also called _____ valve.**
    A. a check   C. a one-pipe
    B. an expansion   D. a pressure

19. **Outdoor reset controls:**
    A. measure the difference between the boiler water temperature and the temperature of the occupied space.
    B. measure the difference between the outside ambient temperature and the temperature of the water in the boiler.
    C. allow for hotter water to be produced when the outdoor temperature is high.
    D. allow for cooler water to be used when the outdoor temperature is low.

20. **In a series-loop hydronic heat piping system:**
    A. the heated water flows in the same direction in parallel pipes.
    B. the heated water flows in the opposite direction in parallel pipes.
    C. all of the heated water flows through all of the terminal units.

21. **How much heat is being supplied by a terminal unit if there is 10 gpm of water flowing through it with a 5°F temperature differential?**
    A. 50 Btu/h
    B. 500 Btu/h
    C. 2500 Btu/h
    D. 25,000 Btu/h

22. **In a two-pipe reverse-return hydronic heat piping system:**
    A. the heated water flows in the same direction in parallel pipes.
    B. the heated water flows in the opposite direction in parallel pipes.
    C. all of the heated water flows through all of the terminal units.
    D. the heated water going into the first terminal unit is the first to be returned to the boiler.

23. **In a series-loop system, the water going through the first baseboard unit will be _____ the water flowing through the last unit.**
    A. cooler than
    B. the same temperature as
    C. warmer than

24. **In a primary-secondary pumping arrangement, the largest circulator pump will be located:**
    A. in the primary loop.
    B. in the zone with the largest terminal unit.
    C. in the zone with the smallest terminal unit.
    D. anywhere in the system as all the pumps are the same size.

25. **The temperature of the water at the outlet of a three-way mixing valve will most likely be:**
    A. equal to the temperature of the water at the hot-water inlet.
    B. lower than the temperature of the water at the cold-water inlet.
    C. higher than the temperature of the water at the hot-water inlet.
    D. higher then the temperature of the water at the cold-water inlet.

Filling a Hot-Water Heating System

Name	Date	Grade

Objectives: Upon completion of this exercise, you will be able to fill an empty hot-water system with fresh water and put it back in operation.

Introduction: You will start with an empty boiler and hot-water system and fill it with water, bleed the air, and put it back into operation.

Text References: Paragraphs 33.1, 33.2, 33.3, 33.4, 33.5.

Tools and Materials: Two pipe wrenches (12"), straight-blade and Phillips screwdrivers, water hose if needed, and a boiler and hot-water heating system.

Safety Precautions: NEVER ALLOW COLD WATER TO ENTER A HOT BOILER. IF YOUR BOILER IS HOT, SHUT IT OFF THE DAY BEFORE THIS EXERCISE OR DO NOT DRAIN THE BOILER AS A PART OF THE EXERCISE. Shut the power off to the boiler when it is not in operation. Lock and tag the electrical panel and keep the single key in your possession.

PROCEDURES

1. Turn the power off and lock out the boiler and water pumps. Drain the entire system at the boiler drain valve if the boiler is at the bottom of the circuit. If the boiler is at the top of the system, drain from the system valve at the lowest point.

2. Make sure the boiler is cool to the touch of your hand.

3. Close all valves in the system and start to fill the water system. CHECK TO MAKE SURE THE SYSTEM HAS A WATER-REGULATING VALVE, such as in Figure 33-1. With this valve in the system, you cannot apply too much pressure to the system. IF THE SYSTEM DOES NOT HAVE AN AUTOMATIC FILL VALVE, WATCH THE PRESSURE GAGE AND DO NOT ALLOW THE PRESSURE TO RISE ABOVE THE RECOMMENDED LIMIT, PROBABLY 12 PSIG.

4. Fill the system until the automatic valve stops feeding or the correct pressure is reached (in which case, shut the water off).

Figure 33-1 An automatic water-regulating valve with a system pressure relief valve.
Courtesy ITT Fluid Handling Division

5. Go to the air-bleed valves and make sure that they allow any air to escape. See Figure 33-2 for examples of manual and automatic types. If they are manual valves, bleed them until a small amount of water is bleeding out.

6. Start the pump and check the air bleed valves again.

7. Start the boiler and wait until it becomes hot to the touch.

(A)

8. Carefully feel the water line leaving the boiler. It should become hot, and the hot water should start leaving the boiler and move toward the system. The pump should be moving the water in a forced-water system, and gravity will move it in a gravity system. You should now be able to follow the hot water throughout the system, with the cooler water going back to the boiler.

9. You should now have a system that is operating. The water level in the expansion tank has not been checked for a system with an air-bleed system at the expansion tank. See Lab 33-2.

10. Clean up the work area; do not leave water on the floor. LEAVE THE BOILER AS YOUR INSTRUCTOR INDICATES.

Maintenance of Workstation and Tools: Return all tools to their respective places.

Summary Statements: Describe the distribution piping system in the unit you were working with.

(B)

Figure 33-2 (A) Manual air vent. (B) Automatic air vent. *Courtesy ITT Fluid Handling Division*

QUESTIONS

1. How does air get into a water system?

2. How can you get air out of a water system?

3. What are the symptoms of air in a water system?

4. Why should cold water not be introduced into a hot boiler?

5. What is inside an automatic air-bleed valve to allow the air to get out?

6. What type of pump is typically used on a small hot-water heating system?

7. What is the purpose of balancing valves in a hot-water system?

8. How is air separated from the water at the boiler in some systems?

9. When it is separated, as in question 8, where does it go?

10. What is the purpose of a system pressure relief valve?

Determining Expansion Tank Pressure

Name	Date	Grade

Objectives: Upon completion of this exercise, you will be able to determine the proper pressure for a bladder-type expansion tank.

Introduction: Most of the time, bladder-type expansion tanks come from the factory with a preset pressure of 12 psig. For most applications, this is sufficient, but the actual pressure that is needed is determined by the height of the piping circuit with respect to the inlet of the expansion tank.

Text Reference: Paragraph 33.3.

Tools and Materials: Tape measure, hot-water hydronic heating system with terminal units on multiple floors if possible.

Safety Precautions: Always be aware that the pipes on a hydronic heating system may be very hot. Be sure to protect yourself by wearing proper personal protection equipment.

PROCEDURES

1. At the boiler, measure the distance from the inlet of the expansion tank to the floor and record that distance here: _____ ft _____ in.

2. Determine the number of floors in the structure that have baseboard heating (above the floor on which the boiler is located) and record that number here: _____

3. Go to the top floor on which there is a baseboard heating unit.

4. Measure the distance from the top of the highest pipe to the floor and record that distance here: _____ ft _____ in.

5. Go to the next highest floor on which there is a baseboard heating unit. If the next highest floor is the level on which the boiler is located, go to step 10.

6. Measure the height of the ceiling and record it here: _____ ft _____ in.

7. Go to the next highest floor on which there is a baseboard heating unit. If the next highest floor is the level on which the boiler is located, go to step 10.

8. Measure the height of the ceiling and record it here: _____ ft _____ in.

9. Measure the height of the ceiling of the boiler room and record it here: _____ ft _____ in.

10. Add the measurements obtained in steps 4, 6, 8, and 9 and record that total here: _____ ft _____ in.

11. Multiply the number in step 2 by 0.50 and enter that number here: _____ ft _____ in. If the number from step 2 is even, the entry will be only in feet. If the number from step 2 is odd, the entry will be the number of whole feet and 6 inches. For example, if the number in step 2 is 3, the result will be 1.5, or 1 foot, 6 inches. If the number in step 2 is 2, the result will be 1.0, or 1 foot, 0 inches.

12. Add the numbers in steps 10 and 11 and record that value here: _____ ft _____ in.

13. Subtract the measurement from step 1 from the value in step 12 and record that value here: _____ ft _____ in.

14. Convert the value in step 13 to the decimal form of feet and record the result here: _____ ft For example, 20 feet, 6 inches is 20.5 feet.

15. Divide the value in step 14 by 2.31 and record that value here: _____ psi

16. Add 5 to the result from step 16 and record the result here: _____ psig

17. The result in step 16 is the required expansion tank pressure for that system.

Maintenance of Workstation and Tools: Make certain that all tools are put back in their proper places and that the hot-water hydronic heating system is left in good working order.

Summary Statement: Compare the value obtained in this exercise to the "normal" factory preset pressure and explain why the manufacturer of expansion tanks typically uses this pressure.

QUESTIONS

1. Explain where the 2.31 conversion factor in step 15 comes from.

2. Explain why we add 5 psig to the end result.

3. What prediction can be made regarding the required expansion tank pressure on systems that are used on structures with more floors?

4. Why is it that only the height of the piping arrangement is factored into the expansion tank pressure calculation?

Near-Boiler Piping Component Identification

Name	Date	Grade

Objectives: Upon completion of this exercise, you will be able to identify the components typically found on and around a typical hot-water hydronic heating system.

Introduction: There are a number of system components that are located on or very close to the boiler. These components enhance system operation and allow it to operate in an efficient manner. This exercise will provide a means to become familiar with these system components.

Text References: Paragraphs 33.1, 33.2, 33.3, 33.4, and 33.5.

Tools and Materials: Hot-water hydronic heating system, pad, pencil, and flashlight.

Safety Precautions: Always be aware that the pipes on a hot water hydronic heating system may be very hot. Protect yourself by wearing proper personal protection equipment.

PROCEDURES

1. Draw a sketch of the boiler and the piping that is connected to it. Include the pipe sizes and the fittings that were used to complete the piping of the unit.

2. Identify all of the system components that are found in the piping arrangement on the sketch.

3. Obtain information from the following system components:

Water pressure reducing valve

    Manufacturer _____ Model _____

Expansion tank

    Manufacturer _____ Model _____

    Circle one: Steel tank   Bladder-type

Air separator

    Manufacturer _____ Model _____

Air vent

    Manufacturer _____ Model _____

Pressure relief valve

    Manufacturer _____ Model _____

Circulator pump 1

    Manufacturer _____ Model _____

Circulator pump 2

    Manufacturer _____ Model _____

Circulator pump 3

    Manufacturer _____ Model _____

Zone valve

    Manufacturer _____ Model _____

Other _____

    Manufacturer _____ Model _____

Other _____

    Manufacturer _____ Model _____

4. Identify the point of no pressure change on the system you are working on and indicate that point on your piping diagram.

5. On your diagram, indicate the direction of water flow as it makes its way through the system.

Maintenance of Workstation and Tools: Make certain that all tools are put back in their proper places and that the hot-water hydronic heating system is left in good working order.

Summary Statement: Explain the importance of properly piping in a boiler and knowing how all of the near-boiler system components operate.

QUESTIONS

1. What is the typical opening pressure on the boiler's pressure relief valve?

2. Explain why the pressure relief valve must be installed directly on the boiler with as few fittings as possible and with no valves in between the component and the boiler.

3. If an expansion tank has a static (resting) pressure of 12 psig, what is the desired fill pressure (water pressure) in the hot water hydronic heating system prior to boiler operation?

4. Of the near-boiler piping components that are found on a typical hot-water hydronic heating system, determine which are intended to be operational controls and which are intended to operate specifically as safety devices.

5. What would happen to a hot-water hydronic heating system if the bladder on the expansion tank ruptured?

System Configuration Identification

Name	Date	Grade

Objectives: Upon completion of this exercise, you will be able to identify various types of hot-water hydronic heating systems.

Introduction: There are many different configurations for hot water hydronic heating systems. To properly troubleshoot and evaluate a hydronic heating system, the technician must first understand what type of system he is dealing with. By evaluating the near-boiler piping and the components that are present, the technician will be able to determine the piping configuration that is being used.

Text Reference: Paragraph 33.6.

Tools and Materials: Hot-water hydronic heating system, pad, pencil, and flashlight.

Safety Precautions: Always be aware that the pipes on a hydronic heating system may be very hot. Protect yourself by wearing proper personal protection equipment.

PROCEDURES

1. Determine which system piping components are installed on the system. Enter information on the following lines:

    Circulator Pump   Yes   No

    How many circulator pumps are installed on this system? _____

    List the different models of the pumps here:

    Zone valves   Yes   No

    Diverter tees   Yes   No

    Mixing valves   Yes   No

    Thermostatic radiator valves   Yes   No

    Balancing valves   Yes   No

2. Based on the information obtained in step 1, compare the system being evaluated to the following:

	Circulator (single or multiple)	Zone Valves	Diverter Tees	Mixing Valves	Thermostatic Radiator Valves	Balancing Valves
Series Loop	Single	No	No	No	No	No
One Pipe	Single	Maybe	Yes	No	Maybe	No
2 Pipe Direct	Single or multiple	Maybe	No	No	Maybe	Yes
2 Pipe Reverse	Single or multiple	Maybe	No	No	Maybe	No
High-temp. primary-secondary	Multiple	Maybe	No	Maybe	Maybe	No
Multi-temp. primary-secondary	Multiple	Maybe	No	Yes	Maybe	No

3. From the information obtained in step 1 and the table in step 2, the system being evaluated is most likely a _____ system.

4. Sketch out the piping circuit, omitting the near-boiler piping if already included in previous lab sketches.

5. Indicate the direction of water flow through the system by appropriate arrows on your sketch.

Maintenance of Workstation and Tools: Make certain that all tools are put back in their proper places and that the hot-water hydronic heating system is left in good working order.

Summary Statement: Explain the importance of properly piping in a boiler and knowing the type of system that is being worked on.

QUESTIONS

1. Why are zone valves, thermostatic radiator valves, and other temperature-controlling valves not found on series loop hydronic heating systems?

2. Why are series loop hydronic heating systems equipped with only one circulator?

3. Explain the function/purpose of the diverter tee.

4. Explain how a thermostatic mixing valve is used to provide different heating zones or areas with water at different temperatures.

Indoor Air Quality

Unit Overview

In the last several years there has been evidence that the air in homes and other buildings has become increasingly polluted. Methods to control indoor air contamination are eliminating the source of the contamination, providing adequate ventilation, and providing means for cleaning the air. Indoor air quality tests and monitoring instruments are available. Ventilation is the process of supplying and removing air by natural or mechanical means to and from any space. There are three types of air cleaners: mechanical filters, electronic air cleaners, and ion generators. Dirty ducts may be a source of pollution and may need to be cleaned. Air may need to be dehumidified or humidified.

Key Terms

Absorbtion
Adsorption
Air Filtration
Asbestos
Biological
  Contaminates
Carbon Monoxide (CO)
Charcoal Air Purifier

Duct Cleaning
Dust Mites
Energy Recovery
  Ventilator (ERV)
Filter Media
High Efficiency
  Particulate Arrestor
  (HEPA)

Humidistat
Indoor Air Pollution
Lead Paint
Minimum Efficiency
  Reporting Value
  (MERV)
Mold
Mold Remediation

Radon
Relative Humidity
Sick Building
  Syndrome (SBS)
Spores
Ventilation
Volatile Organic
  Compounds (VOC)

REVIEW TEST

Name	Date	Grade

Circle the letter that indicates the correct answer.

1. Radon is a (an):
  A. radioactive gas.
  B. refrigerant used for special systems.
  C. oil additive.
  D. filter media.

2. Formaldehyde byproducts may be found in:
  A. old fruits and vegetables.
  B. aerosols.
  C. cleaning supplies.
  D. pressboard and carpets.

3. Dust mites are:
  A. usually found in damp areas in basements.
  B. often found in dust blown through vacuums with HEPA filters.
  C. microscopic spider-like insects.
  D. related to mold spores.

4. Mold:
  A. is usually found in dusty, dry areas.
  B. is usually found in areas where there is moisture and/or high humidity.
  C. is seldom found in areas where it can be seen.
  D. spores are not harmful to humans.

5. The first step with regard to mold cleanup should be to:
  A. remove the source of moisture.
  B. vacuum the area.
  C. check for radioactive activity.
  D. check the filter media in the area.

6. Carbon monoxide:
  A. has a bluish color.
  B. can easily be seen.
  C. is heavier than air.
  D. is produced by incomplete combustion in appliances that use the combustion process.

7. **Auto exhaust produces carbon monoxide.**
   A. true
   B. false

8. **Wood smoke may contain:**
   A. polycyclic organic matter.
   B. nonpolycyclic organic matter.
   C. particles from combustion.
   D. all of the above.

9. **Disturbed materials containing asbestos:**
   A. should be brushed down and swept up by the HVAC technician.
   B. should be removed by a certified or licensed contractor.
   C. may be removed when found as part of a mold remediation program.
   D. should be vacuumed up when found.

10. **Ventilation is:**
    A. a problem to deal with.
    B. generally not recommended.
    C. never used in a residence.
    D. the process of supplying or removing air by natural or mechanical means.

11. **Duct cleaning is accomplished by:**
    A. polishing the outside of the duct.
    B. washing out with a hose and water.
    C. using compressed air.
    D. using a vacuum system and brushing.

12. **All humidifiers must:**
    A. be installed correctly.
    B. be serviced regularly.
    C. use good water.
    D. all of the above.

13. **Air in homes tends to be drier in the:**
    A. summer.
    B. winter.

14. **If the relative humidity is 60%, it means that:**
    A. 40% of air is moisture.
    B. 60% of the air is moisture.
    C. the air has 60% of the moisture it has the capacity to hold.
    D. the air has 40% of the moisture it has the capacity to hold.

15. **Humidification generally relates to a process of causing water to _____ into the air:**
    A. condense
    B. evaporate
    C. be ionized
    D. be filtered

16. **Infrared humidifiers reflect infrared energy:**
    A. into the air.
    B. onto water.
    C. onto an evaporative media.
    D. into a hydronic boiler.

17. **A humidistat has _____ element and is generally used to control a humidifier.**
    A. a moisture-sensitive
    B. a light-sensitive
    C. an infrared-sensing
    D. an air pressure-sensing

18. **Atomizing humidifiers discharge _____ into the air.**
    A. water vapor
    B. electronic ions
    C. trivalent oxygen
    D. tiny water droplets

19. **Two types of atomizing humidifiers are the:**
    A. thermistor and ionization generator.
    B. spray nozzle and centrifugal.
    C. disc and drum types.
    D. particulate and oxidation types.

20. **If reservoir-type humidifiers are not properly maintained, which of the following may collect and be discharged into the air?**
    A. fungi
    B. bacteria
    C. algae
    D. all of the above

21. **In the ionizing section of an electronic air cleaner, the particles in the air are:**
    A. trapped by a filter.
    B. charged with an electrical charge.
    C. washed with a special cleaner.
    D. treated to remove many of the odors.

22. **Extended surface filters often use _____, which provides air cleaning efficiencies up to three times greater than those of fiberglass.**
    A. trivalent oxygen
    B. charcoal
    C. stainless steel
    D. nonwoven cotton

Humidification

Name	Date	Grade

Objectives: Upon completion of this exercise, you will be able to describe the various components and the application of a humidifier.

Introduction: Your instructor will furnish you with a humidifier and you will perform a typical service inspection on the unit.

Text References: Paragraphs 34.9, 34.10, 34.11, and 34.12.

Tools and Materials: Soap, chlorine solution (if possible, to kill the algae), goggles, plastic or rubber gloves, straight-blade and Phillips screwdrivers. $^1/_4$" and $^5/_{16}$" nut drivers, two small (6" or 8") adjustable wrenches, and a pair of slip joint pliers.

Safety Precautions: Turn the power and the water off to the furnace and humidifier (if the humidifier is installed). Lock and tag the electrical panel. DO NOT PUT YOUR HANDS AROUND YOUR FACE WHILE CLEANING A HUMIDIFIER THAT HAS ALGAE IN IT. ALGAE CAN CONTAIN HARMFUL GERMS.

PROCEDURES

1. Put safety goggles on and wear gloves. Set the humidifier on the workbench. If it is installed, remove it from the system using the above cautions.

2. Complete the following information:
   - Manufacturer _____
   - Model number _____
   - Serial number _____
   - Principle of operation _____

3. Remove any necessary panels for service.

4. Clean any surfaces that are dirty. If there is mineral buildup, it may not be removed easily. Do not force the cleaning; if it cannot be removed, let it stay.

5. State the condition of the pads or drum if the unit operates on that principle.

6. Is there algae formed in the water basin (if the unit has a basin)? Algae will be a soft brown, green, or black substance.

7. If there is algae, clean the algae with strong soap. A chlorine solution is preferred to kill the algae. Wear gloves and goggles.

8. Check the float for freedom of movement if the unit has a float.

9. Assemble the unit and place it back in the system if appropriate.

Maintenance of Workstation and Tools: Return all tools to their respective places. Leave your work area clean.

Summary Statement: Describe the complete operation of the humidifier that you worked with, including the type of controls.

QUESTIONS

1. How does humidity added to the air in a residence affect the thermostat setting?

2. What happens to the humidity level in the air if it is heated?

3. How is moisture added to air, using pads and rotating drums in a humidifier?

4. Why is an atomizing type of humidifier not recommended if there is a high mineral content in the water?

5. What is the algae that forms in a humidifier?

6. Why should you not put your hands around your face while handling a humidifier with algae on it?

7. What may be used to remove algae and kill germs?

8. What is the purpose of the drain line on a humidifier with a flat chamber?

9. What control is used to control the humidity in the conditioned space?

10. What is the effect of mineral deposits on the pads or rotating drum of a humidifier?

Comfort and Psychrometrics

Unit Overview

Four factors that can affect our comfort are temperature, humidity, air movement, and air cleanliness. The human body temperature is normally 98.6°F. We are comfortable when the heat level in our body is transferring to the surroundings at the correct rate. Typically, when the body at rest is in surroundings of 75°F, 50% relative humidity with a slight air movement, the body is close to being comfortable during summer conditions. In winter, a different set of conditions apply as we wear different clothing.

The study of air and its properties is called psychrometrics. Air is not totally dry. Surface water and rain keep moisture in the atmosphere. Moisture in the air is called

humidity. The moisture content of air can be checked by using a combination of dry-bulb and wet-bulb temperatures. Dry-bulb temperature is the sensible heat level of air and is taken with an ordinary thermometer. Wet-bulb temperature is taken with a thermometer with a wick on the end that is soaked with distilled water. The difference between the dry-bulb reading and the wet-bulb reading is called the wet-bulb depression.

Dew point temperature is the temperature at which moisture begins to condense out of the air. A glass of ice will begin to condense moisture on the glass when the temperature of the glass falls below the dew point.

Key Terms

Absolute Humidity
Comfort
Dew Point Temperature
Dry-Bulb Temperature
Enthalpy
Generalized Comfort Chart

Humidity
Infiltration
Psychrometric Chart
Psychrometrics
Relative Humidity
Sling Psychrometer

Total Heat
Ventilation
Wet-Bulb Depression
Wet-Bulb Temperature

REVIEW TEST

Name	Date	Grade

Circle the letter that indicates the correct answer.

1. **The human body is comfortable when:**
   A. heat is transferring from the surroundings to the body at the correct rate.
   B. heat is transferring from the body to the surroundings at the correct rate.

2. **To be comfortable in winter, lower temperatures can be offset by:**
   A. lower humidity.
   B. higher humidity.
   C. more air movement.
   D. less activity.

3. **The generalized comfort chart can be used as a basis to:**
   A. determine combinations of temperatures and humidity to produce comfort in summer and winter.
   B. study air and its properties.

C. determine the amount of fresh-air makeup necessary for comfort.
D. compare ventilation with infiltration.

4. **The density of air is the:**
   A. amount of moisture in the air.
   B. amount of movement in the air per cubic foot.
   C. weight of the air per unit of volume.
   D. combination of the temperature and pressure of the air.

5. **Air is made up primarily of:**
   A. oxygen and carbon dioxide.
   B. nitrogen and carbon dioxide.
   C. oxygen and ozone.
   D. nitrogen and oxygen.

6. The dry-bulb temperature is the:
   A. same as the wet-bulb depression.
   B. total of the sensible and latent heat.
   C. sensible heat temperature.
   D. sensible heat loss and humidity.

7. The wet-bulb temperature:
   A. is the same as the wet-bulb depression.
   B. is the total of the sensible and latent heat.
   C. takes into account the humidity.
   D. is always higher than the dry-bulb temperature.

8. The wet-bulb depression is:
   A. the difference between the dry-bulb and the wet-bulb temperature.
   B. the same as the wet-bulb temperature.
   C. the same as the dry-bulb temperature.
   D. when the air is saturated with moisture.

9. The dew point temperature is the:
   A. temperature when the air is saturated with moisture.
   B. temperature when the moisture begins to condense out of the air.
   C. difference between the dry-bulb and the wet-bulb temperatures.
   D. temperature at which the relative humidity is 50%.

10. If the dry-bulb temperature is 76°F and the wet-bulb temperature is 68°F, the wet-bulb depression is:
    A. 6°F.   C. 10°F.
    B. 8°F.   D. 12°F.

11. Fresh-air intake is necessary to keep the indoor air from becoming:
    A. too humid.
    B. superheated.
    C. oxygen starved and stagnant.
    D. too dry.

12. Infiltration is the term used when air comes into a structure:
    A. through an air makeup vent.
    B. around windows and doors.
    C. through a ventilator.

13. Using a sling psychrometer and the wet-bulb depression chart, a technician finds a dry-bulb temperature of 80°F and a wet-bulb temperature of 65°F. The relative humidity is:
    A. 44%.
    B. 28%.
    C. 52%.
    D. 60%.

14. Using a psychrometric chart, the total heat for a wet-bulb reading of 41°F and a dry-bulb reading of 62°F is:
    A. 16.9 Btu/lb.
    B. 14.6 Btu/lb.
    C. 18.6 Btu/lb.
    D. 15.7 Btu/lb.

15. A furnace puts out 122,500 Btu/h sensible heat and the air temperature rise across the furnace is 55°F. The airflow across the furnace is:
    A. 2062 cfm.
    B. 1850 cfm.
    C. 2525 cfm.
    D. 1750 cfm.

16. What is the absolute humidity of an air sample if the sample has the ability to hold 100 grains of moisture and the actual amount of moisture present in the sample is 50 grains?
    A. 50%
    B. 50 grains
    C. 100%
    D. 100 grains

17. What is the relative humidity of an air sample if the sample has the ability to hold 100 grains of moisture and the actual amount of moisture present in the sample is 50 grains?
    A. 50%
    B. 50 grains
    C. 100%
    D. 100 grains

18. As air is heated:
    A. the relative humidity and the absolute humidity both drop.
    B. the relative humidity rises and the absolute humidity decreases.
    C. the relative humidity falls and the absolute humidity increases.
    D. the relative humidity falls and the absolute humidity remains the same.

Comfort Conditions

Name	Date	Grade

Objectives: Upon completion of this exercise, you will be able to take wet-bulb and dry-bulb temperature readings, determine relative humidity from the psychrometric chart, and use this information to determine the level of comfort from the American Society of Heating, Refrigerating and Air-Conditioning Engineers (ASHRAE) generalized comfort chart.

Introduction: With a sling psychrometer, you will take several dry- and wet-bulb temperature readings in a conditioned space. By referring to a psychrometric and generalized comfort chart, you will determine whether the conditioned space is within accepted comfort limits.

Text References: Paragraphs 35.1, 35.3, 35.4, 35.5, 35.6, 35.7, 35.9, and 35.12.

Tools and Materials: A sling psychrometer and the comfort and psychrometric charts. See text, Figure 35-24 and Figure 35-31A.

Safety Precautions: The sling psychrometer is a fragile instrument. Be careful not to break it. Keep away from other objects when using it.

PROCEDURES

1. Review procedures for using the sling psychrometer in different areas, outside and in classrooms.

2. Record the following for ten wet-bulb/dry-bulb readings.

	Dry Bulb	Wet Bulb	Relative Humidity (from psychrometric chart)	Within Generalized Comfort Zone (yes or no)
1.				
2.				
3.				
4.				
5.				
6.				
7.				
8.				
9.				
10.				

Maintenance of Workstation and Tools: Return the sling psychrometer to its storage space.

Summary Statements: Describe the general comfort level of the conditioned space you evaluated. What air-conditioning adjustments can be made to make the space more comfortable?

QUESTIONS

1. What causes the thermometer with the wet sock to read lower than the thermometer with the dry bulb?

2. Why must the sling psychrometer be swung in order to obtain the wet-bulb reading?

3. Name the five factors that can condition air.

4. Name the four comfort factors.

5. How does air movement improve comfort?

6. What is the difference in the summer and winter comfort chart readings?

7. Name three different kinds of air filters.

8. What does it mean when it is said that air is 70% relative humidity (RH)?

9. Name two ways that humidity can be added to air.

10. Name three ways that the human body gives up heat.

Refrigeration Applied to Air Conditioning

Unit Overview

Summer heat leaks into a structure by conduction, infiltration, and radiation (solar load). The summer solar load is greater on the east and west sides because the sun shines for longer periods of time on these parts of the building. Some of the warm air that gets into the structure is infiltrated through the cracks around the windows and doors and leaks in when the doors are opened for entering and leaving. This infiltrated air is different in different parts of the country because the dry-bulb/wet-bulb temperatures will be different, resulting in differing amounts of humidity.

Refrigerated air conditioning is similar to commercial refrigeration because the same components are used to cool the air: the evaporator, the compressor, the condenser, and the metering device. There are two basic types of air-conditioning units: the package type and the split system. The package type has all of the components built into one cabinet. In the split system, the condenser and compressor are located outside, separate from the evaporator, with interconnecting refrigerant lines. The evaporator may be located in the attic, a crawl space, or a closet. The fan to blow the air across the evaporator may be part of the heating equipment.

The evaporator is a refrigeration coil made of aluminum or copper with aluminum fins to give it more surface area for better heat exchange. The refrigerant in the evaporator absorbs heat into the system. It removes both heat and moisture from the air passing over it.

The following types of compressors are used in air-conditioning systems: reciprocating, rotary, scroll, centrifugal, and screw. The centrifugal and screw compressors are used primarily in large commercial and industrial applications. Compressors compress vaporized refrigerant and pump it throughout the system.

The condenser for air-conditioning systems rejects the heat from the system, which causes the vapor refrigerant to condense back to a liquid. Most residential and light commercial air-conditioning condensers are air cooled and reject the heat into the air.

The expansion device meters the refrigerant to the evaporator. The thermostatic expansion valve or the fixed-bore metering device (either a capillary tube or an orifice) are the types most often used.

The air-moving side of the air-conditioning system consists of the supply air and return air systems. The airflow is normally 400 cfm/ton in humid climates. The air should leave the air handler at about 55°F and return at approximately 75°F.

The package air conditioner has the advantage that all equipment is located outside and can be serviced from outside.

R-22 has been one of the leading air-conditioning refrigerants in residential and commercial applications for many years. However, because R-22 is an HCFC refrigerant and contains chlorine, it has a phaseout date for new equipment in 2010 and a total production phaseout in 2020. R-410A and R-407C are two refrigerant blends that will soon become the long-term replacement refrigerant for R-22 in residential and commercial air-conditioning applications. R-410A is designed for new applications, and R-407C can be used in new and retrofit applications. Please refer to Unit 9, Section 9.11 for more detailed information on R-410A and R-407C.

Key Terms

CFM/Ton
Evaporative Cooler
High-Efficiency Condensers
Orifice Metering Device

Package Air Conditioning
R-410A
R-407C
Rotary Compressor

Solar Load
Split-System Air Conditioning

REVIEW TEST

Circle the letter that indicates the correct answer.

1. **The solar load on a structure is the load resulting from the:**
   A. heat from the surrounding air passing through the materials of the structure.
   B. air infiltrating through the windows and doors.
   C. heat from the sun's radiation.
   D. dry-bulb temperature inside the structure.

2. **Infiltrated air of the same dry-bulb temperature may vary from one part of the country to another because of the difference in:**
   A. solar load.
   B. sensible heat.
   C. humidity.
   D. convection currents.

3. **Air blown into a structure through an evaporative cooler tends to _____ than the outside air.**
   A. be less humid
   B. be more humid
   C. have a lower wet-bulb temperature

4. **In a package air-conditioning system, the evaporator will normally be located:**
   A. in a closet.
   B. in the attic.
   C. outside the house.

5. **The fins on an evaporator coil are made of:**
   A. stainless steel.   C. aluminum.
   B. copper.   D. plastic.

6. **More than one evaporator coil circuit may be used when:**
   A. the evaporator is located outside.
   B. there would be excessive refrigerant pressure drop.
   C. a distributor would not be practical.
   D. a slant coil is at an angle greater than 60 degrees.

7. **The cooling for modern reciprocating fully hermetic compressors used in air conditioning is controlled by or directly from:**
   A. the suction gas.
   B. the discharge gas.
   C. a squirrel cage fan.
   D. a water-cooling system.

8. **A compressor that has a stationary vane is the _____ compressor.**
   A. reciprocating
   B. rotary
   C. scroll
   D. centrifugal

9. **High-efficiency condensers:**
   A. can withstand higher pressures than standard condensers.
   B. have more surface area than standard condensers.
   C. have stainless steel coils.
   D. are located in a separate cabinet.

10. **Two types of expansion devices used in residential air conditioning are the thermostatic expansion valve and the:**
    A. automatic expansion valve.
    B. evaporator pressure regulator.
    C. fixed-bore device.
    D. solenoid valve.

11. **The thermostatic expansion valve:**
    A. does not allow the pressure to equalize during the off cycle unless specifically designed to do so.
    B. normally allows the pressure to equalize during the off cycle.
    C. does not normally require the compressor to have a start capacitor.
    D. in an air-conditioning system is designed for the same temperature range as in commercial refrigeration.

12. **The air leaving the air handler in a typical air-conditioning system is approximately _____ .**
    A. 35°F.
    B. 45°F.
    C. 55°F.
    D. 65°F.

13. **The supply air duct in an air-conditioning system installed in an unconditioned space _____ need to be insulated.**
    A. does
    B. does not

14. **Two names for the interconnecting piping in a split air-conditioning system are the:**
    A. discharge and distributor.
    B. fixed-bore and low ambient.
    C. thermostatic and pressure equalizer.
    D. suction (cool gas) and liquid.

15. **In the southwestern United States on a 90°F day, the condensing temperature for a standard air-conditioning system would be approximately _____ .**
    A. 100°F.
    B. 120°F.
    C. 150°F.
    D. 175°F.

16. **A refrigerant blend replacing R-22 in new residential and commercial air-conditioning applications is:**
    A. R-404A.
    B. R-407C.
    C. R-507C.
    D. R-410A.

17. **A refrigerant blend replacing R-22 in residential and commercial air-conditioning applications that can be used in retrofit applications is:**
    A. R-507C.
    B. R-407C.
    C. R-410A.
    D. R-404A.

18. **By the year _____, R-22 can no longer be used in new air-conditioning equipment.**
    A. 2010
    B. 2020
    C. 2030
    D. 2040

19. **The total production phaseout for R-22 will occur in:**
    A. 2010.
    B. 2020.
    C. 2030.
    D. 2040.

Air-Conditioner Component Familiarization

Name	Date	Grade

Objectives: Upon completion of this exercise, you will be able to identify and describe various components in a typical air-conditioning system.

Introduction: You will remove panels and, when necessary, components from an air-conditioning system to identify and record specifications of various components.

Text References: Paragraphs 17.8, 17.9, 17.12, 17.13, 17.14, 17.15, 17.16, 17.17, 17.23, 36.1, 36.4, 36.5, 36.6, 36.8, 36.9, 36.10, 36.11, 36.13, 36.15, 36.16, 36.17, 36.18, 36.19, 36.21, and 36.22.

Tools and Materials: Straight-blade and Phillips screwdrivers, $1/4$" and $5/16$" nut drivers, set of open end wrenches, set of Allen wrenches, flashlight, and an operating split air-conditioning system.

Safety Precautions: Turn the power to the air-conditioning unit off before removing panels. Lock and tag the electrical panel where you have turned the power off and keep the only key in your possession. When loosening or tightening bolts or nuts, be sure to use the correct size wrench.

PROCEDURES

1. Turn the power off and lock the panel. Remove enough panels and/or components to obtain the following information:

Indoor Fan Motor

Horsepower _____
Voltage _____
Locked-rotor amperage _____
Fan wheel width _____
Type of motor _____

Type of motor mount _____
Full-load amperage _____
Fan wheel diameter _____
Shaft diameter _____
Motor speed _____

Indoor Coil

Material coil made of _____
No. of rows in coil _____
Suction line size _____
Liquid-line size _____
Air blow through or draw through coil _____

No. of circuits _____
Type(s) of metering device(s) _____

Type of field piping connection _____

Outdoor Unit

Top or side discharge _____
Type of refrigerant used in this unit _____ Type of field piping connection _____
Compressor suction-line size _____ Discharge line size _____ Unit capacity _____
Type of compressor starting mechanism _____
Type of compressor motor _____ Run capacitor (if there is one) rating _____

Outdoor Fan Motor
Type of motor _____ Size of shaft _____ Motor speed _____
Horsepower _____ Voltage _____ Full-load amperage _____

2. Replace all components you may have removed with the correct fasteners. Replace all panels.

Maintenance of Workstation and Tools: Insure that the air-conditioning system is left ready for its next lab exercise. Replace all tools, equipment, and materials. Leave your workstation neat and orderly.

Summary Statement: State the information that would be necessary to order a replacement fan motor for the indoor and the outdoor unit. Draw a sketch of a typical split air-conditioning system, labeling the components and piping.

QUESTIONS

1. State two advantages of a top air discharge condenser over a side air discharge condenser.

2. Which of the interconnecting refrigerant lines is the largest, the suction or the liquid?

3. What is the purpose of the fins on the indoor and outdoor coils?

4. Why was R-22 the most popular refrigerant for many years for residential air conditioning?

5. Where does the condensate on the evaporator coil come from?

Air Distribution and Balance

Unit Overview

The object of a forced-air heating or cooling system is to deliver the correct quantity of conditioned air to the space to be conditioned. When this occurs, the air mixes with the room air and creates a comfortable atmosphere. Different spaces have different air quantity requirements. The correct amount of air must be delivered to each room or part of a building so that all will be cooled or heated correctly.

The blower or fan provides the pressure difference to move the air. Typically 400 cubic feet of air must be moved per minute per ton of air conditioning. The pressure in a duct system for a residence is measured in inches of water column (in. W.C.) with a water manometer, which uses colored water that rises up a tube.

A duct system is pressurized by three pressures: static pressure, velocity pressure, and total pressure.

Two types of fans used to move air are the propeller fan and the forward curve centrifugal fan, which is also called the squirrel cage fan wheel. The supply duct system distributes air to the registers or diffusers in the conditioned space. A well-designed system will have balancing dampers in the branch ducts to balance the air in the various parts of the structure. When a duct passes through an unconditioned space, it should be insulated if there will be any significant heat exchange. A return air system may have individual room returns, or it may have a central return system. In a central system, larger return air grilles are located so that air from common rooms can easily move back to the common returns.

Air-distribution systems can be zoned so that different areas in the structure can be maintained at different temperatures, based on the comfort requirements of the occupants. Zoning is accomplished by using zone dampers that are located in the ductwork of the air-distribution system. These zone dampers are thermostatically controlled, meaning they open and close based on the sensed temperature in a particular zone.

Key Terms

Air Friction Chart
Balancing Dampers
Diffuser
Forward Curve Centrifugal Fan
  (Squirrel Cage Fan)
Friction Loss
Grille
Inches of Water Column
Pitot Tube
Propeller Fan

Register
Static Pressure
Takeoff Fitting
Total Pressure
Traversing the Duct
Velocity Pressure
Velometer
Water Manometer
Zone Dampers
Zoning

REVIEW TEST

Name	Date	Grade

Circle the letter that indicates the correct answer.

1. In a typical air-conditioning (cooling) system, _____ cfm of air per min. is normally moved per ton of refrigeration.
A. 200
B. 400
C. 600
D. 800

2. The pressure in a duct system is measured in:
A. psi.
B. in. W.C.
C. cfm per min.
D. tons of refrigeration.

3. **Which of the following is typically measured with a water manometer?**
   A. static pressure in a duct system
   B. pressure in a refrigerant cylinder
   C. atmospheric pressure at sea level
   D. suction pressure in an air-conditioning system

4. **Velocity pressure is the:**
   A. same as total pressure.
   B. speed at which air is moving.
   C. speed and pressure at which refrigerant would leave a cylinder.

5. **Two types of fans are the propeller and:**
   A. venturi.
   B. velometer.
   C. plenum.
   D. forward curve centrifugal.

6. **Squirrel cage fans are used in air-condition-ing systems primarily in _____ applications.**
   A. exhaust
   B. duct work
   C. condenser
   D. compressor cooling

7. **An extended plenum system is a type of:**
   A. pressure-measuring system.
   B. duct system.
   C. package air-conditioning system.
   D. refrigerant piping system.

8. **A drive clip is used when installing _____ duct systems.**
   A. prefabricated fiberglass
   B. sheet metal
   C. flexible
   D. spiral metal

9. **The fittings between the main trunk and branch ducts in a duct system are called:**
   A. takeoff fittings.
   B. diffusers.
   C. spiral metal fittings.
   D. boots.

10. **The diffuser:**
    A. records the air velocity as it leaves the terminal unit.
    B. connects the terminal unit to the branch duct.
    C. is a type of balancing damper.
    D. spreads the air from the duct in the desired pattern.

11. **One reason for friction loss in duct systems is the:**
    A. humidity in the air.
    B. lack of humidity in the air.
    C. rubbing action of the air on the walls of the duct.
    D. thickness of the gauge in a sheet metal duct.

12. **Measuring airflow by traversing the duct is to:**
    A. measure the duct to determine the cross-sectional area.
    B. determine the cfm on each side of the duct.
    C. determine the total pressure at the center of the duct.
    D. measure the velocity of the airflow in a pattern across a cross section of the duct and averaging the measurements.

13. **Static pressure in a duct system is:**
    A. equal to the weight of the air standing still.
    B. the force per unit area in a duct.
    C. equal to the velocity and weight of the moving air.
    D. determined with a venturi system.

14. **A common way to fasten a round galvanized steel duct is with:**
    A. S fasteners.
    B. drive clips.
    C. snap-lock connectors.
    D. self-tapping sheet metal screws.

15. **If there is more than a _____ difference between the air inside and the air outside the duct, it should be insulated.**
    A. 5°F.
    B. 15°F.
    C. 25°F.
    D. 30°F.

Evaluating Duct Size

Name	Date	Grade

Objectives: Upon completion of this exercise, you will be able to use a duct chart to evaluate the duct size on a simple residential or commercial duct system for adequate airflow in heating or cooling cycles.

Introduction: You will evaluate a simple residential duct system using the duct evaluation chart in Figure 42-4 in the text.

Text References: Paragraphs 37.1, 37.2, 37.3, 37.4, 37.5, 37.6, 37.29, and 42.4.

Tools and Materials: A flashlight, straight-blade and Phillips screwdrivers, $1/4"$ and $5/16"$ nut drivers, and a tape measure.

Safety Precautions: Care should be taken not to cut your hands on sheet metal edges while taking duct measurements.

PROCEDURES

1. You will be assigned a duct system to evaluate. Find the capacity of the system and record here: heating _____ Btu/h, cooling _____ Btu/h

2. Record the number of supply outlets: _____

3. Record the number of return outlets: _____

4. Size of the supply duct _____ in. Return inlet _____ in.

5. Recommended size from chart, Supply _____ in. Return _____ in.

6. Recommended number and size of supply outlets _____

7. Recommended number of return inlets _____

8. Fan wheel size, Width _____ in. Diameter _____ in.

9. Type of fan motor, direct drive or belt drive _____

10. Draw a line diagram of the duct system you are working with, showing the sizes of the supply and return trunk and the individual branch ducts.

Maintenance of Workstation and Tools: Replace all panels with the correct fasteners.

Summary Statement: Describe the symptoms of a gas furnace that does not have adequate airflow over the heat exchanger.

QUESTIONS

1. Describe static pressure.

2. Describe velocity pressure.

3. Describe total pressure.

4. Name a common gage for measuring static pressure.

5. What unit of pressure is commonly used to describe air pressure in a duct system?

6. Would the static pressure be a positive or negative value in the supply duct?

7. Would the static pressure be a positive or negative value in the return duct?

8. If the velocity of air in a 12" × 20" supply duct is 700 feet per minute, how much air is moving? _____ cfm

Airflow Measurements

Name	Date	Grade

Objectives: Upon completion of this exercise, you will be able to use basic airflow measuring instruments to measure airflow from registers and grilles.

Introduction: You will use a prop-type velometer such as in Figure 37-1 to record the velocity of the air leaving the register. You will then measure the size of the register or use the manufacturer's data to determine in cfm the volume of air leaving the register.

Text References: Paragraphs 37.1, 37.2, 37.3, 37.4, 37.5, 37.6, and 37.29.

Tools and Materials: A prop-type velometer or the equivalent measuring device, and goggles.

Safety Precautions: Do not allow air to blow directly into your eyes because it may contain dust.

Figure 37-1 Prop-type velometer.
Photo by Bill Johnson

PROCEDURES

1. Start the system fan.

2. Take your air velocity measurements from a rectangular air discharge register that has a flat face, such as a high-side wall, floor, or ceiling type.

3. Put on safety goggles. Measure the velocity in several places on the register for an average reading. Record the readings here in fpm, feet per minute: _____ , _____ , _____ , _____ , _____

   • Average the above readings and record here: _____

   • The design volume of air may be found in the manufacturer's literature.

   • The volume of air in cfm may be found by multiplying the velocity of the air in feet per minute (fpm) by the free area of the register in square feet. The free area of the register may be found by the following method:

   > Measure the open area of the front of the register where the air flows out, the fin area. For example, the register may measure 3.5" × 5.5" = 19.25 square inches. Divide by 144 square inches to get the number of square feet (0.134 square feet area register).

   > * The area in the above step represents the face area, not the free area.

   > * The metal blades take up some of the area that you measured. For a register with movable blades, multiply the area × 0.75 (75% free area).

   > For a register with stamped blades, multiply the area × 0.60 (60% free area).

   • _____ fpm × _____ square feet = _____ cfm

4. You may take the above measurements on each outlet in the system and add them together for the total airflow in the system.

5. If the system has one or two return air inlets, you may arrive at the total airflow by using the above method with these grilles. Remember, the air is flowing toward the return air grille.

Maintenance of Workstation and Tools: Return all tools to their respective places.

Summary Statement: Describe what free area in a register or grille means.

QUESTIONS

1. How is a stamped grille different from a register?

2. Which of the above is the most desirable for directing airflow?

3. Which of the above is the most expensive?

4. What is the difference between a register and a grille?

5. What is the typical static pressure behind a supply register that forces air into the room?

6. What problem would excess velocity from the supply registers cause?

7. What would the symptoms be if supply registers (that blow air across the floor) designed for heating were used for air conditioning?

8. A duct that is 12" × 24" has an air velocity of 800 fpm. What is the airflow in cfm in the duct?

9. A supply register with movable blades has a face measurement of 6.5" × 12.5". How much air is flowing out of the register if the average face velocity is 300 fpm? Show your work.

Installation

Unit Overview

Sheet metal duct is fabricated in sections in a sheet metal shop. The duct is rigid, so it is important that all measurements are precise. These sections are then assembled and installed at the job site. Vibration eliminators should be used between the fan section and the duct.

Insulation can be fastened to either the inside or outside of the duct with tabs, glue, or both. When on the inside, it is generally done at the fabrication shop.

Fiberglass ductboard is popular because it can be fabricated in the field using special knives. It can be made into almost any configuration that metal duct can. Ductboard sections can be fastened together with staples and then taped. They must be supported when installed. Ductboard is sound absorbent and does not transmit fan noise.

Flexible duct is round and may be used for both supply and return duct. Sharp bends should be avoided as they can greatly reduce the airflow.

The electrical supply should be installed by a technician licensed to do so. The air-conditioning technician often installs the control voltage supply to the space thermostat if that is permitted in that locality.

The control or low-voltage in air-conditioning and heating equipment is the line voltage, which is reduced through a step-down transformer. The control wire is light duty, typically 18 gauge. It may have four or eight wires in the same sheath and is called 18-4 or 18-8 to describe the wire size (18) and the number of conductors (4) or (8).

Package air-conditioning systems may be air-to-air, air-to-water, water-to-water, or water-to-air. These units should be set on a firm foundation. Roof units should be placed on a roof curb. The unit should be set so that there will be sound isolation.

The evaporator in a split system is normally located near the fan section whether the fan is in a furnace or an air handler.

The condensing unit should be placed where it has power and air circulation, and is convenient for future piping and electrical service.

The refrigerant charge for the system is typically furnished by the manufacturer and stored in the condensing unit. When quick-connect line sets are used, the refrigerant charge for the lines is normally included in the lines. When the manufacturer furnishes the tubing, it is called a line set. The suction line will be insulated and the tubing charged with nitrogen if there are not quick-connect fittings.

Insure that all manufacturer's recommendations are followed when installing all equipment, including piping, and for equipment startup.

Key Terms

Air-to-Air Cooling	Over-and-Under Duct	Spring Isolators
Auxiliary Drain Pan	Connection	Solar Influence
Boot	Precharged Line Set	Transition Fitting
Crankcase Heater	Rubber and Cork Pads	Vibration Eliminator
Drive Clip	S-Fastener	
Line Set	Side-by-Side Duct Connection	

REVIEW TEST

Name	Date	Grade

Circle the letter that indicates the correct answer.

1. **What two fasteners are often used with sheet metal square and rectangular duct?**
   A. S-fasteners and drive clips
   B. staples and tape
   C. self-tapping sheet metal screws and rivets
   D. tabs and glue

2. **A boot in a duct system is used:**
   A. at the plenum.
   B. between the main and branch duct lines.
   C. between the branch duct and register.
   D. to eliminate vibration between the fan and duct.

3. **A good way to fasten insulation to the inside of metal duct is to use:**
   A. tape.
   B. staples.
   C. tabs and glue.
   D. self-tapping sheet metal screws.

4. **Sections of ductboard can be fastened together with:**
   A. staples and tape.
   B. tabs and glue.
   C. rivets.
   D. S-fasteners and drive clips.

5. **The line voltage should be within _____ of the rated voltage on the air-conditioning unit.**
   A. +/−10%   C. +/−20%
   B. +/−15%   D. +/−25%

6. **The control voltage in an air-conditioning system comes directly from:**
   A. a tap on the line voltage.
   B. a terminal on the compressor.
   C. the secondary of a step-down transformer.
   D. the circuit breaker panel.

7. **The standard wire size for the control voltage is _____ gauge.**
   A. 16   C. 24
   B. 18   D. 28

8. **The most common package air-conditioning system used in residential applications is the _____ system.**
   A. air-to-air
   B. air-to-water
   C. water-to-water
   D. water-to-air

9. **Rubber and cork pads are used for:**
   A. electrical insulation.
   B. heat-exchange insulation.
   C. moisture proofing.
   D. vibration isolation.

10. **The trap in the condensate drain line prevents:**
    A. air and foreign materials from being drawn in through the drain line.
    B. the air handler from forcing air out the drain line.
    C. a heat exchange from taking place through the drain line.
    D. excessive noise from traveling through the drain line.

11. **In a split system, the solar influence will have some effect on the:**
    A. condensing unit.
    B. evaporator.
    C. capillary tube.
    D. air handler.

12. **The crankcase heater is used to:**
    A. keep the temperature of the compressor at a predetermined level for better heat transfer.
    B. keep the oil from deteriorating.
    C. keep the motor relay from sticking.
    D. boil the liquid refrigerant out of the crankcase.

Evaluating an Air-Conditioning Installation

Name	Date	Grade

Objectives: Upon completion of this exercise, you will be able to look at an air-conditioning system and evaluate the installation for good workmanship and completeness.

Introduction: You will use this lab exercise as a checklist to ensure a split-system air-conditioning installation is complete and ready for startup.

Text Reference: Unit 38.

Tools and Materials: Straight-blade and Phillips screwdrivers, a VOM, a clamp-on ammeter, $1/4$" and $5/16$" nut drivers, a tape measure, and a flashlight.

Safety Precautions: Turn the power off and verify with the VOM that it is off. Lock and tag the panel or disconnect the box where the power is turned off. You should have a single key. Keep it in your possession while the power is off.

PROCEDURES

Use the following checklist to insure the installation is in good order. Indicate yes or no. Turn off and lock out the power.

1. Evaporator section:

   Air handler level _____ , vibration eliminator installed _____ , duct fastened tightly _____ , connections taped _____ , filter in place _____ , auxiliary drain pan if evaporator in attic _____ , secondary condensate drain line piped correctly _____ , primary drain pan piped correctly and trap with clean-out plug _____ , condensate pump, if needed, to pump the condensate to the drain _____ , electrical connections in the air handler tight, high voltage correct _____ low voltage correct _____ , fan motor and wheel can be removed easily _____

2. Condenser section:

   Condenser level _____ , a firm foundation under the condenser _____ , service panel in correct location _____ , airflow correct and not recirculating _____ , refrigerant piping in good order and insulated _____ , electrical connections tight _____ , high voltage correct _____ , low voltage correct _____ , service valves open, if any _____

3. Condensate drain termination:

   In a dry well _____ , on the ground _____ , in a drain _____

4. Voltage check:

   Line voltage to indoor fan section _____ V, to condenser _____ V, low-voltage power supply voltage _____ V

5. Start the indoor fan motor and verify the voltage and current: Rated voltage _____ V, actual voltage _____ V, rated current _____ A, actual current _____ A

6. Start the condensing unit and verify the voltage and amperage: rated voltage _____ V, actual voltage _____ V, compressor rated current _____ A, actual current _____ A, fan motor rated current _____ A, actual current _____ A

7. Replace all panels with the correct fasteners.

Maintenance of Workstation and Tools: Return all tools to their places.

Summary Statements: Describe the installation you inspected as it compares to the checklist. Describe any improvements you would have made.

QUESTIONS

1. What are vibration isolators for?

2. Describe a dry well for condensate.

3. What is the purpose of insulation on the suction line?

4. What is the purpose of an auxiliary drain pan when the evaporator is in the attic?

5. What is the purpose of the trap in the condensate drain line?

6. Why should water from the roof not drain into the top of a top discharge condenser?

7. How can an evaporator section be suspended from floor joists in a crawl space?

8. Why should air not be allowed to recirculate back into a condenser?

9. What would the symptoms of reduced airflow be at the evaporator?

10. What should be done if the line voltage to the unit is too low?

Using the Halide Leak Detector

Name	Date	Grade

Objectives: Upon completion of this exercise, you will be able to use a halide torch to leak check a system for refrigerant leaks.

Introduction: Using gages, you will check to determine that there is refrigerant pressure in an air-conditioning system. You will then check all field and factory connections with a halide leak detector to determine whether there are leaks. If the leak is from a flare connection, tighten it. If it is from a solder connection, ask your instructor what you should do.

Text References: Paragraphs 5.31 and 38.11.

Tools and Materials: A halide leak detector, a gage manifold, a VOM, straight-blade and Phillips screwdrivers, $1/4$" and $5/16$" nut drivers, goggles, and light gloves. If a leak is found, you will need additional tools to repair it.

Safety Precautions: Turn the electrical power off. Lock and tag the disconnect boxes to both the indoor and outdoor units before beginning the exercise. Check with the VOM to ensure that the power to the system is off. Wear goggles and light gloves when attaching or removing gages. Beware of toxic fumes from the halide torch.

PROCEDURES

You will need instruction in the use of the halide leak detector before beginning this exercise. Beware of toxic fumes.

1. With the power off and locked out and with goggles and gloves on, fasten the gage lines to the high- and low-side gage ports. Check the gage readings to ensure that there is pressure.

2. With the VOM, check the electrical disconnect to both units to ensure that the electrical power is off.

3. With the halide leak detector, check all joints (solder and flare) in the indoor and outdoor piping. Check the joints under the insulation by making a small hole in the insulation with a small screwdriver and holding the sensing tube to the hole.

4. Remove any additional necessary panels and check all factory connections.

5. If a leak is found in a flare joint, tighten it. If one is found in a solder joint, ask for instructions.

6. When you have established that there is no leak, or when you have finished, remove the gages and replace the panels and service valve connection caps.

Maintenance of Workstation and Tools: Return all tools, equipment, and materials to their proper storage places. Clean your workstation.

Summary Statement: Describe the halide leak-check procedure.

QUESTIONS

1. What pulls the refrigerant to the sensor in a halide leak detector?

2. What is the normal color of the gas flame in a halide leak detector?

3. What is the color of the gas flame in a halide leak detector when refrigerant is present?

4. Can a halide leak detector be used to detect natural gas? Why?

5. What does the term "halide" refer to?

6. Name two gases that are commonly used to produce the flame in a halide leak detector.

7. Describe how a halide leak detector discovers leaks.

8. Does the halide leak detector work best in the light or in the dark?

Using Electronic Leak Detectors

Name	Date	Grade

Objectives: Upon completion of this exercise, you will be able to use an electronic leak detector to find refrigerant leaks.

Introduction: You will check an air-conditioning system with gages to determine that there is refrigerant pressure. You will then check all field and factory connections with an electronic leak detector to determine whether there are refrigerant leaks.

Text References: Paragraphs 5.31 and 38.11.

Tools and Materials: An electronic leak detector, a gage manifold, a VOM, straight-blade and Phillips screwdrivers, $1/4$" and $5/16$" nut drivers, goggles, and light gloves. If a leak is found, you will need additional tools to repair it.

Safety Precautions: Turn the electrical power to both the indoor and outdoor unit off before beginning this exercise. Lock and tag the disconnect box and keep the key in your possession. Check with the VOM to ensure that the system is off. Wear goggles and light gloves when attaching or removing gages. Do not allow liquid refrigerant to contact your skin.

PROCEDURES

You will need instruction in the use of the electronic leak detector before beginning this exercise.

1. With the power off and locked out and with goggles and gloves on, fasten the gage lines to the high- and low-side gage ports. Check the gage readings to ensure that there is pressure. If there is no pressure, check with your instructor for the next step.

2. With the VOM, check the electrical disconnect to both units to ensure that the power is off.

3. Using the electronic leak detector, check all connections (flare and solder) at the indoor unit. Check under fittings.

4. Check all connections at the outdoor unit.

5. Remove necessary panels and check the factory connections.

6. Use a small screwdriver to make a small hole in the insulation on the tubing at each joint, and insert the probe in the holes to check for leaks.

7. Tighten any flare connection where you have found a leak. Ask for instructions before repairing solder connections.

8. When you have finished, carefully remove the gage lines and replace all panels with proper fasteners.

Maintenance of Workstation and Tools: Place all tools, equipment, and materials in their proper storage places. Ensure that your workstation is left clean.

Summary Statements: Describe how the electronic leak detector operates. What are some of its good and bad points?

QUESTIONS

1. Can natural gas be checked with an electronic leak detector?

2. How can the refrigerant be moved from the point of the leak to the detector sensor?

3. What can be done to keep wind from affecting the leak detection?

4. What is the power source for an electronic leak detector?

5. How can soap bubbles be used to further pinpoint the leak when it is detected?

6. What is the purpose of the filter under the tip of the sensor on the detector?

7. Does the electronic leak detector work best in a light or dark place?

Start-up and Checkout of an Air-Conditioning System

Name	Date	Grade

Objectives: Upon completion of this exercise, you will be able to check out components of an air-conditioning system for an orderly system start-up.

Introduction: You will start up an air-conditioning system, one component at a time, and check each one to ensure that it is operating correctly.

Text Reference: Paragraph 38.12.

Tools and Materials: A gage manifold, a VOM, a clamp-on ammeter, a thermometer, straight-blade and Phillips screwdrivers, $1/4$" and $5/16$" nut drivers, a flashlight, electrical tape, light gloves, and goggles.

Safety Precautions: Care should be taken while taking voltage and amperage readings on live circuits. Wear gloves and goggles while attaching and removing gages and taking pressure readings. When the power is turned off, lock and tag the panel.

PROCEDURES

With the unit off:

1. Check the line voltage to the indoor and outdoor units: Indoor unit _____ V, outdoor unit _____ V

2. Record the rated current of the indoor fan: _____ A, outdoor fan _____ A, compressor _____ A

3. With goggles and gloves on, fasten the gage lines to the condensing unit gage ports and put one temperature lead in the return air and one in the outdoor air to record the ambient temperature.

4. Disconnect the common wire going to the compressor, and tape the end to insulate it.

5. Set the thermostat to the off position, and turn the power to the fan section and the condenser section on.

6. Turn the fan switch on and start the indoor fan motor. Check and record the voltage and amperage of the motor _____ V, _____ A. Compare with manufacturer's specifications.

7. Verify that air is coming out all registers and that they are open.

8. Turn the thermostat to call for cooling and check the voltage and amperage of the fan motor at the condensing unit _____ V, _____ A. Compare with the manufacturer's specifications.

9. Turn the power off and reconnect the compressor wire.

10. Turn the power on and the compressor will start. Record the voltage and current: _____ V, _____ A. Compare with manufacturer's specifications.

11. Let the unit run and record the following information:
    - Indoor air temperature _____ °F, outdoor _____ °F
    - Suction pressure _____ psig, discharge pressure _____ psig

12. Carefully remove the gages. Fasten the service valve connection caps.

13. Replace all panels with the correct fasteners.

Maintenance of Workstation and Tools: Return all tools to their correct places. Leave your workstation clean.

Summary Statements: Compare the temperature for the suction pressure on the above system in relation to the inlet air temperature. Indicate whether the head pressure was correct for the conditions.

QUESTIONS

1. Why is it wise to disconnect the compressor while going through a start-up?

2. What should be done if the line voltage is too high before start-up?

3. What should be done if the line voltage is too low before start-up?

4. How does the outdoor temperature affect the head pressure?

5. How does the indoor temperature affect the suction pressure?

6. What should be done if there is no refrigerant in the system before start-up?

7. What type of service valve ports did the unit you worked on have, service valves or Schrader valves?

8. What is the typical refrigerant used in residential central air-conditioning systems?

Controls

Unit Overview

The indoor fan, compressor, and outdoor fan must be started and stopped automatically at the correct times to maintain correct air conditions. The indoor fan must operate when the compressor operates. The outdoor fan must operate when the compressor operates unless there is a fan cycle circuit that allows the fan to be cycled off during mild weather.

The low-voltage controls are operated with 24 V produced by a transformer located in the condensing unit or air handler. These low-voltage circuits operate the various power-consuming devices in the control system that start and stop the compressor and fan motors. This transformer is usually a 40 VA transformer. The room thermostat senses and controls the space temperature. It passes power to the fan relay, which starts and stops the indoor fan. It also passes power to the compressor contactor, which starts and stops the compressor and the outdoor fan. In older units, the high-pressure control stops the compressor when a high-pressure condition exists. The low-pressure control stops the compressor from pulling the suction pressure below a predetermined point. These units had overload protection for the common and run circuits of the compressor. This was usually a bimetal device that opened when a current overload produced heating. Short-cycle protection was provided to prevent the compressor from short cycling when a safety or operating control opened the circuit and then closed it in a short period of time.

More modern room thermostats are much smaller. Some use a thermistor as the sensing element and some may be programmed. The modern fan relay may be smaller than older ones, and the newer compressor contactors may have only one set of contacts instead of two. This leaves one circuit with continuous power to provide a small amount of current through the compressor windings to keep the compressor warm during the off cycle. This takes the place of the crankcase heat.

The components used to start and run modern equipment are the run capacitor and possibly a start-assist unit such as the positive temperature coefficient (PTC) device. These devices are used on PSC motors, which are used with metering devices that cause the pressures to equalize during the off cycle. Because of this, the motors require very little starting torque. These motors may have only an internal motor-temperature control that senses the temperature of the motor windings.

Some systems have a loss-of-charge protection in the form of a low-pressure control in the liquid line that may have a setting as low as 5 psig.

Key Terms

40 VA Transformer
Capacitor-Start, Capacitor-Run Motor
Compressor-Internal Relief Valve
Positive Temperature Coefficient (PTC) Device

Potential Start Relay
Solid-State Circuit Board
Thermistor

REVIEW TEST

Name	Date	Grade

Circle the letter that indicates the correct answer.

1. **The 40 VA control transformer can produce up to _____ amperes from the secondary.**
   A. 0.333　　　C. 4.666
   B. 1.666　　　D. 6.333

2. **Most modern compressors turn at _____ rpm.**
   A. 1800　　　C. 4200
   B. 3600　　　D. 4800

3. **Most modern air-conditioning equipment uses _____ refrigerant.**
   A. R-12　　　C. R-124
   B. R-22　　　D. R-134a

4. **The compressor contactor starts and stops the compressor and:**
   A. outdoor fan.　　　B. indoor fan.

5. **A semiconductor that is sensitive to temperature changes and used in some modern thermostats is the:**
   A. PTC device.
   B. triac.
   C. thermistor.
   D. transistor.

6. **Some modern compressor contactors have only one set of contacts instead of two so that:**
   A. power can more easily be disrupted to the compressor.
   B. there will be less pitting on the contacts.
   C. the contactor will be more reliable.
   D. a small amount of power can be fed continuously through the compressor windings to provide heat to the crankcase.

7. **A permanent split-capacitor motor on a modern air-conditioning compressor will have a:**
   A. start capacitor.
   B. run capacitor.
   C. start and run capacitor.
   D. high-pressure control.

8. **When a modern compressor is internally protected and suction gas cooled, what control will shut the compressor off when there is a low refrigerant charge?**
   A. PTC device
   B. compressor internal relief valve
   C. motor-winding thermostat
   D. potential relay

9. **When the condenser is dirty, the compressor head pressure will:**
   A. rise.
   B. go down.
   C. not be affected.

10. **If there is an overcharge of refrigerant that is keeping the compressor motor cool and a high head pressure, which control will function to shut the compressor off?**
    A. motor-winding thermostat
    B. positive temperature coefficient device
    C. run capacitor
    D. compressor internal relief valve

Identifying Controls of a Central Air-Conditioning System

Name	Date	Grade

Objectives: Upon completion of this exercise, you will be able to identify controls for a central air-conditioning system and state their function.

Introduction: You will remove panels from the condensing unit and the indoor air handler, identify each safety control that concerns the airflow for cooling, and describe its function.

Text References: Paragraphs 39.1, 39.2, 39.3, 39.6, 39.7, 39.8, 39.9, 39.10, and 39.11.

Tools and Materials: A flashlight, $1/4$" and $5/16$" nut drivers, straight-blade and Phillips screwdrivers, and a VOM.

Safety Precautions: Turn the power off while working inside the control compartments. Verify with the VOM that the power is off. Lock and tag the panel or disconnect box and keep the key with you.

PROCEDURES

1. What type of air handler are you working with? Gas, oil, electric furnace, or a fan section only?

2. With the power off and locked out, remove the cover from the fan section and read on the motor the type of motor protection. Record the type here: _____ What type of circuit protection is provided for the air handler: fuse or breaker? _____ Record the rating of the circuit protection here: _____ A

3. Replace all parts to the air handler with the correct fasteners.

4. With the power off and locked out, remove the compressor compartment panel. Follow the suction line from where it enters the unit to where it enters the compressor and look for a low-pressure control. Reviewing the wiring diagram may help. Is there a low-pressure control?

5. Follow the discharge line from where it leaves the compressor to where it enters the condenser and look for a high-pressure control. It may be located on the liquid line between the condenser and the line leaving the cabinet. Does the unit have a high-pressure control?

6. Examine the line voltage control for an overload for the compressor. Review the wiring diagram. Does the unit have an overload? Based on the compressor nameplate, does the compressor have internal motor protection? What type of circuit protector does the unit have: fuse or breaker?

7. Replace the panels with the correct fasteners.

Maintenance of Workstation and Tools: Return all tools to their places. Leave the work area clean and orderly.

Summary Statement: Describe how a compressor with an internal overload and internal relief valve has low-charge protection.

QUESTIONS

1. What is the difference between a fuse and a breaker?

2. Why does a fuse only protect the circuit when applied to an air-cooled condensing unit?

3. How is the low-voltage circuit normally protected?

4. What is impedance protection for a motor?

5. How does an internal relief valve work inside a compressor?

6. How would a low-pressure control located in the liquid line in the condenser protect a system when it may be set to shut off at 10 psig?

7. What is the difference between a fuse and a time delay fuse?

8. What is the difference between a thermal motor protector and a current-type motor protector?

9. How is a typical hermetic motor cooled?

10. What can electronic controls do that electromechanical controls cannot do?

Typical Operating Conditions

Unit Overview

Air-conditioning equipment is designed to operate at its rated capacity and efficiency at a specified design condition. This condition is generally accepted to be at an outside temperature of 95°F and an inside temperature of 80°F with a relative humidity of 50%. This rating has been established by the Air-Conditioning and Refrigeration Institute (ARI). However, many homeowners prefer to operate their systems at 75°F.

The evaporator will normally operate at 40°F boiling temperature when operating at the 75°F, 50% relative humidity condition. This will result in a suction pressure of approximately 70 psig for R-22 or 118 psig for R-410A. There will be liquid refrigerant nearly to the end of the coil, and there will be a superheat of approximately 10°F.

There are normally three grades of equipment: economy, standard efficiency, and high efficiency. Some manufacturers offer only one grade; others offer all three. Economy and standard grades have about the same efficiency. The high-efficiency grade has different operating characteristics. A standard condenser will condense the refrigerant at a temperature of approximately 30°F above the outside ambient temperature. In the high-efficiency condenser, the condensing temperature may be as low as 20°F over the outside ambient air.

One of the first things a technician should do is determine whether the equipment is standard or high efficiency. The high-efficiency equipment is often larger than normal.

High-efficiency air-conditioning equipment may utilize a thermostatic expansion valve rather than a fixed-bore metering device. It may also have a larger than normal evaporator. With a larger evaporator, the refrigerant boiling temperature may be as high as 45°F at design conditions. This results in a suction pressure of 76 psig for R-22 or 130 psig for R-410A.

The Air-Conditioning and Refrigeration Institute (ARI) has developed a rating system for equipment. The original system was called the EER, or Energy Efficiency Ratio, which is the output in Btu/h divided by the input in watts of power. The EER rating does not take into account the energy required to bring the system up to peak efficiency from start-up and the energy used in shutting the system down at the end of the cycle. Another rating called the SEER, or Seasonal Energy Efficiency Ratio, has been developed. This rating does take into consideration the start-up and shutdown for each cycle.

There are also typical electrical operating conditions. The first thing a technician should determine is the supply voltage. Residential units will normally be 230 V. Light commercial units will be 208 or 230 V single-phase. Light commercial units may operate from a three-phase power supply. The compressor may be three-phase and the fan motors single-phase obtained from the three-phase. The unit rated voltage must be within ±10% of the supply voltage.

Key Terms

Energy Efficiency Ratio (EER)

Seasonal Energy Efficiency Ratio (SEER)

REVIEW TEST

Name	Date	Grade

Circle the letter that indicates the correct answer.

1. **What design conditions are most air-conditioning systems designed for?**
   A. outside temperature 90°F, inside temperature 70°F, relative humidity 50%
   B. outside temperature 95°F, inside temperature 75°F, relative humidity 50%
   C. outside temperature 95°F, inside temperature 80°F, relative humidity 50%
   D. outside temperature 100°F, inside temperature 80°F, relative humidity 60%

2. **A standard evaporator will operate with a _____ refrigerant boiling temperature with a 75°F inside air temperature and other conditions being standard.**
   A. 40°F   C. 50°F
   B. 45°F   D. 55°F

3. **If the head pressure increases at the condenser, the suction pressure will:**
   A. decrease.
   B. increase.
   C. not be affected.

4. The refrigerant condensing temperature in a standard-efficiency condenser will be _____ above the outside ambient temperature.
   A. 20°F
   B. 25°F
   C. 30°F
   D. 35°F

5. A normal superheat in a standard efficiency system operating under design conditions will be _____.
   A. 10°F.
   B. 15°F.
   C. 20°F.
   D. 25°F.

6. If the head pressure at the condenser is 297 psig, the condensing temperature will be _____ if the system uses R-22. (Use the temperature/pressure relationship chart at the end of this guide.)
   A. 115°F
   B. 120°F
   C. 125°F
   D. 130°F

7. The head pressure for a high-efficiency condenser will be _____ for a standard condenser under the same conditions.
   A. lower than
   B. higher than
   C. about the same as

8. If the space temperature is higher than standard conditions, the suction pressure will be _____ normal.
   A. lower than
   B. higher than
   C. about the same as

9. If the head pressure at a high-efficiency condenser is 243 psig and the system is using R-22, how high will the condensing temperature be above the ambient of 95°F? (Use the temperature/pressure relationship chart at the end of this guide.)
   A. 20°F
   B. 25°F
   C. 30°F
   D. 35°F

10. The more accurate rating system for air-conditioning equipment that takes into consideration operation during the complete cycle is the:
   A. EER.
   B. SEER.
   C. ARI.
   D. NEC.

11. The metering device for a high-efficiency air-conditioning system may be a:
   A. low-side float.
   B. high-side float.
   C. orifice type.
   D. thermostatic expansion valve.

12. Higher efficiency for a condenser may be accomplished by using:
   A. an aluminum condenser.
   B. larger size condenser tubing.
   C. a larger condenser.
   D. a faster compressor.

13. Suppose you were working on a piece of equipment that matched the chart in the text in Figure 40-11, the outdoor temperature was 90°F, and the indoor wet-bulb temperature was 68°F. The superheat at the suction service valve should be:
   A. 19°F.
   B. 14°F.
   C. 9°F.
   D. 16°F.

14. The approximate full-load amperage (using Figure 40-12) for a 5 hp single-phase motor operating on 230 V would be:
   A. 28 amps.
   B. 15.2 amps.
   C. 10 amps.
   D. 18 amps.

15. If the head pressure at the condenser is 475 psig, the condensing temperature will be _____ if the system uses R-410A.
   A. 115°F
   B. 120°F
   C. 125°F
   D. 130°F

16. The low-side pressure of an air-conditioning system reads 118 psig at the evaporator. If the air-conditioning system is using R-410A as its refrigerant, what will the evaporating temperature be?
   A. 30°F
   B. 35°F
   C. 40°F
   D. 45°F

Evaporator Operating Conditions

Name	Date	Grade

Objectives: Upon completion of this exercise, you will be able to state typical evaporator pressures and describe the results of reduced airflow across an evaporator.

Introduction: You will use a gage manifold and thermometer to measure and then record pressures and temperatures under normal conditions in an evaporator in an air-conditioning system. You will then reduce the airflow and measure and record pressures and temperatures again.

Text References: Paragraphs 40.1, 40.2, 40.3, 40.4, and 40.5.

Tools and Materials: A gage manifold, a thermometer, gages, goggles, gloves, cardboard for blocking the evaporator coil, a service valve wrench, straight-blade and Phillips screwdrivers, and $1/4$" and $5/16$" nut drivers.

Safety Precautions: Wear gloves and goggles while connecting and removing gages. Be careful not to come in contact with any electrical connections when cabinet panels are off.

PROCEDURES

1. Place a thermometer sensor in the supply and return air streams.

2. With gloves and goggles on, fasten the low- and high-side gages to the gage ports.

3. Start the unit and record the following after the unit and conditions have stabilized.

   • Suction pressure _____ psig • Refrigerant boiling temperature _____ °F • Inlet air temperature _____°F • Outlet air temperature _____°F • Discharge pressure _____ psig

4. Reduce the airflow across the evaporator to approximately one-fourth by sliding the cardboard into the filter rack. If there is a single return inlet, the cardboard may be placed across three-fourths of the opening.

5. When the unit stabilizes, record the following:

   • Suction pressure _____ psig • Refrigerant boiling temperature _____ °F • Inlet air temperature _____°F • Outlet air temperature _____°F • Discharge pressure _____ psig

6. Remove the air blockage and gages and return the unit to normal operation.

Maintenance of Workstation and Tools: Replace all panels with proper fasteners. Return all tools to their places.

Summary Statement: Describe how the unit suction pressure and air temperature differences responded.

QUESTIONS

1. How does a fixed-bore expansion device respond to a reduction in load, such as when the filter is restricted?

2. How does a thermostatic expansion valve respond to the above condition?

3. What are the results of small amounts of liquid reaching a hermetic compressor?

4. How does the temperature drop across an evaporator respond to a reduced airflow?

5. What happens to the head pressure when a reduced airflow is experienced at the evaporator?

6. List four things that can happen to reduce the airflow across an evaporator.

7. What procedure may be used to thaw a frozen evaporator?

8. What would happen if an air conditioner were operated for a long period of time without air filters?

9. How may an evaporator coil be cleaned?

Condenser Operating Conditions

Name	Date	Grade

Objectives: Upon completion of this exercise, you will be able to describe how a condenser responds to reduced airflow.

Introduction: You will measure and record a typical condenser's refrigerant pressure and temperature, the condenser entering airstream temperature, and the compressor amperage under normal operating conditions. You will then reduce the airflow across the condenser, measure and record the same conditions, and compare the results.

Text References: Paragraphs 40.1, 40.2, 40.3, and 40.9.

Tools and Materials: A thermometer, a clamp-on ammeter, a gage manifold, goggles, gloves, sheet plastic or cardboard to reduce the airflow through the condenser, a service valve wrench, straight-blade and Phillips screwdrivers, and $^{1}/_{4}$" and $^{5}/_{16}$" nut drivers.

Safety Precautions: Wear goggles and gloves while attaching and removing gage lines. Be very careful not to come in contact with any electrical terminals or moving fan blades.

PROCEDURES

1. On an air-conditioning system that uses R-22 refrigerant and with goggles and gloves on, fasten the gage lines to the gage ports.

2. With the unit off, fasten one thermometer lead to the discharge line on the compressor and place the other in the air stream entering the condenser. Replace the compressor compartment panel if the compressor compartment is not isolated from the condenser.

3. Clamp the ammeter on the compressor common terminal wire.

4. Start the unit and allow it to run until stabilized (when the temperature of the discharge line does not change). This may take 15 minutes or more. Record the following information:

   • Suction pressure _____ psig • Refrigerant boiling temperature _____°F • Discharge pressure _____ psig • Refrigerant condensing temperature _____°F • Compressor current _____ A

5. Cover part of the condenser coil with the plastic or cardboard until the discharge pressure rises to about 300 psig for R-22 and allow the unit to run until it stabilizes, about 15 minutes. Then record the following information:

   • Suction pressure _____ psig • Refrigerant boiling temperature _____°F • Discharge pressure _____ psig • Refrigerant condensing temperature _____°F • Compressor current _____ A

6. Remove the cover and allow the unit to run for about 5 minutes to reduce the head pressure.

7. Shut the unit off and remove the gages.

8. Replace all panels with the correct fasteners. Return the unit to its normal condition.

Maintenance of Workstation and Tools: Return all tools to their places.

Summary Statement: Describe how and why the pressures and temperatures responded to the reduced airflow across the condenser.

QUESTIONS

1. Does the compressor work more or less efficiently with an increased head pressure?

2. What does the discharge-line temperature do when an increase in head pressure is experienced?

3. What would a very high discharge-line temperature do to the oil circulating in the system?

4. Name four things that would cause a high discharge pressure.

5. How does the compressor amperage react to an increase in discharge pressure?

6. Would the discharge pressure be more or less for a high-efficiency unit when compared to a standard-efficiency unit?

7. How is high efficiency accomplished?

8. Is a high-efficiency unit more or less expensive to purchase?

9. Why do many manufacturers offer both standard- and high-efficiency units?

Troubleshooting

Unit Overview

The gage manifold indicates the low- and high-side pressures while the air-conditioning system is operating. These pressures can then be converted to the evaporator refrigerant boiling temperature and the refrigerant condensing temperature at the condenser. Two types of pressure connections are used: the Schrader valve and the service valve.

The four-lead electronic thermometer is a useful instrument when taking temperature readings. Temperatures that may need to be taken are the evaporator inlet air, evaporator outlet air, suction line, discharge line, liquid line, and outside ambient air.

The volt-ohm-milliammeter and the clamp-on ammeter are the main instruments used for electrical trouble-shooting. The main power panel is divided into many circuits. For a split-system, there are usually separate circuit breakers in the main panel for the indoor unit and the outdoor unit. For a package system, there is usually one breaker for the unit.

Before performing any troubleshooting, be sure that you are aware of all safety practices and follow them. All safety practices are not repeated in this manual. Follow instructions from your instructor after you have read Unit 41 in the text and are aware of all safety precautions throughout the text.

Key Terms

Gage Manifold
Schrader Valve

Service Valve
Subcooling

REVIEW TEST

Name	Date	Grade

Circle the letter that indicates the correct answer.

1. **The suction or low-side gage is located on the _____ side of the gage manifold.**
   A. left
   B. right

2. **Which of the following temperature equivalent charts are printed on the pressure gages on the gage manifold?**
   A. R-11, R-12, R-123
   B. R-22, R-12, R-134a
   C. R-22, R-123, R-134a
   D. R-22, R-12, R-502

3. **A service valve can be used to:**
   A. check the electrical service to the compressor.
   B. isolate the refrigerant so the system can be serviced.
   C. check the airflow across the evaporator.
   D. check the ambient temperature at the condenser.

4. **On a 95°F day and a standard-efficiency air conditioner, the R-22 refrigerant pressure on the high-side should be about _____ psig. (Use the temperature/pressure chart at the end of the guide.)**
   A. 182   C. 278
   B. 108   D. 243

5. **On an 85°F day, the high-side R-22 refrigerant pressure for a high-efficiency condenser should be about _____ psig.**
   A. 156   C. 243
   B. 211   D. 181

6. **The liquid-line temperature may be used to check the _____ efficiency of the condenser.**
   A. superheat
   B. low-side pressure
   C. suction pressure
   D. subcooling

7. **A capillary tube is _____ metering device.**
   A. an adjustable
   B. a fixed-bore
   C. a thermostatic
   D. a nonpressure equalizing

8. **To accurately add refrigerant to a system, a _____ ambient temperature should be simulated.**
   A. 75°F
   B. 85°F
   C. 95°F
   D. 105°F

9. **If a short in a compressor motor winding is suspected, which of the following instruments would be used to check it?**
   A. ohmmeter   C. voltmeter
   B. ammeter   D. milliammeter

10. **Gage lines should not be connected to a system unless necessary as:**
    A. the Schrader valve may become worn.
    B. the service valve is used only to isolate refrigerant.
    C. most gage lines leak.
    D. a small amount of refrigerant will leak from the system when the gage lines are disconnected.

11. **A Schrader valve is a:**
    A. pressure check port.
    B. low-side valve only.
    C. high-side valve only.
    D. high-efficiency expansion device.

12. **A short service connection may be used:**
    A. to make quick service calls.
    B. to eliminate the loss of large amounts of liquid while checking the liquid line pressure.
    C. for heat pumps only.
    D. only in summer.

13. **Electronic thermometers are normally used for checking superheat and subcooling because:**
    A. they are cheap.
    B. they are easy to use and accurate.
    C. they are all that is available.
    D. none of the above.

14. **The common meter for checking the current flow in the compressor electrical circuit is the:**
    A. VOM.
    B. clamp-on ammeter.
    C. electronic thermometer.
    D. ohmmeter.

Manufacturer's Charging Procedure

Name	Date	Grade

Objectives: Upon completion of this exercise, you will be able to use a manufacturer's charging chart to check for the correct operating charge for a central air-conditioning system.

Introduction: You will use a manufacturer's charging chart for a suggested charge and charging procedures for top efficiency in a central air-conditioning system in outside temperatures above 65°F. This unit may have any type of metering device.

Text References: Paragraphs 41.2, 41.3, 41.4, 41.5, 41.6, 41.7, and 41.8.

Tools and Materials: A thermometer, a psychrometer, a gage manifold, two 8" adjustable wrenches, service valve wrench, straight-blade and Phillips screwdrivers, $1/4$" and $5/16$" nut drivers, light gloves, goggles, tape, insulation (to hold the temperature sensor and insulate it), and an air-conditioning system with manufacturer's charging chart.

Safety Precautions: Wear gloves and goggles while fastening gage lines and removing them from the gage ports. Be very careful around rotating components.

PROCEDURES

1. Review the manufacturer's procedures for charging. It may call for a superheat check. **NOTE:** All manufacturers' literature specifies that the indoor airflow be correct for their charging procedures to work. If the readings you get are off more than 10% and cannot be corrected by adjusting the refrigerant charge, suspect the airflow.

2. With goggles and gloves on, fasten the gage lines to the gage ports (Schrader ports or service valve ports).

3. Start the unit and allow it to run for at least 15 minutes.

4. While the unit is stabilizing, prepare for taking and recording the required readings:

   - Suction pressure reading _____ psig
   - Superheat (suction line temperature − boiling temperature _____°F
   - Return air temperature _____°F
   - Evaporator approach temperature (return air temperature − evaporator temperature) _____°F
   - Discharge pressure _____ psig
   - Condensing temperature (converted from discharge pressure) _____°F
   - Indoor wet-bulb _____°F, dry-bulb _____°F
   - Outdoor dry-bulb _____°F
   - Condenser approach temperature (condensing temperature − outdoor air temperature) _____°F

5. If the unit has the correct charge because it corresponds to the manufacturer's chart, you may shut the unit off and disconnect the gages.

6. If it does not, under your instructor's supervision, adjust the charge to correspond to the chart. Then shut the unit off and disconnect the gages.

7. Replace all panels with the correct fasteners.

Maintenance of Workstation and Tools: Return all tools to their respective places and be sure your workstation is in order.

Summary Statement: Describe how the wet-bulb temperature reading affects the load on the evaporator.

QUESTIONS

1. What would a typical superheat reading be at the outlet of an evaporator on a design temperature day for a central air-conditioning system?

2. How does the outdoor ambient temperature affect the head pressure?

3. What is the recommended airflow per ton of air conditioning for the unit you worked on?

4. What would the result of too little airflow be on the readings you took?

5. What type of metering device did the unit you worked with have?

6. Did you have to adjust the charge on the unit you worked with?

7. How would too much charge affect a unit with a capillary tube?

8. Would the overcharge in question 7 be evident in mild weather?

Charging a System in the Field Using Typical Conditions (Fixed-Bore Metering Device)

Name	Date	Grade

Objectives: Upon completion of this exercise, you will be able to correctly charge a system using a fixed-bore metering device when no manufacturer's literature is available.

Introduction: You will check the charge in a central air-conditioning system that uses R-22 under field conditions when no manufacturer's literature is available. This system must have a fixed-bore metering device. Always use manufacturer's literature when it is available. The following procedure may be done in any outdoor ambient temperature; however, the colder it is, the more condenser cover you will need.

Text References: Paragraphs 41.2, 41.3, 41.4, 41.5, 41.6, 41.7, and 41.8.

Tools and Materials: A thermometer, a gage manifold, tape, insulation, gloves, goggles, two 8" adjustable wrenches, a service valve wrench, plastic or cardboard to restrict airflow to the condenser, straight-blade and Phillips screwdrivers, $1/4$" and $5/16$" nut drivers, a cylinder of R-22, and a central air-conditioning system with a fixed-bore metering device.

Safety Precautions: Wear gloves and goggles when attaching and removing gages. Do not let the head pressure rise above 350 psig. Be careful of rotating parts and electrical terminals and connections.

PROCEDURES

1. With goggles and gloves on, fasten the gages to the gage ports.

2. Start the unit and allow it to run for at least 15 minutes for it to stabilize. During this time, fasten the thermometer to the suction line at the condensing unit and insulate it. See the text, Figure 41-21.

3. Observe the pressures and cover the condenser to increase the head pressure to 275 psig, if necessary. Keep track of it and do not let it rise above 275 psig. Regulate the pressure until it stabilizes at exactly 275 psig. **NOTE:** If it will not rise to 275 psig, the unit may not have enough charge. Check with your instructor. The superheat should be 10°F to 15°F at the condensing unit, with a correct charge and an average length suction line (see example, Figure 41-1).

4. Adjust the charge to the correct level. Do not add refrigerant too fast or you will overcharge the system. It is easier to adjust the charge while adding refrigerant than while removing it.

5. When the charge is correct, record the following:
   - Suction pressure _____ psig
   - Suction-line temperature _____°F
   - Refrigerant boiling temperature (converted from suction pressure) _____°F
   - Superheat (suction-line temperature − refrigerant boiling temperature) _____°F
   - Recommended superheat for line set length _____°F
   - Discharge pressure _____ psig
   - Condensing temperature _____°F

6. When you have adjusted the charge, remove the head pressure control (cardboard or plastic) and allow the unit to stabilize. Then record the following information:
   - Suction pressure _____ psig
   - Suction-line temperature _____°F
   - Refrigerant boiling temperature (converted from suction pressure) _____°F
   - Superheat (suction-line temperature − refrigerant temperature converted from suction pressure) _____°F
   - Superheat (suction-line temperature − refrigerant boiling temperature) _____°F

7. Shut the unit off, remove the temperature sensors and gages, and replace all panels with the correct fasteners.

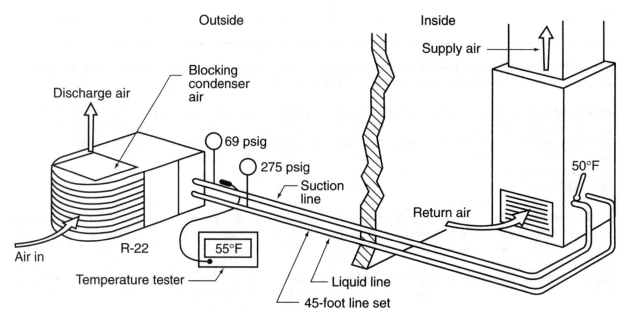

Notice that 5°F superheat was gained between the air handler and the condensing unit in this example.

Figure 41-1

Maintenance of Workstation and Tools: Return all tools, equipment, and materials to their storage places.

Summary Statement: Describe how head pressure affects superheat with a fixed-bore metering device.

QUESTIONS

1. How does line length affect refrigerant charge?

2. Why is the head pressure adjusted to 275 psig instead of some other value?

3. What is the typical refrigerant used for central air conditioning?

Charging a System in the Field Using Typical Conditions (Thermostatic Expansion Valve)

Name	Date	Grade

Objectives: Upon completion of this exercise, you will be able to correctly charge a system using a thermostatic expansion valve metering device when no manufacturer's literature is available.

Introduction: You will use the subcooling method to ensure the correct charge for the unit. This procedure is to be used when there is no manufacturer's literature available and may be performed in any outdoor ambient temperature. A unit with R-22 is suggested because it is the most common refrigerant; however, a unit using another refrigerant may be used.

Text References: Paragraphs 41.2, 41.3, 41.4, 41.5, 41.6, 41.7, and 41.8.

Tools and Materials: A thermometer, a gage manifold, R-22 refrigerant, goggles, light gloves, plastic or cardboard (to restrict airflow to the condenser), two 8" adjustable wrenches, a service valve wrench, straight-blade and Phillips screwdrivers, $1/4$" and $5/16$" nut drivers, tape, and insulation for the temperature sensor.

Safety Precautions: Wear gloves and goggles while fastening and removing gages. Do not let the head pressure rise above 350 psig. Be careful of rotating components and electrical terminals and connections.

PROCEDURES

1. Fasten the gage lines to the test ports (Schrader or service valve ports). Locate the sight glass in the liquid line if there is one.

2. Start the unit. While it is stabilizing, fasten the thermometer to the liquid line at the condensing unit and insulate it. See the text, Figure 41-22.

3. Place the condenser airflow restricter in place and allow the head pressure to rise to 275 psig. Watch closely, particularly if the weather is warm, to make sure that the head pressure stays at 275 psig. Keep adjusting the plastic or cardboard until 275 psig is steady. If it will not reach 275 psig, look to see whether there are bubbles in the sight glass. The unit is probably undercharged, Figure 41-2.

4. Record the following information:

   - Suction pressure _____ psig
   - Discharge pressure _____ psig
   - Condensing temperature (converted from head pressure) _____ °F
   - Liquid-line temperature _____ °F
   - Subcooling (condensing temperature − liquid line temperature) _____ °F

5. If the subcooling is 10°F to 15°F, the unit has the correct charge. If it is more than 15°F, the unit has too much refrigerant. If less than 10°F, the unit charge is too low.

6. Adjust the charge to between 10°F and 15°F and record the following again:
   - Suction pressure _____ psig
   - Discharge pressure _____ psig
   - Condensing temperature (converted from head pressure) _____°F
   - Liquid-line temperature _____°F
   - Subcooling (condensing temperature − liquid line temperature) _____°F

7. When the charge is correct, remove the temperature sensor and gages, and replace all panels with the correct fasteners.

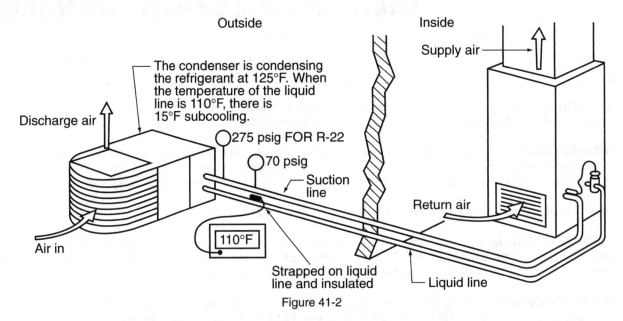

Figure 41-2

Maintenance of Workstation and Tools: Return all tools, equipment, and materials to their storage places and make sure your workstation is clean and in order.

Summary Statement: Describe subcooling and where it takes place.

QUESTIONS

1. If the unit you worked on had a sight glass, did it show bubbles when the charge was correct? What caused the bubbles?

2. Where is the correct location for a sight glass in a liquid line in order to check the charge?

3. How is subcooling different from superheat?

4. Does superheat ever occur in the condenser?

5. Why do you raise the head pressure to 275 psig to check the subcooling of the condenser in the above exercise?

6. Why do you insulate the temperature lead when checking refrigerant line temperatures?

Troubleshooting Exercise: Central Air Conditioning with Fixed-Bore Metering Device

Name	Date	Grade

Objectives: Upon completion of this exercise, you will be able to troubleshoot a basic mechanical problem in a typical central air-conditioning system.

Introduction: This exercise should have a specific problem introduced by your instructor. You will locate the problem using a logical approach and instruments. Instruments may be used to arrive at the final diagnosis; however, you should be able to determine the problem without instruments. THE INTENT IS NOT FOR YOU TO MAKE THE REPAIR BUT TO LOCATE THE PROBLEM AND ASK YOUR INSTRUCTOR FOR DIRECTION BEFORE COMPLETION.

Text Reference: Unit 41.

Tools and Materials: Straight-blade and Phillips screwdrivers, $1/4"$ and $5/16"$ nut drivers, a thermometer, two 8" adjustable wrenches, goggles, gloves, and a set of gages.

Safety Precautions: Use care while connecting gage lines to the liquid line. Wear goggles and gloves.

PROCEDURES

1. Start the system and allow it to run for 15 minutes.

2. Touch test the suction line leaving the evaporator. HOLD THE LINE TIGHT FOR TOUCH TESTING. Record your opinion of the suction line condition here:

3. Touch test the suction line entering the condensing unit and record your impression here:

4. Remove the compressor compartment door. MAKE SURE THAT AIR IS STILL FLOWING OVER THE CONDENSER. IF AIR PULLS IN OVER THE COMPRESSOR WITHOUT A PARTITION, HIGH HEAD PRESSURE WILL OCCUR VERY QUICKLY. COMPLETE THE FOLLOWING VERY QUICKLY AND SHUT THE DOOR. Touch test the compressor body all over. Record the following:

    A. The temperature of the top and sides of the compressor in relation to hand temperature _____
    B. The temperature of the compressor crankcase compared to your hand temperature _____

5. All of these touch test points should be much colder than hand temperature if the system has the correct problem. Record your opinion of the system difficulty here: _____

6. Fasten gages to the gage ports.

7. Fasten a thermometer lead to the suction line at the condensing unit.

8. Record the superheat here: _____°F

9. Place a thermometer lead in the inlet and outlet air streams to the evaporator and record their readings here: Inlet air temperature _____°F, outlet air temperature _____°F, temperature difference _____°F

10. What should the temperature difference be? _____°F

11. Describe your opinion of the problem and the recommended repair.

12. Ask your instructor before making the repair, then do as instructed.

Maintenance of Workstation and Tools: Leave the unit as directed and return all tools to their respective places.

Summary Statement: Describe how to determine the problem with this unit by the touch test.

QUESTIONS

1. How should the touch test of the evaporator suction line feel when a unit has an overcharge of refrigerant?

2. How would the compressor housing feel when a unit has an overcharge of refrigerant?

3. How would the suction line feel when a unit has a restricted airflow?

4. How would the compressor housing feel when a unit has a restricted airflow?

5. How can you tell the difference in an overcharge and a restricted airflow by the touch test?

6. Why should gages not be fastened to the system until the touch test is completed?

Troubleshooting Exercise: Central Air Conditioning with Fixed-Bore Metering Device

Name	Date	Grade

Objectives: Upon completion of this exercise, you will be able to troubleshoot a mechanical problem in a central air-conditioning system with a fixed-bore metering device.

Introduction: This exercise should have a specific problem introduced by your instructor. You will locate the problem using a logical approach and instruments. The problem in this exercise will require instruments to properly solve and diagnose. THE INTENT IS NOT FOR YOU TO MAKE THE REPAIR BUT TO LOCATE THE PROBLEM AND ASK YOUR INSTRUCTOR FOR DIRECTION BEFORE COMPLETION.

Text Reference: Unit 41.

Tools and Materials: Straight-blade and Phillips screwdrivers, $1/4$" and $5/16$" nut drivers, gloves, goggles, two 8" adjustable wrenches, a thermometer, a cylinder of refrigerant, and a gage manifold.

Safety Precautions: Use care while fastening gage lines to the liquid line. Wear gloves and goggles.

PROCEDURES

1. Start the system and allow it to run for 15 minutes.

2. Touch test the suction line leaving the evaporator. HOLD THE LINE TIGHTLY IN YOUR HAND. Record the temperature of the line in comparison to hand temperature:

3. Touch test the suction line entering the condensing unit. Record the temperature of the line in comparison to hand temperature:

4. Remove the compressor compartment door for the purpose of touch testing the compressor. **NOTE:** IF THE COMPRESSOR COMPARTMENT IS NOT ISOLATED, THIS TEST MUST BE PERFORMED QUICKLY OR THE HEAD PRESSURE WILL BECOME TOO HIGH DUE TO AIR NOT FLOWING OVER THE CONDENSER. LOOK THE SYSTEM OVER BEFORE LEAVING THE DOOR OFF TOO LONG. Touch test the compressor and record:

   A. The top and sides of the compressor housing compared to hand temperature.
   B. The bottom, crankcase, of the compressor compared to hand temperature. Replace the door quickly if air is bypassing the condenser.

5. Using the touch test for diagnosis, what is your opinion of the problem with this unit?

6. Fasten gages to the gage ports.

7. Place a thermometer lead on the suction line entering the condensing unit and record the superheat here: _____ °F

8. Take the temperature difference across the evaporator by placing a thermometer lead in the inlet and out-let air. Inlet air temperature _____°F, outlet air temperature _____°F, temperature difference _____°F

9. What should the temperature difference be? _____°F

10. Record your opinion of the unit problem and your recommended repair:

12. Consult your instructor as to what to do next.

Maintenance of Workstation and Tools: Return the unit to the condition to which your instructor directs and return all tools to their respective places.

Summary Statement: Describe how overcharge affects the capacity of a fixed-bore metering device system.

QUESTIONS

1. How does liquid refrigerant affect the oil in the crankcase of a compressor?

2. What is the difference between liquid slugging in a compressor and a small amount of liquid refrigerant entering the compressor crankcase?

3. Do most compressors have oil pumps?

4. How is a compressor lubricated if it does not have an oil pump?

5. Describe how a suction line feels if liquid refrigerant is moving toward the compressor?

6. Do all central air-conditioning systems use fixed-bore metering devices?

7. Name two types of fixed-bore metering devices.

Changing a Hermetic Compressor

Name	Date	Grade

Objectives: Upon completion of this exercise, you will be able to change a compressor in a typical air-conditioning system.

Introduction: You will recover the refrigerant from a system, then completely remove a hermetic compressor from the air-conditioning unit and replace it as though you had a new compressor. You will then leak check, evacuate, and charge the unit with the correct charge.

Text References: Paragraphs 7.5, 7.7, 7.8, 7.9, 7.10, 7.11, 8.6, 8.7, 8.10, 8.11, 10.1, 10.2, 10.3, 10.4, 10.5, and 41.8.

Tools and Materials: Straight-blade and Phillips screwdrivers, an adjustable wrench, $1/4$" and $5/16$" nut drivers, torch setup suitable for this exercise, solder, flux, slip joint pliers, gage manifold, refrigerant recovery equipment, a vacuum pump, a nitrogen tank and regulator, a VOM, an ammeter, refrigerant, a leak detector, gloves and goggles, a set of socket wrenches, and needle-nose pliers.

Safety Precautions: Use care while working with liquid refrigerant. Turn all power off before starting to remove the compressor. Lock and tag the panel or disconnect box where the power is turned off. There should be a single key; keep it on your person.

PROCEDURES

1. Put on safety goggles. Recover the refrigerant from the system using an approved recovery system.

2. With the refrigerant out of the system, TURN OFF AND LOCK OUT THE POWER.

3. Remove the door to the compressor compartment.

4. Remove the compressor mounting bolts or nuts.

5. Mark all electrical connections before removing. USE CARE REMOVING THE COMPRESSOR SPADE TERMINALS AND DO NOT PULL ON THE WIRE. USE THE NEEDLE-NOSE PLIERS AND ONLY HOLD THE CONNECTOR.

6. Disconnect the compressor refrigerant lines. You would normally use the torch to heat the connections and remove them. Your instructor may instruct you to cut the compressor suction and discharge lines, leaving some stubs for later exercises.

7. Completely remove the compressor from the condenser section. You now have the equivalent of a new compressor, ready for replacement.

8. Place the compressor on the mounts.

9. Reconnect the compressor suction and discharge lines in the way your instructor tells you.

10. When the refrigerant lines are reconnected, pressure the system enough to leak check. **NOTE:** YOUR IN-STRUCTOR MAY WANT YOU TO ADD A FILTER-DRIER. THIS WOULD BE NORMAL PRACTICE WITH A REAL CHANGEOUT.

11. Evacuate the system. While evacuation is taking place, fasten the motor terminal connectors and the motor mounts.

12. When evacuation is complete, add the correct charge, either by weight or volume.

13. Place the ammeter on the common wire of the compressor and the voltmeter across the line side of the contactor and turn the power on.

14. Ask your instructor to look at the system, then start the compressor and observe the amperage and voltage. Compare with those listed on the nameplate.

15. Let the unit run for 15 minutes and check the charge using the touch test to be sure refrigerant is not flooding over into the compressor.

16. When you are satisfied all is well, turn the unit off and remove the gages.

Maintenance of Workstation and Tools: Replace all panels and return the tools to their places.

Summary Statement: Describe the three steps of triple evacuation.

QUESTIONS

1. Why is it necessary to recover the refrigerant rather than just remove it to the atmosphere?

2. Why should you always check the voltage and amperage after a compressor changeout?

3. Should you always add a filter-drier when changing a compressor?

4. What type of compressor overload did this unit have?

5. What method did you use to measure the charge back into the system?

6. Why is refrigerant recovery necessary?

7. What solder or brazing material did you use to replace the compressor?

Changing an Evaporator in an Air-Conditioning System

| Name | Date | Grade |

Objectives: Upon completion of this exercise, you will be able to change an evaporator on a central air-conditioning system.

Introduction: You will recover the refrigerant from a central air-conditioning system and change the evaporator coil. You will pressure check and evacuate the system, then add the correct charge.

Text References: Paragraphs 7.5, 7.7, 7.8, 7.9, 7.10, 8.6, 8.7, 8.10, 8.11, 10.1, 10.2, 10.3, 10.4, 10.5, and 41.8.

Tools and Materials: Straight-blade and Phillips screwdrivers, a refrigerant recovery system, a vacuum pump, a nitrogen tank and regulator, $1/4$" and $5/16$" nut drivers, two 8" adjustable wrenches, a set of sockets, gloves, goggles, a liquid-line drier, an adjustable wrench, a cylinder of refrigerant, and gages.

Safety Precautions: Wear goggles and gloves while transferring liquid refrigerant.

PROCEDURES

1. Put on your goggles and gloves, fasten gages to the gage ports, and purge the gage lines.

2. Recover the refrigerant into an approved recovery system.

3. Turn the electrical power off and lock and tag the disconnect box. With the system open to the atmosphere, sweat or loosen the fitting connecting the suction line to the evaporator. Then loosen the liquid line.

4. Remove the condensate drain line. Plan how this is done so it can be replaced leak free.

5. Remove the evaporator door or cover.

6. Slide the evaporator out.

7. Examine the evaporator for any problems. Look at the tube turns. Clean it if needed. IF YOU CLEAN IT, BE CAREFUL NOT TO ALLOW WATER TO ENTER THE COIL. TAPE THE ENDS.

8. Examine the condensate drain pan for rust and possible leakage. You may want to clean and seal the drain pan while it is out.

9. Examine the coil housing for air leaks. Seal them from the inside if the coil housing is in a positive pressure to be sure it is leak free.

10. When you are sure all is well, slide the coil back into its housing.

11. Reconnect the suction and liquid line.

12. Purge the system with nitrogen, then install the liquid-line drier.

13. Pressure the system for leak checking purposes and leak check all connections.

14. Evacuate the system.

15. While evacuation is taking place, install the condensate drain line and replace the coil cover.

16. When evacuation is complete, charge the system using weight or volume.

17. With your instructor's permission, start the system.

18. Allow the system to run for at least 15 minutes, then check the charge using the touch test.

19. Turn the system off.

Maintenance of Workstation and Tools: Return the system to the condition to which the instructor directs and return all tools to their places.

Summary Statement: Describe why the system was purged before adding the liquid-line drier.

QUESTIONS

1. What will a liquid-line filter-drier remove from the system?

2. What will the vacuum pump remove from the system?

3. Why is it recommended to measure the charge into the system?

4. Why would an evaporator ever need changing?

5. Is it necessary to remove an evaporator to clean it properly if it is really dirty?

6. Name three styles of evaporator.

7. What type of tubes did the evaporator have, copper or aluminum?

8. What material was the condensate drain line made of?

Situational Service Ticket

Name	Date	Grade

NOTE: Your instructor must have placed the correct service problem in the system for you to successfully complete this exercise.

Technician's Name: _____ Date: _____

CUSTOMER COMPLAINT: No cooling.

CUSTOMER COMMENTS: Unit stopped running in the night. The indoor unit will not run on "Fan on."

TYPE OF SYSTEM:

OUTDOOR UNIT:
MANUFACTURER_____   MODEL NUMBER_____   SERIAL NUMBER_____

INDOOR UNIT:
MANUFACTURER_____   MODEL NUMBER_____   SERIAL NUMBER_____

COMPRESSOR (where applicable):
MANUFACTURER_____   MODEL NUMBER_____   SERIAL NUMBER_____

TECHNICIAN'S REPORT

1. SYMPTOMS: _____

_____

2. DIAGNOSIS: _____

_____

3. ESTIMATED MATERIALS FOR REPAIR: _____

_____

4. ESTIMATED TIME TO COMPLETE THE REPAIR: _____

SERVICE TIP FOR THIS CALL

1. Use your VOM to check all power supplies.

Situational Service Ticket

Name	Date	Grade

NOTE: Your instructor must have placed the correct service problem in the system for you to successfully complete this exercise.

Technician's Name:_____ Date: _____

CUSTOMER COMPLAINT: No cooling.

CUSTOMER COMMENTS: We had a bad electrical storm in the night and the unit stopped. It will not restart. The indoor fan will run.

TYPE OF SYSTEM:

OUTDOOR UNIT:
MANUFACTURER_____   MODEL NUMBER_____   SERIAL NUMBER_____

INDOOR UNIT:
MANUFACTURER_____   MODEL NUMBER_____   SERIAL NUMBER_____

COMPRESSOR (where applicable):
MANUFACTURER_____   MODEL NUMBER_____   SERIAL NUMBER_____

TECHNICIAN'S REPORT

1. SYMPTOMS: _____

    _____

2. DIAGNOSIS: _____

    _____

3. ESTIMATED MATERIALS FOR REPAIR: _____

    _____

4. ESTIMATED TIME TO COMPLETE THE REPAIR: _____

SERVICE TIP FOR THIS CALL

1. Use your VOM to check all power supplies.

Electric, Gas, and Oil Heat with Electric Air Conditioning

Unit Overview

Frequently a heating system is installed and then an air-conditioning (cooling) system is installed at a later date. Air-conditioning systems must have the correct air circulation, which may be more than what is needed for a forced-air heating system. The furnace on the existing system must be able to furnish the required airflow, and the duct system should be sized for the airflow. If the duct is installed outside the conditioned space, it should be insulated or it will sweat and absorb heat from the surrounding air.

A study should also be made of the grilles and registers. If the original purpose of the installation was heating only, the registers may not provide the correct air pattern for cooling. Baseboard heating registers are designed to keep the air near the floor. These will not work well with air-conditioning because the cool air will stay near the floor.

If arrangements are made to provide enough airflow for the summer, dampers can be installed with a summer and a winter position. Someone would need to change the damper positions each summer and winter.

The control system must be capable of operating both the heating and cooling equipment. Some thermostats have a manual changeover from heat to cool and some have an automatic changeover. When air conditioning is added, a second low-voltage transformer may have to be added. Most furnaces can operate on a 25 VA transformer. If two transformers are used and wired in parallel, they must be in phase. Most furnace fans are not controlled with a fan relay. If air conditioning is added, a separate fan relay must be added.

Key Terms

Add-On Air Conditioning
All-Weather System
Conditioned Space

Forced-Air Systems
Transformer Relay Package
(Fan Center)

Transformers in Phase

REVIEW TEST

Name	Date	Grade

Circle the letter that indicates the correct answer.

1. **Typically, air-conditioning (cooling) systems require an airflow of _____ cfm per ton.**
   A. 200   C. 600
   B. 400   D. 800

2. **If air conditioning is added to a heating system, the supply duct should be insulated to keep it from sweating and to keep the air in the duct from _____ in the cooling season.**
   A. absorbing heat
   B. giving up heat
   C. becoming too dry

3. **A heating system typically requires _____ a cooling system.**
   A. less airflow than
   B. greater airflow than
   C. about the same airflow as

4. **In evaluating an air distribution system, the terminal units called the _____ are an important factor.**
   A. electric baseboard heaters and radiators
   B. evaporator and condenser
   C. grilles and registers
   D. blower and damper

5. When two low-voltage transformers are wired in parallel, care must be taken to ensure that they are wired:
   A. in phase.
   B. in a circuit with the same resistance.
   C. so that they produce the same amperage.

6. When adding an air-conditioning system to an existing furnace, frequently a _____ must be added to the control circuits.
   A. thermocouple and limit switch
   B. glow coil and static pressure disc
   C. transformer and fan relay
   D. stack switch and cad cell

7. When designing an all-weather system, the ductwork is designed around _____ system.
   A. the cooling
   B. the heating
   C. either the heating or cooling

8. Too much airflow across the heat exchanger in a gas or oil heating system can cause _____ problems.
   A. venting   C. intermittent ignition
   B. poor combustion D. flame retention

9. When adding a control transformer for air conditioning to an existing heating system, the second transformer can be wired in parallel with the original or:
   A. in a series with the original.
   B. as a separate circuit.
   C. in parallel with the fan relay.
   D. in parallel with the compressor contactor.

10. You need to install a 3-ton air-conditioning system on an existing furnace. Using the chart in the text, Figure 42-4, which of the following will meet the fan needs?
    A. $1/4$ HP and a fan wheel that is 10" in diameter and 8" wide.
    B. $1/3$ HP and a fan wheel that is 10" in diameter and 8" wide.
    C. $1/3$ HP and a fan wheel that is 9" in diameter and 9" wide.
    D. $1/4$ HP and a fan wheel that is 9" in diameter and 8" wide.

11. A $2^{1}/_{2}$-ton cooling system must circulate approximately _____ cfm of air.
    A. 800
    B. 1000
    C. 1400
    D. 2000

Control Checkout for an All-Weather System

Name	Date	Grade

Objectives: Upon completion of this exercise, you will be able to perform a simple control checkout on an all-weather system for heating and cooling.

Introduction: You will be assigned a system that has gas, oil, or electric heat and electric air conditioning, and perform a low-voltage control checkout, causing each component to function.

Text References: Paragraphs 42.1, 42.7, 42.8, 42.10, 42.11, 42.12, 42.13, 42.14, and 42.15.

Tools and Materials: Straight-blade and Phillips screwdrivers, $1/4$" and $5/16$" nut drivers, a VOM with clip-on alligator clips, electrical tape, and a flashlight.

Safety Precautions: Extreme care should be taken while working with live electrical circuits. When turning the electrical power off, lock and tag the panel or disconnect box. Keep the key with you.

PROCEDURES

1. Turn the power off to the outdoor unit you are working with. Lock and tag the disconnect box.

2. Remove the compressor compartment door and the control panel door.

3. Verify with the VOM that the power is off, then remove the wire to the common terminal on the compressor. Insulate the end with tape. This allows you to adjust the thermostat and control circuits without starting and stopping, possibly causing harm to the compressor.

4. Turn the power to the condensing unit on and go to the indoor unit.

5. Turn the power off to this unit and lock the panel. Turn your meter range switch so that you can record 24 V. Fasten your VOM leads to the 24 V control that activates the heat: the sequencer or relay for electric heat, gas valve for gas heat, or a relay for oil heat.

6. Turn the power on to the indoor unit. Turn the room thermostat to the heat position and set the temperature adjustment to call for heat. Go to the indoor unit and record the meter reading: _____ V. Did the heat start up?

7. If you had no voltage and the heat did not start, check to be sure there is 24 V low voltage at the secondary of the power supply.

8. If the low-voltage power supply is energized, proceed to the room thermostat and remove it from its base.

9. Using a jumper, jump from the hot wire to the heat terminal and check to see whether the meter at the heating unit has power. If you can get power to the meter by jumping the terminals and cannot get power by adjusting the thermostat, the thermostat is defective.

10. When you are assured that the heating components operate, remove your meter from the indoor terminal block and place it on the 24 V contactor coil at the condensing unit.

11. Turn the room thermostat to cool and set the adjustment lever to call for cooling. The indoor fan should start.

12. If the indoor fan does not start, check the fan relay for power. If it is energized and the fan does not turn, there is a fan motor problem.

13. Check for power at the outdoor unit 24 V coil and record the voltage. _____ V. If you have power, the unit should start (the condenser fan only will run). If it does not, the problem is in the condenser control circuit. Try to find the problem. Your instructor may have to help you.

14. Be sure that the system will both heat and cool before you stop.

15. Turn the power off to the outdoor unit, lock and tag the box cover, and replace the common wire to the compressor.

16. Replace all panels with the correct fasteners.

Maintenance of Workstation and Tools: Leave the system you are working with ready for the next exercise. Return all tools to their places.

Summary Statement: Describe how the low-voltage thermostat distributes the electrical power to the control components.

QUESTIONS

1. Why is 24 V used for the control circuits on residential and light commercial heating and air conditioning?

2. What is the purpose of the heat anticipator in the room thermostat?

3. What type of resistor is the heat anticipator?

4. Is the heat anticipator in series or parallel with the thermostat heating contacts?

5. What is the purpose of the cooling anticipator?

6. Why does cold air distribute better from the ceiling?

7. What would be the minimum blower motor for a system with 3 tons of cooling?

8. What would be the minimum round supply duct for a 2-ton system?

Air Source Heat Pumps

Unit Overview

Heat pumps are refrigeration systems that can pump heat two ways, either from indoors to outdoors or from outdoors to indoors. They are used to provide both heating and cooling.

The heat pump removes heat from the outside air in the winter and rejects it into the conditioned space of the house. This is an air-to-air heat pump. In the summer, the heat pump acts like an air conditioner, absorbing heat from the inside of the house and rejecting it into the air outside. The heat absorbed into the system is contained and concentrated in the discharge gas. The four-way reversing valve diverts the discharge gas and the heat in the proper direction to either heat or cool the conditioned space.

Air is not the only source from which a heat pump can absorb heat, but it is the most popular. Other sources are water and the earth. Water may come from a lake, well, or a loop of pipe buried in the ground.

The air-to-air heat pump resembles the central air-conditioning system. There are indoor and outdoor system components. The coil that serves the inside of the house is called the indoor unit. The unit outside the house is called the outdoor unit. The indoor coil is a condenser in the heating mode and an evaporator in the cooling mode. The outdoor coil is a condenser in the cooling mode and an evaporator in the heating mode. The large diameter pipe of the interconnecting piping is called the gas line because it always contains gas. The smaller line is called the liquid line because it contains liquid during both heating and cooling cycles. The direction of the liquid flow is toward the inside unit in the summer and toward the outside unit in the winter.

There is a metering device at both the indoor and outdoor units. Various combinations of metering devices are used. Thermostatic expansion valves, capillary tubes, electronic expansion valves, and orifice metering devices are used.

Auxiliary heat is provided on air-to-air systems because the heat pump system may not have the capacity to heat a structure in colder temperatures. The balance point is when the heat pump is providing as much heat as the structure is leaking out. At this point the heat pump will heat the building but will run continuously. Auxiliary heat is required when the temperature is below the balance point to provide the heat needed to keep the structure up to the desired temperature. This auxiliary heat may be electric, oil, or gas. Electric heat has a coefficient of performance (COP) of 1 to 1. A heat pump under certain conditions may have a COP of 4 to 1, or it may be 400% more efficient than electric resistance heat. This higher COP only occurs during higher winter temperatures. The water-to-air heat pump may not need auxiliary heat because the heat source (the water) can be a constant temperature all winter.

A defrost cycle is used to defrost the ice from the outside coil during winter operation. The outside coil operates below freezing any time the outdoor temperature is below 45°F. The outdoor coil (evaporator in winter) operates 20 to 25°F colder than the outdoor temperature. Defrost is accomplished with air-to-air systems by reversing the system to the cooling mode and stopping the outdoor fan. The coil will become quite warm even in the coldest weather. The indoor coil will then be in the cooling mode and blow cold air throughout the house. One stage of the auxiliary heat is normally energized during defrost to prevent the cool air from being noticed. This is not efficient and must operate only when necessary. There are various ways to start and stop defrost cycles.

Key Terms

Air Source Heat Pump
Air-to-Air Heat Pump
Auxiliary Heat
Balance Point
Biflow Filter-Drier
Check Valve
Coefficient of Performance (COP)
Defrost Time and Temperature Initiated
Defrost Time or Temperature Terminated
Demand Defrost

Electric Strip Heaters
Emergency Heat
Four-Way Valve
Gas Line
Indoor Coil
Liquid Line
Outdoor Thermostat
Outdoor Unit
Water Source Heat Pump
Water-to-Air Heat Pump

REVIEW TEST

Name	Date	Grade

Circle the letter that indicates the correct answer.

1. A heat pump that absorbs heat from surrounding air and rejects it into air in a different place is called _____ heat pump.
   A. a solar-assist
   B. an air-to-air
   C. a water-to-air
   D. a groundwater

2. A heat pump has about the same components as an electric air-conditioning system except it has:
   A. an indoor coil.
   B. an outdoor coil.
   C. a four-way valve.
   D. a capillary tube.

3. The larger diameter interconnecting tubing in a heat pump system is called the _____ line.
   A. liquid
   B. suction
   C. discharge
   D. gas

4. The indoor coil operates as the _____ in the heating cycle.
   A. condenser
   B. evaporator
   C. compressor
   D. accumulator

5. Metering devices in most heat pumps are piped in parallel with:
   A. a check valve.
   B. the evaporator.
   C. the outdoor coil.
   D. the compressor.

6. Filter-driers installed in heat pump systems should be of the biflow type or installed:
   A. with thermistors.
   B. with an electrical disconnect.
   C. in parallel with the condensate line.
   D. with check valves.

7. The capillary tube is used in many heat pumps as the:
   A. liquid line.
   B. gas line.
   C. check valve.
   D. metering device.

8. When orifice metering devices are used in heat pumps, the one at the indoor coil has _____ the one at the outdoor coil.
   A. the same diameter bore as
   B. a larger bore than
   C. a smaller bore than

9. When a heat pump is pumping in exactly as much heat as the building is leaking out and it is running continuously, it is said to be at the:
   A. auxiliary heat capacity.
   B. coefficient of performance.
   C. beginning of the defrost cycle.
   D. balance point.

10. The coefficient of performance for electric resistance heat is:
    A. 1 to 1.
    B. 2 to 1.
    C. 3 to 1.
    D. 4 to 1.

11. The outdoor coil operates as the _____ in the heating cycle.
    A. evaporator
    B. condenser
    C. accumulator
    D. auxiliary heat

12. When electric resistance auxiliary heat is used in a heat pump, the indoor refrigerant coil must be located in the airstream _____ the electric heating coil.
    A. before
    B. after
    C. either before or after

13. When oil or gas auxiliary heat is used, the indoor refrigerant coil must be located _____ of the furnace.
    A. in the outlet airstream
    B. in the return air duct
    C. in either the outlet airstream or the return air duct

14. Most heat pumps require a minimum of _____ cfm/ton of capacity in the cooling mode.
    A. 200
    B. 300
    C. 400
    D. 500

15. The temperature of the air in the airstream from the heat pump under typical heat-pump operating conditions is approximately _____ or less.
    A. 80°F
    B. 100°F
    C. 120°F
    D. 130°F

16. **For the greatest efficiency, the leaving air from the heat pump should be directed through the registers to the:**
    A. inside walls.
    B. outside walls.
    C. ceiling.
    D. any of the above.

17. **The larger diameter interconnecting pipe may reach temperatures as high as _____ in the winter.**
    A. 120°F
    B. 150°F
    C. 170°F
    D. 200°F

18. **In the cooling cycle, which of the following components start first?**
    A. compressor
    B. outdoor fan
    C. indoor fan
    D. all three start together

19. **In the heating cycle, the outdoor coil operates approximately _____ colder than the outside air temperature.**
    A. 5°F to 10°F
    B. 20°F to 25°F
    C. 30°F to 35°F
    D. 40°F to 45°F

20. **The defrost cycle in a heat pump is used to melt the ice that forms on the:**
    A. indoor coil.
    B. outdoor coil.
    C. compressor.
    D. both the indoor and outdoor coil.

Familiarization with Air-to-Air Heat Pump Components

Name	Date	Grade

Objectives: Upon completion of this exercise, you will be able to identify and describe the typical components in an air-to-air heat pump system.

Introduction: You will inspect the components listed below and record the specifications called for. By referring to your text, you should be able to recognize all components listed.

Text References: Paragraphs 43.1, 43.3, 43.7, 43.8, 43.9, 43.10, 43.11, 43.12, 43.14, 43.17, 43.18, 43.20, 43.21, 43.24, 43.25, 43.26, 43.27, and 43.28.

Tools and Materials: Straight-blade and Phillips screwdrivers, $1/4$" and $5/16$" nut drivers, a combination wrench set, flashlight, rags, and a VOM.

Safety Precautions: Ensure that power is off to both indoor and outdoor units. To make sure, check the voltage with the VOM before starting this exercise. Lock and tag the electrical power panel and keep the key with you.

PROCEDURES

1. Remove the panels on the indoor unit and record the following:

   Motor _____ hp _____ V _____ A

   Motor type _____ Shaft size _____

   Shaft rotation _____ Number of speeds _____

   Electric heat _____ kw _____ A _____ V

   Resistance of heaters _____

   Number of heat stages _____ Factory wire size _____

   Refrigerant coil material _____ Number of refrigerant circuits in coil _____ Type of metering device _____
   Size of gas line _____ Size of liquid line _____
   Type of coil (A, H, or slant) _____ Number of tube passes _____ Size of tubes _____

2. Remove the panels on the outdoor unit and record the following:

   Fan motor _____ hp _____ V _____ A

   Motor type _____

   Shaft size _____ Type of blade _____

   Shaft rotation _____

   Compressor motor type _____ A _____ V _____

   Suction-line size _____

Discharge line size _____ Number of motor terminals _____

Coil material _____ Number of refrigerant circuits in coil _____ Coil tubing size _____ Number of tube passes _____ Four-way valve voltage _____ Line sizes _____

Maintenance of Workstation and Tools: Replace any components you may have removed or disturbed. Have your instructor inspect your system to ensure it is left in proper working order. Replace all panels with appropriate fasteners. Replace all tools, equipment, and materials to their proper places. Clean your work area.

Summary Statement: Describe in detail the air and refrigerant flow in the air-to-air heat pump when it is in the heating cycle.

QUESTIONS

1. What are typical condenser tubes made of?

2. What are typical evaporator tubes made of?

3. When is the four-way valve energized in the above heat pump, summer or winter?

4. What is used to start the compressor motor?

5. What does kW mean?

6. What component is used to change the heat pump from cooling to heating?

7. What does "air-to-air" mean when referring to heat pumps?

8. When and where does water or frost collect on the air-to-air heat pump?

9. How is the frost buildup removed?

10. Name two conditions that are normally met before defrost can begin.

Using Manufacturer's Information to Check Heat Pump Performance

Name	Date	Grade

Objectives: Upon completion of this exercise, you will be able to check system performance and compare it to the system performance chart furnished in the manufacturer's literature.

Introduction: You will check the indoor and outdoor wet-bulb and dry-bulb temperatures for the existing weather conditions for the season of the year, as called for in the manufacturer's performance chart. Then chart the suction and discharge pressures and compare with the pressures indicated on the chart.

Text References: Paragraphs 35.9, 43.4, 43.7, 43.8, 43.9, 43.31, and 43.33.

Tools and Materials: A thermometer with leads set up to check wet-bulb and dry-bulb temperatures, a gage manifold, straight-blade and Phillips screwdrivers, $1/4$" and $5/16$" nut drivers, light gloves, goggles, and an operating heat pump system with accompanying manufacturer's performance chart.

Safety Precautions: Wear light gloves and goggles when attaching gage lines to gage ports. Do not allow liquid refrigerant to get on your bare skin.

PROCEDURES

1. Check the manufacturer's literature to determine on which lines to install the gages of the gage manifold. This will depend on the weather or the season. With the system off, and wearing gloves and goggles, install the gages as indicated.

2. Set up your thermometer to measure the temperatures the manufacturer recommends. Use their literature.

3. Start the system in the cooling or heating mode, depending on the mode for which you have installed the gages.

4. Allow the unit to run until the conditions are stable and pressures and temperatures are not changing.

5. Measure the temperatures and pressures as called for below and record.

   For summer or cooling: Indoor WB _____ °F, DB _____ °F, Outdoor DB _____ °F

   Suction pressure _____ psig, Discharge pressure _____ psig

   For winter or heating: Indoor DB _____ °F, Outdoor WB _____ °F, DB _____ °F

   Suction pressure _____ psig

   Discharge pressure _____ psig

6. Compare with the manufacturer's performance chart. **NOTE:** In the cooling mode, the charge can sometimes be adjusted, but in the heating mode the charge cannot normally be adjusted using the chart.

7. Turn the system off and wait for the pressure to equalize. Wearing gloves and goggles, remove the gages.

Maintenance of Workstation and Tools: With the system off, replace all panels with proper fasteners as instructed. Return all tools and equipment to their proper places. Leave your workstation in order.

Summary Statement: Describe how the heat pump will operate under a low-charge condition in the winter heating cycle.

QUESTIONS

1. Why is the wet-bulb reading necessary when checking the charge?

2. Where is the wet-bulb temperature measured (indoors or outdoors) when checking the charge in the winter or heating cycle?

3. Where is the wet-bulb temperature measured (indoor or outdoors) when checking the charge in the summer or cooling cycle?

4. Wet-bulb temperature is related to which kind of heat, sensible or latent?

5. Would a high indoor wet-bulb temperature result in a higher or lower suction pressure in the cooling mode?

6. Why should the heat pump not be operated above 65°F for long periods of time in the heating mode?

7. Why should the technician not simply adjust the charge in a heat pump according to his or her best judgment?

8. Many heat pumps have a critical charge to the nearest _____ oz.

Heat Pump Performance Check in Cooling Mode

Name	Date	Grade

Objectives: Upon completion of this exercise, you will be able to make a heat pump performance check on a heat pump with fixed-bore metering devices (cap tube and orifice) while it is in the cooling mode, and adjust refrigerant charge for desired performance.

Introduction: You will be taking superheat readings on the suction line under a standard condition. You will install gages on the gas and liquid refrigerant lines to check pressures and to add or remove refrigerant. The head pressure will be increased to 275 psig in the heat pump and used as a standard condition. Refrigerant will be added to or removed from the system until the desired superheat is achieved. **NOTE:** This method of adjusting the charge is recommended only when manufacturer's data is not available. It may be performed in the winter while operating in the summer mode.

Text References: Paragraphs 43.10, 43.11, 43.13, and 43.33.

Tools and Materials: An electric thermometer, a gage manifold, straight-blade and Phillips screwdrivers, $1/4$" and $5/16$" nut drivers, R-22, light gloves, goggles, and an R-22 heat pump system with gage ports and fixed-bore metering devices.

Safety Precautions: Wear goggles and light gloves when attaching gages to or removing from refrigerant ports. Liquid refrigerant can cause frostbite. Watch the low-side gage to ensure that the system does not go into a vacuum. Turn the system off if it appears that this is happening; it is low on refrigerant. If the compressor makes any unusual sound or if it acts like it is under a strain, shut it off and check with your instructor.

PROCEDURES

1. With the unit off, fasten one lead of the thermometer to the gas line just before the outdoor unit service valve. Insulate the thermometer sensor.

2. Carefully screw the low-pressure gage line onto the gas (large line) port and the high-pressure gage line onto the liquid (small line) port. Wear light gloves and goggles, and keep your fingers to the side.

3. Switch the thermostat to the cooling cycle and start the heat pump. Allow the unit to operate for about 15 minutes to stabilize. Check the gages. If it appears to be going into a vacuum, shut it off. This indicates it is very low on refrigerant. Ask your instructor for directions.

4. After the system has stabilized, use a plastic bag, a piece of cardboard, or heavy paper to partially block the airflow over the condenser. Adjust until the head pressure has stabilized at 275 psig for 15 minutes. (Very slight adjustments in the plastic or cardboard can be made until this condition is reached.)

5. Record the temperature reading on the gas line: _____°F. Determine the superheat and record: _____°F.

6. The superheat should be 12°F ± 2°F. If the superheat is above this range, add refrigerant until it is within range. If it is below, recover refrigerant until the superheat is within this range.

7. After any refrigerant charge adjustment, let the unit operate for at least 15 minutes with a stable 10°F to 14°F superheat before concluding that the charge is correct.

8. When this condition is met, turn the unit off. Carefully remove the gage lines and the thermometer leads.

Maintenance of Workstation and Tools: Remove the plastic bag or cardboard from around the condenser area. Replace all panels on the heat pump unit with proper fasteners. Return tools, equipment, and materials to their proper places.

Summary Statement: Describe why the heat pump can be checked for performance and refrigerant charge more easily in the summer or cooling cycle.

QUESTIONS

1. Name three metering devices that can be found on typical heat pumps.

2. What type of drier is used in the liquid line on heat pumps?

3. Where is the only true suction line?

4. Where is the only true discharge line?

5. Where is the suction-line accumulator located?

6. Why did you simulate summer conditions when checking the charge?

7. Could the above method of charge adjustment be used on a heat pump with a thermostatic expansion device?

8. Why should you shut the system off to remove the liquid-line gage line?

Checking the Control Circuit in the Emergency and Auxiliary Heat Modes

Name	Date	Grade

Objectives: Upon completion of this exercise, you will be able to follow the control circuits in the auxiliary and emergency heat modes in a heat pump system and measure the voltage at each terminal.

Introduction: You will draw a wiring diagram for the low-voltage control circuits in the auxiliary and emergency heat modes to help you become familiar with these circuits. After tracing each circuit on the diagram, you will locate each terminal on the heat pump system and check the voltage at each terminal.

Text References: Paragraphs 43.14 and 43.22.

Tools and Materials: Straight-blade and Phillips screwdrivers, a flashlight, $1/4$" and $5/16$" nut drivers, and a VOM.

Safety Precautions: Turn the power off, and lock and tag the disconnect box while becoming familiar with the control circuits and components for the auxiliary and emergency heat systems. Use the VOM to verify that the power is off. The line voltage should remain off at the outdoor unit while you are taking voltage readings. Be careful not to come in contact with the high voltage at the indoor unit.

PROCEDURES

1. Study the wiring diagrams shown in Figures 43-40 through 43-50 in your text.

2. Draw a wiring diagram of the complete low-voltage control circuits for auxiliary and emergency heat modes. Study this diagram until you understand the function of each component.

3. Find each component and terminal in the actual system.

4. Turn the power to the indoor unit on.

5. Set the room thermostat to call for heat.

6. Check and record the voltage for each circuit at the indoor unit terminal block. Place one meter lead on the common terminal and record the voltage at the terminal, controlling the following:

   • Sequencer coil voltage _____ V
   • Fan relay coil voltage _____ V
   • Compressor contactor coil voltage _____ V

7. Switch the thermostat to the emergency heat mode at the thermostat and record the following:

   • Sequencer coil voltage _____ V
   • Fan relay coil voltage _____ V
   • Compressor contactor coil voltage _____ V

   NOTE: The difference between the two circuits above is caused by the switching action in the thermostat from auxiliary to emergency heat.

Maintenance of Workstation and Tools: Replace all panels on the heat pump system. Return all tools and equipment to their proper places. Leave your workstation in a neat and orderly condition.

Summary Statement: Describe the difference between auxiliary and emergency heat.

QUESTIONS

1. What is the COP of the emergency heat?

2. What is the COP of the auxiliary heat?

3. When should emergency heat be used?

4. When would auxiliary heat normally operate?

5. Where are the electric heating elements located in relation to the heat pump refrigerant coil in the airstream?

6. How often does a heat pump normally defrost?

7. How long does a unit normally stay in defrost?

8. Where does the heat come from for defrost?

9. What keeps the unit from blowing cold air in the house during defrost?

Checking the Four-Way Valve During the Heating and Cooling Cycles

Name	Date	Grade

Objectives: Upon completion of this exercise, you will be able to check the four-way valve for internal gas leaking during the cooling mode.

Introduction: Using an electronic thermometer, you will measure the temperature of the permanent suction line and the indoor coil line at the four-way valve while the unit is in the cooling mode. You will compare the two readings. A difference of more than 3°F indicates hot gas is passing into the suction line from the discharge line and the valve is defective.

Text References: Paragraphs 43.3 and 43.29.

Tools and Materials: A gage manifold, an electronic thermometer, straight-blade and Phillips screwdrivers, $^1/_4$" and $^5/_{16}$" nut drivers, insulation material for attaching thermometer sensors to tubing, light gloves, and goggles.

Safety Precautions: If there is an indication the heat pump may go into a vacuum, stop the unit and check with your instructor. Wear gloves and goggles when attaching gage lines to and removing them from gage ports. Shut the power off to the outdoor unit while fastening thermometer sensors to refrigerant lines to avoid electrical contact. Lock and tag the panel and keep the key with you.

PROCEDURES

1. With the power to the outdoor unit off and locked out, fasten one thermometer sensor to the permanent suction line and one to the indoor coil line. Fasten them at least 4" from the four-way valve. Insulate the sensors.

2. With goggles and gloves on, fasten the high- and low-side gages to the gage ports and start the unit in the cooling mode. Check to ensure that the low side does not go into a vacuum.

3. Let the unit run for 15 minutes and then record the following:

   • Suction pressure _____ psig
   • Discharge pressure _____ psig
   • Permanent suction-line temperature _____ °F
   • Indoor coil line temperature _____ °F

4. Stop the unit. Turn the power off to the outdoor unit and lock the disconnect box. Remove the thermometer sensors. With your gloves and goggles on, remove gage lines.

   NOTE: For a complete test of the four-way valve, the same test may be performed in the heating mode by moving the lead from the indoor coil line to the outdoor coil line. **CAUTION:** DO NOT RUN THE UNIT FOR MORE THAN 15 MINUTES IN THE HEATING CYCLE IF THE OUTDOOR TEMPERATURE IS ABOVE 65°F.

Maintenance of Workstation and Tools: Replace all panels on the system as you are instructed. Return all tools, equipment, and materials to their respective places.

Summary Statement: Explain what the symptoms would be if the four-way valve were leaking across the various ports.

QUESTIONS

1. How much temperature difference should be expected between the indoor line and the permanent suction line during the cooling cycle?

2. How much temperature difference should be expected between the outdoor line and the permanent suction line during the heating cycle?

3. If the permanent suction line were 10°F warmer than the indoor coil line at the four-way valve, where could the heat come from?

4. Why should you place the thermometer sensor at least 4" from the four-way valve when checking the refrigerant line temperature?

5. What cautions should be taken when soldering a new four-way valve in the line?

6. What force actually changes the four-way valve over from one position to another?

7. What would the symptoms of a four-way valve that is stuck in the exact midposition be?

8. What terminal on the thermostat typically energizes the four-way valve coil?

Checking Low-Voltage Field Control Circuits at the Individual Components

Name	Date	Grade

Objectives: Upon completion of this exercise, you will be able to identify and take low-voltage readings at each individual component.

Introduction: To become familiar with the control wiring for the heat pump you are working with, you will make a pictorial drawing of the wiring to the three terminal blocks. You will then operate the system in the cooling cycle, the 1st- and 2nd-stage heat cycles, and the emergency heat cycle, while tracing the current flow on your diagram and checking the voltage at the various components.

Text References: Paragraphs 43.22, 43.23, 43.24, 43.25, and 43.27.

Tools and Materials: Straight-blade and Phillips screwdrivers, a flashlight, $1/4$" and $5/16$" nut drivers, a VOM, a split-system heat pump system and wiring diagram, and four different colored pencils.

Safety Precautions: Turn the power off to the outdoor unit when working with electrical wiring and components in the unit. Lock and tag the disconnect box and keep the key with you. Always take all precautions when working around electricity.

PROCEDURES

1. Draw the following.

   - In the center of the page, draw the terminal block for the indoor unit, indicating the terminal designations.
   - On the right side, draw the terminal block for the outdoor unit.
   - On the left side of the page, draw the terminals for the room thermostat.

2. Refer to the system wiring diagram and draw in the wiring from terminal to terminal on your drawing.

3. Turn off and lock out the power to the outdoor unit. Verify with the VOM that the power is off. Remove the common wire that runs from the load side of the contactor to the common terminal on the compressor. Tape it to insulate it. This will allow you to turn the system on and off without starting and stopping the compressor.

4. Turn the power to the outdoor unit on. Switch the room thermostat to the cool cycle and adjust to call for cooling. **NOTE:** The outdoor fan will start and stop each time the compressor would normally start and stop. Check to see whether the indoor and outdoor fans are on. Trace the control circuitry for cooling on your diagram. Make this circuit a specific color. With your VOM, check the voltage at the following components:

   • Fan relay coil voltage _____ V
   • Compressor contactor coil voltage _____ V
   • Sequencer coil voltage _____ V
   • Four-way valve coil voltage _____ V

5. Set the thermostat to call for 1st-stage heat. Check the indoor and outdoor units to ensure that they are operating. Trace the control wiring for this circuit on your diagram. Color it a different color. With your VOM, check the voltage at the following components:

   • Fan relay coil voltage _____ V
   • Compressor contactor coil voltage _____ V
   • Sequencer coil voltage _____ V
   • Four-way valve coil voltage _____ V

6. Set the thermostat to call for 2nd-stage heat. Check the indoor and outdoor units to ensure that they are operating. Follow the control wiring for this circuit on your diagram. Make it a different color. With your VOM, check the voltage at the following components:

   • Fan relay coil voltage _____ V
   • Compressor contactor coil voltage _____ V
   • Sequencer coil voltage _____ V
   • Four-way valve coil voltage _____ V

7. Turn the thermostat to the emergency heat cycle. Check to see that the indoor unit is on. Follow the control wiring for this circuit on your diagram and make it a different color. With your VOM, check the voltage at the following components:

   • Fan relay coil voltage _____ V
   • Compressor contactor coil voltage _____ V
   • Sequencer coil voltage _____ V
   • Four-way valve coil voltage _____ V

8. Turn off and lock out the power to the outdoor unit. Replace the wire that you previously removed from the common terminal to the contactor. Turn the power back on.

Maintenance of Workstation and Tools: Replace the panels with proper fasteners as instructed. Return all tools and equipment to their proper places.

Summary Statements: Describe all functions of the control system from the time the thermostat calls for cooling until the thermostat is satisfied.

Describe all functions of the control system from the time the thermostat calls for heat, including 2nd-stage heat, until the thermostat is satisfied.

Describe all functions of the control circuit from the time the system is switched to emergency heat until the thermostat is satisfied.

QUESTIONS

1. What terminals in the thermostat are energized in the cooling cycle?

2. What terminals in the thermostat are energized in the heating cycle, 1st stage?

3. What terminals in the thermostat are energized in the heating cycle, 2nd stage?

4. What terminals in the thermostat are energized in the emergency heat mode?

5. What would the symptoms be under the following conditions?

- Dirty filter in the heating cycle?

- Dirty outdoor coil in the cooling cycle?

- Dirty indoor coil in the cooling cycle?

- Overcharge in the heating cycle?

- Overcharge in the cooling cycle?

- The four-way valve stuck in the midposition?

Troubleshooting Exercise: Heat Pump

Name	Date	Grade

Objectives: Upon completion of this exercise, you will be able to troubleshoot basic defrost problems in a heat pump.

Introduction: This unit should have a specified problem introduced by your instructor. You will locate the problem using a logical approach and instruments. THE INTENT IS NOT FOR YOU TO MAKE THE REPAIR BUT TO LOCATE THE PROBLEM AND ASK YOUR INSTRUCTOR FOR DIRECTION BEFORE COMPLETION.

Text Reference: Unit 43.

Tools and Materials: Straight-blade and Phillips screwdrivers, $1/4$" and $5/16$" nut drivers, a VOM, and needle-nose pliers.

Safety Precautions: Use care while checking live electrical circuits.

PROCEDURES

1. Start the unit in the heating mode and allow it to run for 15 minutes. If the outdoor air temperature is above 60°F, you should either block the airflow to the outdoor coil or disconnect the outdoor fan to keep from overloading the compressor. Frost should begin to form on the outdoor coil in a short period of time.

2. With the unit running, remove the control compartment door and locate the method for simulating defrost. If the unit has a circuit board, the manufacturer's directions will have to be followed. If the unit has time and temperature defrost, you may follow the methods outlined in the text.

3. Describe the type of defrost this unit has.

4. Wait for the coil to cover with frost before trying to defrost using the time and temperature method. RE-MEMBER, THE SENSING ELEMENT IS LARGE AND REQUIRES SOME TIME FOR IT TO COOL DOWN TO COIL TEMPERATURE.

5. With the coil covered with frost, force the defrost.
   - Did the reversing valve reverse to cooling?
   - Did the outdoor fan motor stop?
   - Did the first stage of strip heat energize?

6. Using the VOM, is the defrost relay coil energized? Is it supposed to be energized during defrost?

7. Is the defrost relay coil responding as it should?

8. Describe what you think the problem may be.

9. Describe any materials you think it would take to repair the equipment.

10. Ask your instructor whether the unit should be repaired or shut down and used for the next student.

Maintenance of Workstation and Tools: Leave the workstation as your instructor directs you, and return all tools to their places.

Summary Statement: Describe the entire defrost cycle for this unit.

QUESTIONS

1. Why must a heat pump have a defrost cycle?

2. Is the electric heat energized during the defrost cycle for all heat pumps?

3. Why is the electric heat energized during the defrost cycle on any heat pump?

4. Do all heat pumps have a defrost cycle? If not, which ones do not?

5. What is the purpose of a circuit board on a heat pump?

6. What is the purpose of the air switch used for defrost on some heat pumps?

7. What type of metering devices did this heat pump have?

8. How could a heat pump be safely defrosted if the technician were to arrive and find the coil a solid bank of ice?

9. How high should a heat pump be installed above the snow line?

10. What is the purpose of emergency heat on a heat pump system?

Troubleshooting Exercise: Heat Pump

Name	Date	Grade

Objectives: Upon completion of this exercise, you will be able to troubleshoot an electrical problem with the changing from cool to heat.

Introduction: This unit should have a specified problem introduced by your instructor. You will locate the problem using a logical approach and instruments. THE INTENT IS NOT FOR YOU TO MAKE THE REPAIR BUT TO LOCATE THE PROBLEM AND ASK YOUR INSTRUCTOR FOR DIRECTION BEFORE COMPLETION.

Text Reference: Unit 43.

Tools and Materials: Straight-blade and Phillips screwdrivers, $1/4$" and $5/16$" nut drivers, needle-nose pliers, and a VOM.

Safety Precautions: Use care while checking energized electrical connections.

PROCEDURES

1. Study the wiring diagram and determine whether the four-way valve should be energized in the cooling or heating mode.

2. Start the unit in the cooling mode and allow it to run for 15 minutes. Is it cooling or heating the conditioned space? _____ If it is heating and is supposed to be cooling, block the evaporator coil to reduce the airflow to the outdoor unit, or it may overload the compressor.

3. Locate the low-voltage terminal board at the unit and check the voltage from the common terminal to the terminal that energizes the four-way valve. **NOTE:** This is the "O" terminal for many manufacturers. Record the voltage here: _____V

4. Is the correct voltage supplied?

5. Check the voltage at the four-way valve coil. **NOTE:** THIS MAY BE FULL-LINE VOLTAGE ON SOME UNITS. IF SO, THE VOLTAGE IS USUALLY SWITCHED AT THE DEFROST RELAY.

6. Is the correct voltage being supplied to the four-way valve coil?

7. You should have discovered the problem with the unit at this time. Describe what the problem may be.

8. Describe what the repair should be, including parts.

9. Consult your instructor before making a repair.

Maintenance of Workstation and Tools: Leave the unit as your instructor directs you, and return all tools to their respective places.

Summary Statement: Using the four-way valve as an example, describe how a pilot valve works.

QUESTIONS

1. Why would a manufacturer use line voltage for a four-way valve coil when 24 V could be used?

2. Why is the four-way valve solenoid known as a pilot-operated solenoid valve?

3. Name the four refrigerant line connections on the four-way valve.

4. Can a four-way valve stick in a midposition?

5. Does the four-way valve electrical circuit run through the defrost relay?

6. What is the purpose of stopping the fan during the defrost cycle?

7. What would happen if the fan did not stop during defrost?

8. What would the complaint be if the electric heat were not energized during the defrost cycle?

9. Which is larger on a heat pump, the indoor coil or the outdoor coil?

10. Did this unit have a low-pressure control?

Changing a Four-Way Valve on a Heat Pump

Name	Date	Grade

Objectives: Upon completion of this exercise, you will be able to change a four-way valve on a heat pump.

Introduction: You will recover the refrigerant, change a four-way valve, pressure check and leak check the system, charge the system, and start it. You will then check the system in the heating and cooling modes.

Text References: Units 7, 8, 9, 10, and paragraph 43.3.

Tools and Materials: Straight-blade and Phillips screwdrivers, two pairs of slip joint pliers, $1/4$" and $5/16$" nut drivers, soldering or brazing torch setup, solder and flux (ask your instructor what is preferred), some strips of cloth and water, a leak detector, gloves, goggles, a two-way liquid-line drier, dry nitrogen, and refrigerant.

Safety Precautions: Wear goggles and gloves while fastening gages and transferring liquid refrigerant. Remove the four-way valve solenoid coil before soldering.

PROCEDURES

1. Put goggles and gloves on. Fasten gages to the heat pump. Purge the gage lines.

2. Recover the refrigerant from the unit with an approved recovery system.

3. With all of the refrigerant recovered, raise the system pressure to the atmosphere, using the gage manifold and dry nitrogen.

4. Turn the power off. Lock and tag the power panel. Keep the key with you. Remove enough panels to comfortably work on the four-way valve. Remove the control wires from the valve solenoid.

5. The text recommends that you cut the old valve out of the system, leaving stubs to fasten the new valve to. Ask your instructor what is recommended in your case and remove the valve using your instructor's directions. Prevent filings from entering the system.

6. With the old valve out, prepare the new valve and piping stubs for installation. Clean all connections thoroughly.

7. With all connections clean, apply flux as recommended and place the new valve in place. WRAP THE VALVE BODY WITH THE STRIPS OF CLOTH AND DAMPEN THEM. HAVE SOME WATER IN RESERVE.

8. Solder the new valve into the system, using care not to overheat the valve body. Point the torch tip away from the body. If the body begins to heat, water can be added from the reserve water.

9. When the valve body cools, pressure the system with nitrogen with trace refrigerant R-22, and leak check the valve installation.

10. When you are satisfied with the valve installation, remove the nitrogen and refrigerant from the system and install a liquid-line two-way drier.

11. Pressure the system and leak check the drier connections.

12. When you are satisfied the system is leak free, evacuate the system and prepare to charge it.

13. After evacuation, measure the charge into the system.

14. Ask your instructor to look at the installation. If approved, replace all panels and turn the power on.

15. Start the system in the cooling mode and observe the pressures. Allow the system to run for about 15 minutes, then shut the system off.

16. Allow the system to stay off for 5 minutes to allow the pressure to equalize, then start the system in the heating mode. If the temperature for the outdoor coil is above 65°F, do not allow the unit to run for more than 15 minutes or the compressor may overload.

Maintenance of Workstation and Tools: Leave the unit as your instructor directs you and return all tools to their places.

Summary Statement: Describe how you could tell by touching the lines when the heat pump is in the cooling mode versus the heating mode.

QUESTIONS

1. Why does it hurt the four-way valve to be overheated?

2. What is the seat in the four-way valve made of?

3. Why is the four-way valve known as a pilot-operated valve?

4. What was the coil voltage of the valve on this system?

5. Describe the permanent discharge line.

6. Which line would have to be used if a suction-line drier had to be installed?

7. What is the purpose of a suction-line accumulator?

Wiring a Split System Heat Pump, Low-Voltage Control Wiring

Name	Date	Grade

Objectives: Upon completion of this exercise, you will be able to follow a wiring diagram and wire the field control wiring for a split system heat pump.

Introduction: You will draw a wiring diagram of all low-voltage terminal boards and draw the interconnecting wiring. You will connect these wires on a real system, then start the system and perform a control checkout.

Text Reference: Unit 43.

Tools and Materials: Straight-blade and Phillips screwdrivers, $1/4$" and $5/16$" nut drivers, VOM, a pencil and paper, and a flashlight.

Safety Precautions: Turn all power off before starting this exercise. Lock and tag the panel where you turned off the power. Keep the key in your possession.

PROCEDURES

1. Turn off and lock out the power and remove all panels to low-voltage connections. With a split system, this should be the indoor unit, the outdoor unit, and the thermostat (remove the thermostat from the subbase).

2. Using your paper, draw the terminal block for the outdoor unit on the left side of the page, the terminal block for the indoor unit in the middle of the page, and the terminal subbase on the right-hand side of the page.

3. Using the manufacturer's diagram, draw the interconnecting diagram.

4. Connect the wires to the indoor coil using your diagram.

5. Connect the wires to the outdoor coil using your diagram.

6. Connect the wires to the thermostat subbase using your diagram.

7. Ask your instructor to look over the connections, and get permission to start the system.

8. Replace the panels on the units and start the system in the cooling mode. Allow it to run for about 15 minutes. Verify that it is in the cooling mode. If not, remove the panels to the indoor unit and check for the correct voltage at the proper terminals.

9. After you are satisfied with cooling, shut the unit off and leave it off for 5 minutes, allowing the pressures to equalize, then start it in the heating mode. Verify that it is heating, then shut the system off.

10. Start the system in the emergency heat mode and verify that it is operating correctly.

Maintenance of Workstation and Tools: Leave the unit as your instructor directs you and return all tools to their places.

Summary Statement: Describe the complete function of the thermostat.

QUESTIONS

1. Is a heat pump compressor the same as an air-conditioning compressor?

2. Did this heat pump have more than one stage of auxiliary heat?

3. Did this heat pump have outdoor thermostats for the auxiliary heat?

4. What was the terminal designation for emergency heat?

5. What was the terminal designation for the four-way valve?

6. What was the coil voltage for the four-way valve?

7. Did the thermostat have any indicator lights, for example, for auxiliary heat? If so, name them all.

8. What gage wire was the field low-voltage wire?

9. How many wires were in the cable from the outdoor unit to the indoor unit?

10. How many wires were in the cable from the indoor unit to the room thermostat?

Situational Service Ticket

Name	Date	Grade

NOTE: Your instructor must have placed the correct service problem in the system for you to successfully complete this exercise.

Technician's Name: _____ Date: _____

CUSTOMER COMPLAINT: No heat with heat pump.

CUSTOMER COMMENTS: Indoor fan runs and emergency heat works, but heat pump will not run.

TYPE OF SYSTEM:

OUTDOOR UNIT:
MANUFACTURER_____   MODEL NUMBER_____   SERIAL NUMBER_____

INDOOR UNIT:
MANUFACTURER_____   MODEL NUMBER_____   SERIAL NUMBER_____

COMPRESSOR (where applicable):
MANUFACTURER_____   MODEL NUMBER_____   SERIAL NUMBER_____

TECHNICIAN'S REPORT

1. SYMPTOMS: _____

_____

2. DIAGNOSIS: _____

_____

3. ESTIMATED MATERIALS FOR REPAIR: _____

_____

4. ESTIMATED TIME TO COMPLETE THE REPAIR: _____

SERVICE TIP FOR THIS CALL

1. Use a VOM to check all components for heat cycles that pertain to the heat pump.

Situational Service Ticket

Name	Date	Grade

NOTE: Your instructor must have placed the correct service problem in the system for you to successfully complete this exercise.

Technician's Name: _____ Date: _____

CUSTOMER COMPLAINT: Outdoor coil is frosting all the time.

CUSTOMER COMMENTS: Unit cannot be operating efficiently with outdoor coil frosting all the time.

TYPE OF SYSTEM:

OUTDOOR UNIT:
MANUFACTURER_____   MODEL NUMBER_____   SERIAL NUMBER_____

INDOOR UNIT:
MANUFACTURER_____   MODEL NUMBER_____   SERIAL NUMBER_____

COMPRESSOR (where applicable):
MANUFACTURER_____   MODEL NUMBER_____   SERIAL NUMBER_____

TECHNICIAN'S REPORT

1. SYMPTOMS: _____

_____

2. DIAGNOSIS: _____

_____

3. ESTIMATED MATERIALS FOR REPAIR: _____

_____

4. ESTIMATED TIME TO COMPLETE THE REPAIR: _____

SERVICE TIP FOR THIS CALL

1. Fasten gages to the system and check pressures and superheat on the outdoor coil in heating mode.

Geothermal Heat Pumps

Unit Overview

Geothermal heat pumps are much like air-to-air heat pumps. They move heat from one place to another. Unlike air-to-air heat pumps, which move heat from outdoors to indoors in the winter, they move heat from the ground or lake water to the indoors. In the summer, they move heat from the indoors to the ground or lake water. The four-way valve is used to direct the refrigerant gases to the correct coils to accomplish the task of heat movement.

Open-loop systems either take water from a well in the ground or a lake. The water is circulated through a coaxial heat exchanger that operates as either an evaporator or condenser, depending on the four-way valve position. The circulated water either adds heat to the geothermal heat exchanger or removes heat from the geothermal heat exchanger. Different methods of returning water to a well dictate the use of several alternate open-loop system configurations. Open-loop systems are subject to mineral deposits on the tube surfaces that reduce heat exchange efficiency.

Closed-loop systems use plastic pipe buried in the ground in different configurations depending on the land area and soil type. Heat is transferred from water or the ground through the pipe to a fluid inside the pipe in winter or from the fluid in the pipe to the water or ground in summer.

The waterless, earth-coupled, closed-loop geothermal heat pump system consists of small $3/8$- to $7/16$-inch polyethylene plastic-coated copper pipes buried in the ground, with refrigerant flowing through them. The copper pipes act as the evaporator, where the refrigerant experiences a phase change from liquid to vapor, and are often referred to as the waterless earth loop, phase-change loop, or refrigerant loop. In the heating mode, heat is absorbed from the earth "directly" to the vaporizing refrigerant. The refrigerant vapor then travels through the reversing valve to the compressor, where it is compressed and superheated to a higher temperature. The superheated, heat-laden hot gas from the compressor then travels to the condenser. The condenser is a refrigerant-to-air, finned-tube heat exchanger located in the ductwork of the conditioned space or building. It is often referred to as the air coil. Heat is then rejected to the air as the refrigerant condenses at a higher temperature than the air. A fan delivers the heated air to the conditioned space. The condensed liquid then travels through the expansion valve and vaporizes in the evaporator or waterless earth loop buried in the ground.

Water-to-water heat pumps use the heat contained in groundwater to add heat to other water that is circulating throughout the occupied space. This type of heating, referred to as radiant heating, is gaining popularity, and geothermal heat pump systems are a popular equipment choice for supplying the water at the desired temperature.

R-410A is a leading alternate refrigerant to R-22 in air-conditioning and heat pump systems. However, other alternate refrigerants and refrigerant blends have entered the market and may also be substitutes for R-22. Please refer to UNIT 3, "Refrigeration and Refrigerants," for more detailed information on alternate refrigerants and refrigerant blends.

Key Terms

Antifreeze Solutions
Closed-Loop
Coaxial Heat Exchanger
Counterflow
Cupro-Nickel
Delta-T
Dry Well
Earth-Coupled
Four-Way Valve
Geothermal Heat Pumps

Horizontal Loop
Open Loop
Parallel Series Flow
Radiant Heating Systems
Slinky Loop
Vertical Systems
Water Quality
Water Source Heat Pump
Water-to-Water Heat Pump

REVIEW TEST

Name	Date	Grade

Circle the letter that indicates the correct answer.

1. The component that makes a heat pump different from any other refrigeration equipment is the:
   A. defroster.
   B. compressor type.
   C. compressor rotation device.
   D. four-way valve.

2. The geothermal heat pump uses which of the following as a heat source?
   A. the earth
   B. chemicals
   C. air
   D. coal

3. The liquid heat exchanger in a geothermal heat pump is also called a:
   A. coaxial heat exchanger.
   B. four-way valve.
   C. delta-T.
   D. reverse acting valve.

4. Water from wells or lakes is used in:
   A. open-loop systems.
   B. closed-loop horizontal systems.
   C. slinky systems.
   D. air-to-air systems.

5. Water for open-loop systems may contain which of the following?
   A. chlorine
   B. fluoride
   C. sand and clay
   D. algae

6. When there are possibilities of minerals, the heat exchanger tubes should be specified as:
   A. steel.
   B. cupro-nickel.
   C. copper.
   D. brass.

7. A closed-loop geothermal heat pump system may use which of the following circulating fluids?
   A. antifreeze
   B. sugar solution
   C. vinegar solution
   D. fluoride water

8. When a coaxial heat exchanger becomes coated with minerals, which of the following is used to clean it?
   A. acid circulation
   B. scrubbing
   C. brushes
   D. airflow

9. When a coaxial heat exchanger becomes coated with minerals, the cooling cycle symptom would be:
   A. low suction pressure.
   B. low head pressure.
   C. low amperage.
   D. high head pressure.

10. When a coaxial heat exchanger becomes coated with minerals, the heating cycle symptom would be:
    A. low suction pressure.
    B. low head pressure.
    C. high amperage.
    D. high head pressure.

11. In a waterless, earth-coupled, closed-loop geothermal heat pump system, what is the fluid flowing in the polyethylene plastic coated copper pipes buried in the ground?
    A. glycol
    B. water
    C. refrigerant
    D. antifreeze

12. The waterless, earth-coupled, closed-loop geothermal heat pump system extracts heat directly from the earth and is often referred to as a(n) _____ heat exchange with the earth.
    A. direct
    B. indirect
    C. exact
    D. differential

13. The purpose of a buffer tank on a water-to-water heat pump is to:
    A. keep the head pressure of the system as high as possible.
    B. maintain water temperature at a constant level.
    C. act as the water supply tank for a radiant heating system.
    D. both B and C.

Familiarization with Geothermal Heat Pumps

Name	Date	Grade

Objectives: Upon completion of this exercise, you will remove the necessary panels to study and record the various components of the geothermal heat pump.

Introduction: You will remove the necessary panels to study and record the various components of the geothermal heat pump.

Text Reference: Unit 44.

Tools and Materials: Straight-blade and Phillips screwdrivers, and $\frac{1}{4}$" and $\frac{5}{16}$" nut drivers.

Safety Precautions: Turn the power off and lock it out before removing the electrical panels.

PROCEDURES

1. After turning the power off, remove all of the panels to expose the fan, compressor, controls and heat exchanger. Record the following:

2. Fan motor information:

   - Fan motor full-load amperage _____ A
   - Type of fan motor _____
   - Diameter of fan motor _____ in.
   - Diameter of fan motor shaft _____ in.
   - Motor rotation (looking at the fan shaft, if the motor has two shafts, look at the motor lead end) _____
   - Fan wheel diameter _____ in.
   - Width _____ in.
   - Number of motor speeds _____
   - High speed rpm _____ Low speed rpm _____
   - Location of the return air inlet _____
   - Location of the supply air discharge _____

3. Nameplate information:

   - Manufacturer _____
   - Model number _____
   - Serial number _____
   - Capacity in tons of cooling _____
   - Heating capacity _____ Btu/h
   - Voltage _____ Amperage _____
   - Type of pressure controls _____ High _____ Low

4. Identify the coaxial heat exchanger:

- Inlet pipe size _____ in.
- Outlet pipe size _____ in.
- If available, what is the water-side pipe made of, copper or cupro-nickel? _____
- Gas line size at the heat exchanger _____ in.
- Liquid-line size at the heat exchanger _____ in.
- Type of metering device used for the indoor coil _____
- Type of metering device used on the coaxial heat exchanger coil _____

5. Control identification:

- Line voltage to the unit _____ V
- Control voltage to the unit _____ V
- Is there a control voltage terminal block? _____
- When is the four-way valve energized, in cooling or heating? _____

Maintenance of Workstation and Tools: Replace all panels with the correct fasteners and return all tools to their places.

Summary Statement: Describe how the four-way valve directs the refrigerant gas flow to determine how the pump heats or cools.

QUESTIONS

1. Why is there no defrost cycle on a geothermal heat pump?

2. Name a disadvantage of the open-loop geothermal heat pump system.

3. Why are cupro-nickel heat exchangers often used for geothermal heat pumps?

4. What type of fan was used in the geothermal heat pump you worked on?

5. What type of refrigerant was used in the geothermal heat pump you worked on?

6. How much refrigerant did it hold?

7. Why is there no auxiliary heat for geothermal heat pumps?

8. Can an auxiliary water heater be operated from a geothermal heat pump?

9. Name three different fluids that may be circulated in a closed-loop system.

10. Describe a dry well.

11. Describe a waterless, earth-coupled, closed-loop geothermal heat pump system.

Checking the Water Flow, Geothermal Heat Pumps

Name	Date	Grade

Objectives: Upon completion of this exercise, you will be able to check for the correct waterflow through a coaxial heat exchanger on a geothermal heat pump.

Introduction: You will use pressure gages, thermometers, and manufacturers' data to check for the correct water flow through a coaxial heat exchanger on a geothermal heat pump.

Text Reference: Paragraph 44.10.

Tools and Materials: A geothermal heat pump with Pete's test ports, a thermometer and pressure gage for Pete's plug-type access ports (see text, Figure 44-24), a temperature tester, and manufacturer's literature with pressure drop charts for the coaxial heat exchanger coil.

Safety Precautions: None.

PROCEDURES

1. Place one lead of the temperature tester in the return air inlet of the air handler and the other lead in the supply air outlet.

2. Start the unit cooling cycle and wait for the air temperatures to settle down until there is no change, at least 15 minutes.

3. Measure the pressure drop across the coaxial heat exchanger using the pressure gage at the Pete's plug connection at first the inlet, and then the outlet. **NOTE:** Using the same gage gives a more accurate reading.

   - Pressure in _____ psig
   - Pressure out _____ psig
   - Pressure difference _____ psig

4. Measure the temperature difference (delta-T) across the coaxial heat exchanger using the Pete's plug connection at first the inlet, and then the outlet. Again, using the same thermometer for temperature difference will give the best results.

   - Temperature in _____ °F
   - Temperature out _____ °F
   - Temperature difference _____ °F

5. Record the air temperatures:

   - Temperature in _____ °F
   - Temperature out _____ °F
   - Temperature difference _____ °F

6. Compare the pressure and temperature differences to the manufacturer's pressure chart and record the estimated flow rate here: _____ gallons per minute (gpm)

Maintenance of Workstation and Tools: Remove all gages or temperature probes and cap the Pete's plug ports. Replace all panels with the correct fasteners and return all tools to their correct places.

Summary Statement: Describe how the pressure drop can give an accurate measure of the amount of water flowing through a liquid heat exchanger.

QUESTIONS

1. What type of fluid flowed in the system you worked on, water or antifreeze?

2. Where is water ordinarily exhausted to in an open-loop geothermal heat pump system?

3. What material is used for the coaxial heat exchanger when there may be excess minerals present?

4. How can a coaxial heat exchanger be cleaned should it become fouled with minerals or scum?

5. What is the predicted water source temperature from a well used for a geothermal heat pump in your location? See the map in the text, Figure 44-4.

6. In what increment did the pressure gage you used read?

7. What would the temperature rise across a coaxial heat exchanger do in the cooling cycle if the water flow were to be reduced?

8. How would this affect the head pressure?

9. What would the temperature drop across a coaxial heat exchanger do in the heating cycle if the water flow were reduced?

10. How would this affect the suction pressure?

Domestic Refrigerators

Unit Overview

A small fan circulates air inside the refrigerator across a cold refrigerated coil. The air gives up sensible heat to the coil and the air temperature is lowered. The air gives up latent heat (from moisture in the air) to the coil, and dehumidification takes place, causing frost to form on the coil. When the air has given up heat to the coil, it is distributed back to the box so that it can absorb more heat and humidity. This process continues until the desired temperature in the box is reached.

A typical household refrigerator has two compartments, one for frozen food and one for fresh food such as vegetables and dairy products. This refrigerator maintains both compartments with one compressor that operates under the conditions for the lowest box temperature. The evaporator must operate at the low-temperature condition as well as maintain the fresh-food compartment. This may be accomplished by allowing part of the air from the frozen-food compartment to flow into the fresh-food compartment. It may also be accomplished with two evaporators piped in series, one for the frozen food and one for the fresh food. In either case, frost will form on the evaporator. The air movement over the evaporators may be natural draft or forced draft. Most use forced-draft coils utilizing a small fan.

Natural-draft evaporators are normally the flat-plate type with the refrigerant passages stamped into the plate. They may have automatic or manual defrost.

There are two fans on forced-draft refrigerators, one for the evaporator and one for the condenser. They are usually of the prop type, although a small squirrel case fan may be used for the evaporator. The evaporator fan may run all the time except when the box is in defrost.

Most domestic refrigerated devices do not have gage ports. The system is hermetically sealed at the factory and the use of gages may never be needed. It is a poor practice to routinely apply gages. Gages should be applied as a last resort, and when applied, great care should be taken.

Key Terms

Altitude Adjustment
Automatic Defrost
Compressor Oil Coolers

Defrost Condensate
Forced-Draft Evaporator
Heat-Laden Refrigerant

Line Tap Valve
Manual Defrost
Natural-Draft Evaporator

REVIEW TEST

Name	Date	Grade

Circle the letter that indicates the correct answer.

1. **A household refrigerator may have more than one evaporator.**
   A. true B. false

2. **One of the best ways to remove ice from the evaporator in a manual-defrost system is to scrape it with a sharp knife.**
   A. true B. false

3. **An accumulator at the outlet of the evaporator is there to:**
   A. accumulate moisture in the system.
   B. provide for the optimum amount of liquid refrigerant in the evaporator.
   C. provide space for refrigerant vapor to collect.
   D. accumulate foreign particles and other contaminants.

4. **The condensate from the evaporator coil is typically evaporated using heat from the compressor discharge line or by:**
   A. heated air from the condenser.
   B. air from the evaporator fan.
   C. heat from the evaporator defrost cycle.
   D. heat from the crankcase fins.

5. **Two means for providing heat for the automatic defrost cycle are with electric heaters embedded in the evaporator fins and:**
   A. with heated air from the condenser.
   B. through a closed piping circuit from the compressor crankcase.
   C. with a hot gas discharge from the compressor.
   D. from a heat exchange with the suction line.

6. The two types of compressors used most frequently in domestic refrigerators are the reciprocating and:
   A. scroll.
   B. centrifugal.
   C. screw.
   D. rotary.

7. Natural-draft condensers are normally located at the _____ of the refrigerator.
   A. top
   B. bottom
   C. back

8. The capillary tube metering device on a domestic refrigerator is generally fastened to the _____ for a heat exchange.
   A. liquid line
   B. compressor discharge line
   C. condenser coil
   D. suction line

9. On some household refrigerators, the _____ is routed through the compressor crankcase to cool the crankcase oil.
   A. condenser tubing
   B. suction line
   C. suction process tube
   D. capillary tube

10. The heaters located near the refrigerator door to keep the temperature above the dew point are called _____ heaters.
    A. defrost cycle
    B. mullion or panel
    C. condensate
    D. defrost limiter

11. The adjustment device used for low atmospheric pressure conditions is called the _____ thermostatic adjustment control.
    A. current relay
    B. PTC
    C. altitude
    D. mullion

12. The compressor may contain one of the following for start-assist: current-type start relay, potential relay and start capacitor, or:
    A. contactor.
    B. dual-voltage winding.
    C. centrifugal switch.
    D. positive temperature coefficient device.

13. If the compressor is sweating around the suction line, there is:
    A. too much refrigerant.
    B. not enough refrigerant.
    C. probably a refrigerant leak.

14. If a suction-line filter-drier is added to a system, it _____ require an extra charge of refrigerant.
    A. will
    B. will not

15. With a forced-draft condenser, it is important that:
    A. all cardboard partitions, baffles, and backs be in place.
    B. the refrigerator not be located under a cabinet.
    C. the refrigerant pressures be checked when servicing the refrigerator.
    D. the system be swept with nitrogen on a regular basis.

Domestic Refrigerator Familiarization

Name	Date	Grade

Objectives: Upon completion of this exercise, you will be able to identify and describe various parts of a domestic refrigerator.

Introduction: You will move a refrigerator from the wall and remove enough panels to have access to the compressor compartment and identify all components.

Text References: Paragraphs 45.1 through 45.10.

Tools and Materials: Straight-blade and Phillips screwdrivers, $1/4$" and $5/16$" nut drivers, and a flashlight.

Safety Precautions: Make sure the unit is unplugged before opening the compressor compartment.

PROCEDURES

1. Unplug the refrigerator and pull it far enough from the wall to remove the covers from the compressor compartment and the evaporator. Remove the cover, and fill in the following information:

   Refrigerator Nameplate:
   - Manufacturer _____
   - Model number _____
   - Serial number _____
   - Operating voltage _____ V
   - Full-load current _____ A
   - Power (if available) _____ W

   Compressor Compartment:
   - Discharge line size _____ in.
   - Suction line size _____ in.
   - Type of start assist for the compressor (relay, capacitor, or PTC device) _____.

   Refrigerator Components and Features:
   - Side by side, over and under _____
   - Evaporator location _____
   - Forced draft or natural draft _____
   - Evaporator coil material _____
   - Condenser coil material _____
   - Type of condenser (forced air or natural draft) _____
   - If forced draft, where does the air enter the condenser? _____
   - Leave the condenser? _____

Defrost Method:

- Type of defrost (manual or automatic) _____
- If automatic, what type? (hot gas or electric) _____
- How is the condensate evaporated? _____
- Where is the defrost timer located? _____

Maintenance of Workstation and Tools: Replace all panels with the correct fasteners and move the refrigerator back to its correct location. Put all tools in their proper places. Restart the refrigerator if your instructor advises.

Summary Statements: Describe the flow of refrigerant through the entire cycle for the refrigerator that you worked on. Make a simple drawing showing all refrigeration components, such as the compressor, evaporator, condenser, and expansion device, in their respective locations.

QUESTIONS

1. What type of metering device was used on the refrigerator you inspected?

2. Does the domestic refrigerator have an air- or water-cooled system?

3. Does the domestic refrigerator ever have two separate evaporators?

4. Did the refrigerator you inspected have a natural or forced-draft condenser?

5. How are compressors in domestic refrigerators typically mounted, internally or externally?

6. How is heat from the compressor used to evaporate condensate?

7. What does the heat exchange between the capillary tube and the suction line cause the temperature to do in the suction line?

8. What is full-load amperage (FLA)?

9. Name two types of wiring diagrams.

10. What may happen at the evaporator if there is a low refrigerant charge?

Refrigerator Temperature Conditions During Pulldown

Name	Date	Grade

Objectives: Upon completion of this exercise, you will be able to state the characteristics of the refrigerated space temperature and compressor amperage during a pulldown of a warm refrigerated box.

Introduction: You will start with a refrigerator with the inside at room temperature and place a thermometer in the fresh-food and frozen-food compartments. Then start the refrigeration system and observe the temperatures and compressor amperage during the temperature reduction of the box.

Text References: Paragraphs 45.1 through 45.3.

Tools and Materials: An electric thermometer with at least two leads or two glass stem thermometers and a clamp-on ammeter.

Safety Precautions: Use care while fastening the clamp-on ammeter and do not pull any connections loose. TAKE CARE TO ENSURE CORRECT AIRFLOW ACROSS A FORCED-DRAFT CONDENSER.

PROCEDURES

1. Make sure the refrigerator is unplugged and move it from the wall far enough to remove the compressor panel. Fasten the clamp-on ammeter to the common line to the compressor. **NOTE:** IF THE PANEL IS VITAL FOR CORRECT AIRFLOW ACROSS A FORCED-DRAFT CONDENSER, GENTLY REROUTE THE WIRE TO OUTSIDE THE PANEL AND REPLACE IT.

2. Place a thermometer lead in the frozen-food compartment and suspend the lead in the air space, not touching the coil.

3. Place another lead in the fresh-food compartment on a rack in the middle of the box.

4. Shut both doors and record the following temperatures before plugging the box into the power supply:
   - Frozen-food box temperature   _____°F
   - Fresh-food box temperature   _____°F

5. Plug the box in and record the following at 15-minute intervals:
   - Frozen-food box temperature   _____°F
   - Fresh-food box temperature   _____°F
   - Compressor amperage   _____A
   - Frozen-food box temperature   _____°F
   - Fresh-food box temperature   _____°F
   - Compressor amperage   _____A
   - Frozen-food box temperature   _____°F
   - Fresh-food box temperature   _____°F
   - Compressor amperage   _____A

- Frozen-food box temperature   _____°F
- Fresh-food box temperature   _____°F
- Compressor amperage   _____A
- Frozen-food box temperature   _____°F
- Fresh-food box temperature   _____°F
- Compressor amperage   _____A
- Frozen-food box temperature   _____°F
- Fresh-food box temperature   _____°F
- Compressor amperage   _____A

6. If the box temperature has pulled down to 35°F, you may want to stop the exercise at this point.

7. When you stop the exercise, place the refrigerator where the instructor tells you.

Maintenance of Workstation and Tools: Remove the thermometers and ammeter. Replace all panels and return the box to its original location.

Summary Statement: Describe the temperature pulldown. If the box did not pull down and shut off, describe how long it would likely take.

QUESTIONS

1. Why should all panels be in place when the refrigerator is operating?

2. Where is the warm air from a forced-draft condenser normally routed?

3. How is the flow of refrigerant to the evaporator controlled?

4. Why is there often frost accumulation around a defective refrigerator door gasket?

5. What happens to the temperature of the refrigerant in the discharge line when the head pressure increases?

6. What is the temperature in a crisper compartment compared to the rest of the refrigerator?

Charging a Refrigerator Using the Frost-Line Method

Name	Date	Grade

Objectives: Upon completion of this exercise, you will be able to establish the correct operating charge in a refrigerator by using the frost-line method.

Introduction: You will fasten gages to a refrigerator and charge the unit to the correct level if refrigerant is needed.

Text References: Paragraphs 45.9, 45.21, 45.22, and 45.23.

Tools and Materials: A refrigerator with pressure taps if available (if not, the instructor's choice of pressure taps), a thermometer, a gage manifold, R-12 or R-134a, and straight-blade and Phillips screwdrivers.

Safety Precautions: Gloves and goggles should be worn when transferring refrigerant.

PROCEDURES

1. Unplug the refrigerator and pull it from the wall.

2. Place a thermometer sensor in the box. With goggles and gloves on, fasten a high-pressure and a low-pressure gage to the proper gage ports. If the refrigerator does not have gage ports, the instructor will provide taps and/or instructions for installing taps or ports.

3. Purge the center gage line from both gage connections to remove any air from the gage lines.

4. Replace any panels that may cause air not to pass over the condenser properly if the condenser is forced air.

5. Locate the point on the suction line just as it leaves the evaporator and before the capillary tube, Figure 45-1.

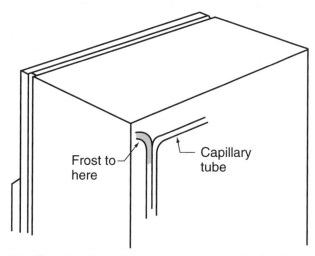

Frost to here

Capillary tube

Figure 45-1 Checking the refrigerator charge using the frost-line method.

6. Turn the thermostat to the off position so the unit will not start, and plug the refrigerator into the power supply.

7. Grip the refrigerant line between your thumb and first finger at the point indicated in step 5 so that you can feel the suction-line temperature.

8. Start the compressor while holding the line. **NOTE:** IF THE LINE FLASHES COLD FOR A FEW SECONDS JUST AFTER THE COMPRESSOR STARTS, THE SYSTEM CHARGE IS CLOSE TO CORRECT.

9. Let the system run long enough so that the box begins to cool inside, to about 50°F, and observe the frost line leaving the evaporator. As the box continues to cool towards 40°F, this line should sweat and turn to frost.

10. The charge is correct when this part of the line has frost at the end of the cycle, at the time the thermostat shuts the unit off. **NOTE:** THE FROST SHOULD NOT EXTEND DOWN THE LINE WHERE THE CAPILLARY TUBE IS FASTENED. IF IT DOES, THE REFRIGERATOR HAS AN OVERCHARGE OF REFRIGERANT, AND SOME SHOULD BE RECOVERED.

11. Shut the compressor off and let the system stand for 10 minutes for the pressures to equalize in the system. Repeat step 8 to get the feel of a suction line on a fully charged system at start-up.

12. When the charge is established as correct, unplug the box and remove the gages. Remove the thermometer sensor and replace all panels. Move the box to the proper location.

Maintenance of Workstation and Tools: Make sure the workstation is clean and in good order. Put all tools in their proper places.

Summary Statement: Describe how the refrigerant line felt when the refrigerator was started with a complete charge.

QUESTIONS

1. Why is it necessary for all panels to be in place on a refrigerator with a forced-draft condenser?

2. Will the refrigerant head pressure be higher or lower than normal if the ambient temperature is below 65°F?

3. If part of the condenser airflow is blocked, will the head pressure be higher or lower?

4. What is the refrigerant most commonly used in household refrigerators?

5. Why should gages be kept under pressure with clean refrigerant when they are not being used?

6. Why is it good practice to remove Schrader valve depressors furnished in the gage lines before starting the evacuation procedure?

7. How many fans would be included in a refrigerator with a forced-draft condenser?

8. Where would these fans be located?

Pressure and Temperature Readings While Refrigerated Box Is Pulling Down in Temperature

Name	Date	Grade

Objectives: Upon completion of this exercise, you will be able to state the refrigerated box fresh-food and frozen-food compartment temperatures, suction and discharge pressures, and compressor amperage at various intervals during a pulldown from room temperature.

Introduction: You will install a gage manifold on the suction and discharge lines, an ammeter on the compressor common line, and thermometer sensors or individual thermometers in the fresh-food and frozen-food compartments of a refrigerator at room temperature. You will record data from these instruments at various intervals as the refrigerator is cooling down.

Text References: Paragraphs 45.21 and 45.22.

Tools and Materials: An electronic thermometer (or two glass stem thermometers), straight-blade and Phillips screwdrivers, a clamp-on ammeter, a gage manifold, and a correctly charged refrigerator with service ports.

Safety Precautions: Goggles and gloves should be worn while making gage connections. Unplug the refrigerator while attaching the clamp-on ammeter. Be sure to have the correct airflow over the condenser.

PROCEDURES

1. Unplug and move a refrigerated box with the inside at room temperature, far enough from the wall to fasten gages and a clamp-on ammeter. **NOTE:** BE SURE TO HAVE THE CORRECT AIRFLOW OVER THE CONDENSER.

2. Place a thermometer in the air space in the freezer and in the fresh-food compartment.

3. Before starting the refrigerator, record the following:
   - Suction pressure _____ psig
   - Discharge pressure _____ psig
   - Temperature of the fresh-food compartment _____°F
   - Temperature of the frozen-food compartment _____°F

4. Start the compressor and record the following every 15 minutes:
   - Suction pressure _____ psig _____ psig _____ psig _____ psig _____ psig _____ psig
   - Discharge pressure _____ psig _____ psig _____ psig _____ psig _____ psig _____ psig
   - Temperature of the fresh-food compartment _____°F _____°F _____°F _____°F _____°F _____°F
   - Temperature of the frozen-food compartment _____°F _____°F _____°F _____°F _____°F _____°F
   - Compressor amperage _____ A _____ A _____ A _____ A _____ A _____ A

5. If the box temperature has not reached 35°F, do not be concerned. You may continue to monitor the box or stop the exercise. It may take several hours for the box to reach 35°F.

6. Unplug the box, remove the ammeter and gages, and replace the box in its proper location.

Maintenance of Workstation and Tools: Return all tools to their correct places and make sure the workstation is clean.

Summary Statement: Describe the relationship of the evaporator temperature converted from suction pressure to box temperature as the box cooled down.

QUESTIONS

1. What would the temperature of the refrigerant in the evaporator be if the refrigerant pressure (R-12) were 26.1 psig or R-134A 23.7 psig?

2. What would the temperature of the refrigerant (R-12) in the condenser be if the refrigerant pressure were 126.6 psig or R-134A 135 psig?

3. What are four ways of determining a refrigerant leak?

4. What might happen if a low-side line tap valve were installed when the system was operating in a vacuum?

5. Why should gages be installed only when absolutely necessary on a domestic refrigerator?

Testing a Compressor Electrically

Name	Date	Grade

Objectives: Upon completion of this exercise, you will be able to identify common, run, and start terminals on a compressor and determine whether the compressor is grounded.

Introduction: You will use an ohmmeter to determine the resistances across the run and start windings and from the compressor to ground.

Text References: Paragraphs 17.7 and 20.6 through 20.9.

Tools and Materials: A refrigerator compressor (it may be installed in a refrigerator), an ohmmeter with R × 1 and R × 1000 scale, needle-nose pliers, straight-blade and Phillips screwdrivers, and $\frac{1}{4}$" and $\frac{5}{16}$" nut drivers.

Safety Precautions: Make sure the unit is unplugged before starting this exercise. Properly discharge the capacitor if there is one.

PROCEDURES

1. Unplug the refrigerator if the compressor is installed in it, and move it from the wall far enough to reach comfortably into the compressor compartment.

2. Remove the cover over the compressor compartment. Properly discharge the capacitor if there is one.

3. Remove the compressor terminal cover.

4. Using needle-nose pliers, remove the terminal connectors from the compressor terminals, one at a time, and LABEL THE WIRES. DO NOT PULL ON THEM. Two of the terminals may be covered by a plug in a current relay type. Ensure the wires are labeled so that you will not make a mistake when replacing them.

5. Set the ohmmeter selector switch to R × 1 and zero the meter.

6. Draw a diagram of the terminal layout using Figure 45-2 as an example. Record the ohms resistance from terminal to terminal as indicated.

   • 1 to 2 _____ ohms
   • 1 to 3 _____ ohms
   • 2 to 3 _____ ohms

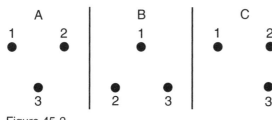

1 to 2 _____ ohms

1 to 3 _____ ohms

2 to 3 _____ ohms

Figure 45-2

7. Based on your resistance readings, identify the start windings: terminal _____ to _____. The run windings: terminal _____ to _____.

8. Set the ohmmeter to the R × 1000 scale and touch one lead to a copper line and the other to each compressor terminal, one by one, and record the readings: _____ ohms

9. Replace the motor terminal wires or relay to the compressor and fasten the terminal cover back in place.

10. Replace the compressor compartment cover with the proper fasteners and return the refrigerator to its proper location.

Maintenance of Workstation and Tools: Turn the ohmmeter to the off position to prevent battery drain and return all tools to their places. Make sure the workstation is clean and orderly.

Summary Statement: Describe what the symptoms would be if the compressor windings to ground reading were 500 ohms.

QUESTIONS

1. Which has the most resistance, the start or run winding?

2. Do the start windings have a smaller or larger diameter wire than the run windings?

3. What is the purpose of the start windings?

4. At what approximate percentage of the normal run speed is the start winding disconnected from the circuit?

5. What is the purpose of the current relay in the start circuit?

6. What two types of compressors are used in household refrigerators?

7. What would be the pressure in a system using R-12 or R-134a if the temperature of the refrigerant were −3°F?

Troubleshooting Exercise: Refrigerator

Name	Date	Grade

Objectives: Upon completion of this exercise, you will be able to troubleshoot the compressor starting circuit.

Introduction: This unit should have a specified problem introduced by your instructor. You will locate the problem, using a logical approach and instruments. THE INTENT IS NOT FOR YOU TO MAKE THE REPAIR BUT TO LOCATE THE PROBLEM AND ASK YOUR INSTRUCTOR FOR DIRECTION BEFORE COMPLETION.

Text References: Units 19, 20, and 45.

Tools and Materials: Straight-blade and Phillips screwdrivers, $1/4$" and $5/16$" nut drivers, safety goggles, trouble light, needle-nose pliers, an electrical test cord for checking the compressor start circuit, an ammeter (with power cord splitter for measuring power in a power cord), and a VOM.

Safety Precautions: Use care while working with energized electrical circuits. Place a shield over the wiring connections at the compressor when attempting to start it with a test cord.

PROCEDURES

1. Put on your safety goggles. Fasten the cord splitter on the end of the power cord and place the ammeter in the splitter.

2. Set the ammeter on the 20-ampere scale.

3. Plug the power cord in and observe the ampere reading. Record the current with the power cord plugged in: _____ A

4. Did the compressor start and stay on? This is evident with a high current on start-up and then the current should drop back to less than 10 amperes. Did this occur? Describe the ammeter reading.

5. The compressor should not have started. Unplug the refrigerator and remove the back panel where you can access the compressor terminal box.

6. Remove the compressor terminal box cover and label the compressor wires.

7. Remove the compressor wires and fasten the wires from the test cord to the compressor terminals. PLACE A SHIELD OVER THE WIRING CONNECTIONS. **NOTE:** THE SHIELD IS VERY IMPORTANT BECAUSE COMPRESSOR TERMINALS MAY BLOW OUT ON A QUESTIONABLE COMPRESSOR. THE OIL MAY IGNITE FROM THE HEAT. YOU HAVE THE POTENTIAL FOR A BLOW TORCH-TYPE FIRE.

8. Try starting the compressor, using the test cord and watching the ammeter. Did the compressor start?

9. Describe your opinion of the problem.

10. What materials or parts do you think are necessary for making a repair?

11. Ask your instructor what you should do regarding making a repair, then proceed.

Maintenance of Workstation and Tools: Leave the unit as your instructor directs you. Return all tools to their places.

Summary Statement: Describe a current relay and how it functions.

QUESTIONS

1. Why is the wire so large in the coil of a current relay?

2. What were the terminal designations on the current relay for this refrigerator?

3. Did this refrigerator have a start capacitor?

4. What is the purpose of a start capacitor?

5. What would the symptoms be for a current relay if the contacts were stuck shut?

6. What would the symptoms be for a current relay if the contacts were stuck open?

7. Where was the relay located on this refrigerator? On the terminals (plug-in type) or in a separate box?

8. Do all refrigerators have start relays?

9. How many wires were in the power cord to this refrigerator?

10. What is the purpose of the green wire in a three-wire power cord?

Troubleshooting Exercise: Refrigerator

Name	Date	Grade

Objectives: Upon completion of this exercise, you will be able to troubleshoot the defrost circuit of a typical household refrigerator.

Introduction: This unit should have a specified problem introduced by your instructor. You will locate the problem, using a logical approach and instruments. THE INTENT IS NOT FOR YOU TO MAKE THE REPAIR BUT TO LOCATE THE PROBLEM AND ASK YOUR INSTRUCTOR FOR DIRECTION BEFORE COMPLETION.

Text Reference: Unit 45.

Tools and Materials: Straight-blade and Phillips screwdrivers, safety goggles, $1/4$" and $5/16$" nut drivers, a VOM, and a troublelight.

Safety Precautions: Use care while working with energized electrical circuits.

PROCEDURES

1. The refrigerator should be running and the evaporator frozen solid. Name at least two things that can cause a frozen evaporator.

2. Put on your safety goggles. Let's assume we have a defrost timer problem. Locate the timer. You may have to follow the wiring diagram to find it. **NOTE:** IF YOU HAVE TO REMOVE THE COMPRESSOR PANEL IN THE BACK, DO NOT ALLOW AIR TO BYPASS THE CONDENSER, OR HIGH HEAD PRESSURE WILL OCCUR AND OVERLOAD THE COMPRESSOR. Describe the location of the defrost timer.

3. Examine the timer. Does it have a viewing window to show the timer turning? If not, it should have some sort of dial that turns over once every hour. If it has a dial that turns over once per hour, mark the dial with a pencil and wait 15 minutes to see whether it is turning. You may measure the voltage to the timer coil during the wait. Is the timer turning?

4. If it is not turning, does it have voltage to the coil?

5. Describe your opinion of the problem.

6. What materials do you think it would take to make the repair?

7. Ask your instructor for directions at this point.

Maintenance of Workstation and Tools: Either repair the unit if your instructor directs you to do so, or leave the unit as directed. Return all tools to their places.

Summary Statement: Describe the complete defrost cycle, including the electrical circuit.

QUESTIONS

1. Draw a wiring diagram of just the electrical circuit for the defrost cycle.

2. What causes the timer motor to turn?

3. Does this refrigerator have hot gas or electric defrost?

4. Did this refrigerator have an ice maker?

5. If yes, was it still making ice with a frozen evaporator?

6. Were the wires color-coded or numbered on this refrigerator?

7. Which do you like best, color-coded or numbered?

8. Was the wiring diagram easy to read?

9. What type of condenser did this refrigerator have, static or forced air?

10. How would high head pressure affect the operation of this refrigerator?

Removing the Moisture from a Refrigerator After an Evaporator Leak

Name	Date	Grade

Objectives: Upon completion of this exercise, you will be able to remove moisture from a refrigerator (or freezer) after an evaporator leak.

Introduction: You will use a high vacuum pump and the correct procedures to remove moisture from a refrigerator. You will then add driers, pressure check, evacuate again, and charge the system.

Text References: Paragraphs 8.6, 8.7, 8.8, 8.10, 8.11, 8.12, 8.13, 8.14, 8.15, 10.1, 10.2, 10.4, 10.5, 18.13, 18.14, 18.15, 45.24, 45.25, 45.26, 45.27, and 45.29.

Tools and Materials: Straight-blade and Phillips screwdrivers, a Schrader valve tool, a nitrogen tank and regulators, $1/4$" and $5/16$" nut drivers, two 100-watt trouble lamps and a 150-watt spotlight lamp, a torch and solder, two $1/4$" tees, two $1/4$" Schrader valve fittings, a liquid-line drier, a suction-line drier, a gage manifold, safety goggles, a valve wrench, a vacuum pump, refrigerant, and a method of measuring refrigerant into the system (charging cylinder or scales).

Safety Precautions: Use care with any heat lamps applied to the refrigerator or the plastic will melt. DO NOT START THE REFRIGERATOR COMPRESSOR WHILE THE SYSTEM IS IN A DEEP VACUUM. DAMAGE TO THE REFRIGERATOR COMPRESSOR MOTOR MAY OCCUR.

PROCEDURES

1. After the evaporator has been repaired and leak checked, you may begin evacuation.

2. Open the refrigerant circuit to the atmosphere by cutting the suction line to the compressor.

3. This procedure is illustrated in the text, Figure 45-118. Cut the liquid line from the condenser.

4. Install the tees and Schrader valve connectors into the tees.

5. Remove the Schrader valve cores from the connectors.

6. Connect the gage manifold to the Schrader ports. Remove the Schrader depressors from the gage manifold or use copper tubing for gage lines for your gage manifold for the best vacuum. It will reduce the vacuum time by at least half.

7. Start the vacuum pump.

8. Place a lightbulb in each compartment of the refrigerator. LEAVE THE DOOR CRACKED ABOUT 2" OR THE PLASTIC WILL MELT, Figure 45-3.

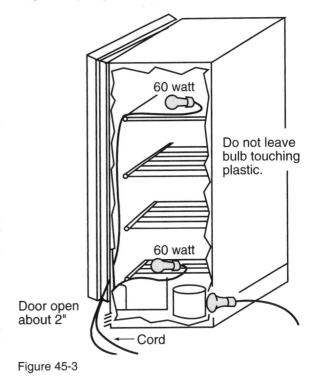

Figure 45-3

9. Place the 150 W spotlight against the compressor shell. This will boil any moisture out of the oil.

10. THE VACUUM PUMP SHOULD RUN FOR SEVERAL HOURS LIKE THIS. IF THERE IS A GREAT DEAL OF WATER IN THE REFRIGERANT CIRCUIT, IT WILL BEGIN TO CONDENSE IN THE VACUUM PUMP CRANKCASE. YOU WILL SEE THE VACUUM PUMP OIL LEVEL RISE AND THE OIL WILL TURN MILKY COLORED. CHANGE THE VACUUM PUMP OIL WHEN THE COLOR CHANGES. YOU MAY HAVE TO DO THIS SEVERAL TIMES FOR A LARGE AMOUNT OF WATER. If the vacuum pump is left on overnight, make sure that the power supply will not be interrupted.

11. The vacuum should be verified with an electronic vacuum gage. When the system is in a deep vacuum, the vacuum pump will stop making pumping sounds. The vacuum may then be broken with nitrogen to atmospheric pressure. IT IS GOOD POLICY TO START THE COMPRESSOR FOR ABOUT 20 SECONDS AT THIS TIME. IF ANY MOISTURE IS TRAPPED IN THE COMPRESSOR CYLINDERS, IT WILL BE PUMPED OUT. Evacuate the system again.

12. After the second vacuum, break the vacuum to atmospheric pressure with nitrogen and install the driers, one in the liquid line and one in the suction line. DO NOT ALLOW THE DRIERS TO BE OPEN TO THE ATMOSPHERE FOR MORE THAN 5 MINUTES. **NOTE:** The liquid-line drier should not be oversized. The suction-line drier will handle any excess moisture.

13. Leak check the system with refrigerant and nitrogen and evacuate for the last time to a deep vacuum.

14. When a deep vacuum is reached, break the vacuum with the nitrogen to about 5 psig.

15. Remove the gage lines, one at a time, and replace the valve cores and depressors.

16. Prepare to measure the refrigerant into the system, start the vacuum pump, and pull one more vacuum. Remember, all refrigerant should be measured into the system from a vacuum. If the volume of the liquid-line drier has not been changed, the charge on the nameplate should be correct.

17. Charge the refrigerant into the system and start the box.

18. IT IS GOOD PRACTICE TO ALLOW A BOX THAT HAS HAD MOISTURE IN THE REFRIGERANT CIRCUIT TO RUN IN THE SHOP FOR SEVERAL DAYS BEFORE RETURNING IT TO THE CUSTOMER. THIS MAY PREVENT A CALLBACK.

Maintenance of Workstation and Tools: Return all tools to their places.

Summary Statement: Describe triple evacuation.

QUESTION

1. Why does it help to start the compressor after the first vacuum?

Changing a Compressor on a Refrigerator

Name	Date	Grade

Objectives: Upon completion of this exercise, you will be able to change a compressor on a household refrigerator.

Introduction: You will recover the refrigerant from a refrigerator, change the compressor, leak check, evacuate, and charge the refrigerator.

Text References: Units 8, 9, and 10. Paragraph 45.28.

Tools and Materials: Straight-blade and Phillips screwdrivers, $1/4$" and $5/16$" nut drivers, a liquid-line drier, one $1/4$" tee, one $1/4$" Schrader fitting, a solder, torch, two 6" adjustable wrenches, a set of sockets, a line tap valve, a charging cylinder or scales, slip joint pliers, an ammeter, a gage manifold, a refrigerant recovery system, a vacuum pump, refrigerant, goggles, and gloves.

Safety Precautions: Use care while using the torch. Wear goggles and gloves when attaching gages to a system and transferring refrigerant.

PROCEDURES

1. Put on your goggles and gloves. Fasten the line tap valve to the suction line and attach the gage manifold suction line to it.

2. Fasten the center port of the gage line to the recovery system and recover the refrigerant.

3. With the refrigerant removed from the system, open the system to the atmosphere.

4. You may want to set the refrigerator on a table for removing the compressor.

5. Mark the compressor terminal wiring and remove the wiring.

6. Sweat the compressor suction and discharge lines loose. Use pliers to gently pull them from their connectors when they are hot enough. USE A HEAT SHIELD TO PREVENT OVERHEATING ADJACENT PARTS.

7. Remove the compressor mounting bolts and set the compressor out of the compartment. This compressor may be reinstalled as if it were a new one.

8. Clean the suction and discharge lines on the compressor. They may be steel and must be perfectly clean. There can be no rust, paint, or pits in the lines, or reconnection will be impossible. Clean the pipe stubs on the refrigerator the same way. A small file is sometimes necessary to clean the pits out and the small solder balls off the pipe.

9. When all is cleaned correctly, replace the compressor into the compartment. There will likely be solder balls inside the fittings that cannot be removed. They will have to be heated and the pipe connection made hot. USE A HEAT SHIELD TO PREVENT OVERHEATING ADJACENT PARTS.

10. With the compressor soldered back in place, cut the suction line and solder a $1/4$" tee with a Schrader connector in the side port. We are only using one gage port for evacuation and charging.

11. Cut the liquid line and sweat the liquid-line drier in place. If the same size liquid-line drier is used, the charge will not need to be adjusted.

12. Pressure the system enough for a leak check and perform a leak check on the system.

13. Triple evacuate the system and on the third evacuation, prepare to charge the system with scales or charging cylinder. **NOTE:** WE ARE EVACUATING ONLY FROM THE SUCTION SIDE OF THE SYSTEM. THE EVACUATION SHOULD BE GIVEN PLENTY OF TIME BECAUSE WE ARE EVACUATING THROUGH THE SCHRADER VALVES AND MUST EVACUATE THE HIGH-PRESSURE SIDE OF THE SYSTEM THROUGH THE CAPILLARY TUBE. IT MAY SAVE TIME FOR YOU TO SWEAT A TEE AND SCHRADER VALVE IN THE HIGH SIDE. While the system is being evacuated, reconnect the wiring and fasten the compressor mounting bolts.

14. With the system evacuated, add the correct charge to the system.

15. When the correct charge is in the system, start the system and observe its operation. Be sure to check the amperage.

Maintenance of Workstation and Tools: Leave the refrigerator as your instructor directs you. Return all tools to their places.

Summary Statement: Describe why it takes longer to evacuate from the low-pressure side of the system only.

QUESTIONS

1. Describe the gas ballast on a vacuum pump.

2. Describe a two-stage rotary vacuum pump and state how it works.

3. Why is a rotary vacuum pump better than utilizing a reciprocating compressor?

Situational Service Ticket

Name	Date	Grade

NOTE: Your instructor must have placed the correct service problem in the system for you to successfully complete this exercise.

Technician's Name:_____ Date: _____

CUSTOMER COMPLAINT: Fresh-food compartment is not cold.

CUSTOMER COMMENTS: Unit runs all the time and the fresh-food compartment is not cold. The milk is not cold.

TYPE OF SYSTEM:

MANUFACTURER_____   MODEL NUMBER_____   SERIAL NUMBER_____

COMPRESSOR (where applicable):
MANUFACTURER_____   MODEL NUMBER_____   SERIAL NUMBER_____

TECHNICIAN'S REPORT

1. SYMPTOMS: _____

   _____

2. DIAGNOSIS: _____

   _____

3. ESTIMATED MATERIALS FOR REPAIR: _____

   _____

4. ESTIMATED TIME TO COMPLETE THE REPAIR: _____

SERVICE TIP FOR THIS CALL

1. Let the box run and monitor the on and off times. Check for excess load.

Situational Service Ticket

Name	Date	Grade

NOTE: Your instructor must have placed the correct service problem in the system for you to successfully complete this exercise.

Technician's Name:_____ Date _____

CUSTOMER COMPLAINT: No cooling.

CUSTOMER COMMENTS: Unit stopped in the middle of the night last night for no apparent reason.

TYPE OF SYSTEM:

MANUFACTURER_____   MODEL NUMBER_____   SERIAL NUMBER_____

COMPRESSOR (where applicable):
MANUFACTURER_____   MODEL NUMBER_____   SERIAL NUMBER_____

TECHNICIAN'S REPORT

1. SYMPTOMS: _____

   _____

2. DIAGNOSIS: _____

   _____

3. ESTIMATED MATERIALS FOR REPAIR: _____

   _____

4. ESTIMATED TIME TO COMPLETE THE REPAIR: _____

SERVICE TIP FOR THIS CALL

1. Use your VOM and follow the power cord.

Domestic Freezers

Unit Overview

The domestic freezer is different from the refrigerator in that it is a single low-temperature system. Food to be frozen must be packaged properly or dehydration will occur, causing freezer burn. Moisture leaves the food, changing from a solid state to a vapor state collecting on the coil. The change from a solid state directly to a vapor is called sublimation.

The freezer box may be a chest type with a lid that raises or an upright with a door that swings out. The upright does not take up as much floor space but may not be as efficient as cold air falls out the bottom of the door area every time the door is opened. In the chest type, the cold air stays in the box.

The evaporator must operate at a temperature that will lower the air temperature to at least 0°F, and in some instances to −10°F. In a plate-type evaporator with air at 0°F, the coil temperature may be as low as −18°F. With a forced-draft type and 0°F air temperature, the coil temperature will be approximately −11°F.

The compressor is very hot to the touch on a freezer that is operating normally. The compressors are air or refrigerant cooled. When air cooled, there are fins on the compressor. When refrigerant cooled, suction gas cools the compressor.

Key Terms

Chimney-Type Condenser
Freezer Burn

Instigate
Sublimation

REVIEW TEST

Name	Date	Grade

Circle the letter that indicates the correct answer.

1. **When a substance changes state from a solid to a vapor, it is known as:**
   A. freezer burn.   C. sublimation.
   B. specific heat.   D. superheat.

2. **A plate- or tube-type evaporator is generally used with:**
   A. natural draft.   B. forced-air draft.

3. **Reciprocating or _____ compressors are used in domestic freezers.**
   A. scroll   C. rotary
   B. centrifugal   D. screw

4. **An air-cooled chimney-type condenser is generally a _____ draft type.**
   A. natural   B. forced-air

5. **Forced-draft condensers are typically used with units that have:**
   A. manual defrost   B. automatic defrost

6. **The type of metering device most commonly used in household freezers is the:**
   A. capillary tube.
   B. thermostatic expansion valve.
   C. electronic expansion valve.
   D. automatic expansion valve.

7. **When the air temperature in a domestic freezer is 0°F, the plate-type evaporator coil temperature may be as low as:**
   A. 0°F.
   B. −5°F.
   C. −12°F.
   D. −18°F.

8. **Compressor running time may be used as a factor to _____ automatic defrost.**
   A. instigate
   B. terminate

9. If a gage reading were taken on the low-pressure side and found to be 1.3 psig, what would the evaporator temperature be on a domestic freezer using R-12? (Use the temperature/pressure relationship chart at the end of this guide.)
A. 18°F
B. 10°F
C. −12°F
D. −18°F

10. What word would best describe the temperature of a compressor operating normally in a household freezer?
A. cold
B. cool
C. warm
D. hot

11. Quick freezing food:
A. makes it mushy.
B. is not recommended.
C. is not cost effective.
D. saves the flavor and texture.

12. The blood and water you find when you thaw out a steak:
A. is caused by quick freezing.
B. is caused by freezing too slowly.
C. is normal.
D. is good.

13. The best method for defrosting a plate-type evaporator that has a heavy ice buildup is to:
A. use a sharp knife.
B. use a meat cleaver.
C. use a putty knife.
D. use external heat or a fan.

14. When moving a heavy freezer in a house, you should:
A. use a fork lift.
B. use refrigerator hand trucks.
C. use four people, one on each corner.
D. all of the above.

15. Ice buildup on the inside of a chest-type freezer is due to:
A. the thermostat being set too low.
B. a defective gasket.
C. food stacked too closely together.
D. the unit running all the time.

Domestic Freezer Familiarization

Name	Date	Grade

Objectives: Upon completion of this exercise, you will be able to recognize the different features of a domestic freezer.

Introduction: You will remove enough panels to be able to identify the compressor, evaporator, and condenser of a freezer.

Text References: Paragraphs 46.1 through 46.7.

Tools and Materials: Straight-blade and Phillips screwdrivers, $1/4$" and $5/16$" nut drivers, and a flashlight.

Safety Precautions: Be sure the unit is unplugged before removing panels.

PROCEDURES

1. Unplug a freezer and pull it far enough from the wall to be able to remove the compressor compartment cover.

2. Remove the compressor compartment cover and fill in the following information:
   - Compressor suction line size _____ in. • Discharge-line size _____ in.
   - Suction-line material _____ • Discharge-line material _____
   - Type of compressor relay (potential or current, see text paragraphs 17.18 and 17.19) _____
   - Pressure ports? _____ • Type of condenser (natural or forced draft) _____
   - Where is the condenser located? (in the bottom of the box, at the back, or in the sides) _____
   - Condenser material _____ • Does the compressor have an oil cooler circuit? _____ • How is the compressor mounted, on springs or rubber feet? _____

   - Type of freezer (upright or chest) _____ • Type of evaporator (forced, shelf, or wall) _____ • Manual or automatic defrost _____ • Where is the thermostat located? _____
   - Does the power cord have a ground leg? _____ Does the box have a lock on the door? _____ What holds the door closed, latch or magnetic gasket? _____

   - Does the box have a wiring diagram on the back? _____
   - Are there any heaters in the diagram? _____ • Is there a lightbulb and, if so, how is it controlled? _____ • If the unit has electric defrost, where is the timer located? _____
   - What is the full load amperage draw of the freezer? _____ A • The voltage rating _____ V

3. Replace all panels with the correct fasteners and move the freezer back to the correct location.

Maintenance of Workstation and Tools: Return all tools to their correct places and make sure the workstation is clean and neat.

Summary Statement: What would the best location for this freezer be, and why?

QUESTIONS

1. What is the primary problem with forced-draft condensers?

2. Between what temperatures should a freezer operate to keep ice cream hard?

3. What three types of refrigerant are used in a domestic freezer?

4. How can the heat from the compressor be used to evaporate condensate in the condensate pan?

5. A high-efficiency condenser (forced draft) will condense refrigerant at how many degrees above the ambient temperature.

6. How does the compressor that is operating normally feel to the touch?

7. What conditions may produce a heavy load on a freezer?

8. What are two controls located on a freezer that does not have automatic defrost?

9. Describe two ways to speed defrost in a manual defrost freezer?

10. What substance may be used to keep food frozen temporarily while a freezer is being serviced?

Charging a Freezer by the Frost-Line Method

Name	Date	Grade

Objectives: Upon completion of this exercise, you will be able to charge a domestic freezer using the frost-line method.

Introduction: You will be provided a unit from which some refrigerant has been removed so the charge is low. You will then add refrigerant until the charge is correct according to the frost-line method.

Text References: Paragraphs 46.8 and 46.9.

Tools and Materials: Straight-blade and Phillips screwdrivers, a cylinder of refrigerant (ensure that it is the same type as in the freezer you will be working with), a gage manifold, a thermometer, and a freezer with gage ports.

Safety Precautions: Gloves and goggles should be worn while working with refrigerant.

PROCEDURES

1. Unplug the freezer and pull it far enough from the wall to access the compressor compartment. Remove the compressor compartment cover.

2. Put on your goggles and gloves. Fasten gages to the suction and discharge line. Replace any panels that may affect condenser airflow if the unit is a forced-draft unit. Suspend the thermometer in the air in the frozen-food compartment.

3. Plug in the unit and start the compressor. If the box was plugged in and cold, you may need to open the door for a few minutes or turn the thermostat to the lowest point to start the compressor.

4. Purge the suction and discharge line through the center gage line and fasten it to the refrigerant cylinder. If the unit has a correct charge, you may recover some refrigerant from the unit to reduce the charge in the unit.

5. Locate a point on the suction line where it leaves the evaporator before the capillary tube heat exchange. If the box is a type with the evaporator in the wall of the box, you will probably not be able to find this point. In this case, locate the evaporator in the wall so that you can follow the frost pattern as it develops in the evaporator. You will need to monitor the point on the suction line or the frost pattern of the evaporator in the wall. The suction line leaving the evaporator may be inside the box also. In these cases, the door must remain closed except for brief periods when you are checking the frost line. **NOTE:** YOU WILL HAVE TO MONITOR THIS SPOT. IT MAY BE INSIDE THE BOX. THE DOOR MUST REMAIN CLOSED UNLESS YOU ARE CHECKING THE FROST LINE.

6. Record the suction and discharge pressure. **NOTE:** The suction pressure will be in a vacuum if the charge is low. Suction pressure _____ psig. Discharge pressure _____ psig.

7. Slowly add vapor in short bursts with about 15 minutes between them until there is frost on the suction line leaving the evaporator, or the pattern of frost is toward the end of the evaporator on a type with the evaporator in the wall. If the space temperature is near 0°F, the suction pressure should rise when refrigerant is added. As the space temperature drops, the suction pressure will drop. THE COMPRESSOR SHOULD SHUT OFF WHEN THE SPACE TEMPERATURE IS AT ABOUT 0 to −5°F, AND THE SUCTION PRESSURE SHOULD BE ABOUT 2 PSIG WITH R-12 REFRIGERANT OR 0.7 WITH R-134a.

8. When the frost line is correct on the suction line or when the evaporator is frosted on the wall unit, and the suction pressure is near correct, you have the correct charge. Shut the unit off and remove the gages. Replace all panels with the correct fasteners and return the box to its assigned location.

Maintenance of Workstation and Tools: Return all tools to their respective places and clean the workstation.

Summary Statement: Describe why the suction pressure rises when adding refrigerant and falls as the box temperature falls.

QUESTIONS

1. How is a forced-draft evaporator fan in a freezer wired with respect to the compressor?

2. Is there ever a fan switch? If so, where is it located?

3. What is typically wrong with a fan motor when it does not run?

4. What may happen if an owner becomes impatient with the manual defrost and an attempt is made to chip away the frost and ice?

5. What will the ohmmeter read when checking an open motor winding?

6. What would happen if the compressor were sweating down the side with an overcharge of refrigerant?

Charging a Freezer Using a Measured Charge

Name	Date	Grade

Objectives: Upon completion of this exercise, you will be able to charge a freezer using a measured charge.

Introduction: You will evacuate a freezer to a deep vacuum and make the correct charge by measurement using scales or a graduated cylinder.

Text References: Paragraphs 8.3, 8.4, 8.5, 8.6, 8.11, 8.12, 10.1, 10.2, 10.4, and 10.5.

Tools and Materials: Straight-blade and Phillips screwdrivers, $1/4$" and $5/16$" nut drivers, a refrigerant recovery system, a cylinder of refrigerant containing the type of refrigerant used in the freezer you are working with, a gage manifold, a vacuum pump, a charging cylinder, scales, gloves, goggles, and a freezer with at least a suction gage port.

Safety Precautions: Gloves and goggles should be worn while transferring refrigerant.

PROCEDURES

1. Move the freezer to where you can access the compressor compartment.

2. Put your gloves and goggles on. Fasten the gage manifold to the service port or ports. If there is any refrigerant in the system, recover it the way your instructor advises.

3. When the system pressure is reduced to the extent indicated by your instructor, fasten the center gage line to the vacuum pump and start it.

4. Open the low-side manifold gage valve to the low-side gage line. This will start the vacuum pump removing the rest of the refrigerant from the system. Allow the vacuum pump to run while you charge the charging cylinder or get set up with scales, whichever you use. Write the refrigerant charge from the freezer nameplate here in ounces: _____

5. When the vacuum pump has lowered the system pressure to below 28 in. Hg on the gage manifold, you are ready to add refrigerant to the system. **NOTE:** We are not evacuating the system for the purpose of removing moisture, but for the purpose of removing all the refrigerant in the system.

6. Shut the gage manifold valve and transfer the center gage line to the refrigerant cylinder. Open the valve on the refrigerant cylinder and allow refrigerant to enter the center gage line. **NOTE:** PURGE THE CENTER LINE WITH REFRIGERANT FROM THE CYLINDER TO REMOVE ATMOSPHERE THAT ENTERED THE GAGE LINE WHEN DISCONNECTED FROM THE VACUUM PUMP.

7. Recheck your initial scale or graduated cylinder reading and enter here in pounds and ounces: _____ lb _____ oz. This is a total of _____ oz. The amount of refrigerant that should be left in the cylinder after charging the refrigerator should be a total of _____ oz, or _____ lb and _____ oz.

8. Allow the vapor refrigerant to flow into the refrigeration system by opening the suction-gage manifold valve. When refrigerant has stopped flowing, you may start the compressor. Refrigerant will start to flow again. You will have to watch your scale or graduated cylinder reading and shut off the gage manifold valve when the correct weight has been reached. **NOTE:** IF FLOW HAS STOPPED BEFORE THE COMPLETE CHARGE HAS ENTERED THE SYSTEM, IT MAY BE BECAUSE THE VAPOR BOILING FROM YOUR STORAGE CYLINDER HAS LOWERED AND IS AS LOW AS THE SYSTEM PRESSURE. ASK YOUR INSTRUCTOR FOR ADVICE.

9. When the correct charge has been charged into the system, remove the gage from the gage port. This should be done while the system is running in order to keep the correct charge in the system.

10. Cover all gage ports with the correct caps and replace all panels. Move the freezer to its correct location.

Maintenance of Workstation and Tools: Return all tools to their correct storage places.

Summary Statement: Explain why it is necessary to measure the charge placed in a freezer.

QUESTIONS

1. Why is it necessary to use a charging cylinder or accurate scales when measuring a charge of refrigerant into a freezer?

2. What is the advantage of using electronic scales?

3. Can you see the refrigerant in the charging cylinder?

4. Why must the temperature of the refrigerant in the charging cylinder be known?

5. How is the temperature registered in the charging cylinder?

6. Why do some graduated cylinders have heaters?

7. How is automatic defrost normally started in a domestic freezer?

8. What type of component is often used to accomplish automatic defrost?

Freezer Gasket Examination

Name	Date	Grade

Objectives: Upon completion of this exercise, you will be able to determine whether the gaskets on a freezer are sealing correctly.

Introduction: Depending on the type of gasket on the freezer you are working with, you will use one of two different procedures to determine whether the gasket is sealing correctly to prevent air and moisture from entering the freezer. Most late model upright freezers will have magnetic gaskets and will be checked with a light. Chest-type freezers may have compressible gaskets and will be checked with a crisp piece of paper.

Text References: Paragraphs 46.2 and 46.3.

Tools and Materials: Straight-blade and Phillips screwdrivers, tape, a troublelight, and a crisp piece of paper such as a dollar bill.

Safety Precautions: Do not place the troublelight in the freezer next to any plastic parts because it may cause them to melt. Use caution while shutting the door on the troublelight cord.

PROCEDURES

Choose a freezer for the gasket test. Most new upright freezers have magnetic gaskets while chest-type freezers may have compressible-type gaskets. One of each type should be used.

FREEZER WITH MAGNETIC GASKETS

1. Place the troublelight inside the box and turn it on. **NOTE:** DO NOT LAY THE LIGHTBULB NEXT TO THE PLASTIC.

2. Gently shut the door on the cord. If too much pressure must be applied to the door, move the cord to another location.

3. Turn the lights off and examine the door gasket all around its perimeter for light leaks. Light will leak around the cord, so it should be moved to a different location and the gasket checked where the cord was first placed.

4. If leaks are found, make a note and mark the places with tape.

5. Open the door and examine any leak spots closely to see whether the gasket is dirty or worn.

FREEZER WITH COMPRESSIBLE GASKETS

1. Close the lid on the piece of paper (a dollar bill) repeatedly around the perimeter of the door and gently pull it out. The paper should have some drag as it is pulled out. If it has no drag, air will leak in. Mark any places that may leak with tape.

2. After you have proceeded around the perimeter of the door and marked any questionable places, open the lid or door and examine each place for dirt or deterioration.

3. Remove all the tape and return each freezer to its permanent location. Your instructor may want you to make a repair on any gaskets found defective.

Maintenance of Workstation and Tools: Return all tools to their designated places.

Summary Statement: Describe the repair or replacement procedures for a magnetic gasket that is defective.

QUESTIONS

1. What type of material is used as insulation in the walls of a modern freezer?

2. What type of conditions will a worn gasket cause?

3. How should a refrigerator or freezer be prepared for storage, and how should it be placed in the storage area?

4. What type of gasket is used in most modern upright refrigerators and freezers?

5. What will be the condition of the outside of the compressor if liquid refrigerant is returning to it through the suction line?

6. Sketch a wiring diagram of a freezer with automatic defrost.

Room Air Conditioners

Unit Overview

Single room air conditioners are package units. They may be window or through-the-wall units, front discharge or top-air discharge. Most units range from 4000 Btu/h to 24,000 Btu/h. A special application is a roof-mount design for travel trailers and motor homes. Many units have electric resistance heating and some are heat pumps.

The refrigeration cycle components are the same four major components described in other units. The evaporator absorbs heat into the system. The compressor compresses and pumps the refrigerant through the system. The condenser rejects the heat, and the metering device controls the flow of the refrigerant.

Room units are designed for window installation or wall installation. Window units can be removed and the window used as it was originally intended. A hole must be cut in the wall for installation of wall units, so these are installed more permanently.

Controls for room units are mounted within the cabinet. A typical room unit will have a selector switch to control the fan speed and provide power to the compressor circuit. The power cord is wired straight to the selector switch and will contain a hot, a neutral, and a ground wire for a 120 V unit. A 208/230 V unit will have two hot wires and a ground. The neutral wire for the 120 V unit is routed straight to the power-consuming devices. On 208/230 V units, one hot wire is wired through the selector switch to the power-consuming devices; the other is connected directly to the other side of the power-consuming devices. These units have only one fan motor, so a reduction in fan speed slows both the indoor and outdoor fan. This reduces the capacity and noise level.

Key Terms

Condensate
Dew Point Temperature
Four-Way Valve
Hermetically Sealed Compressor

Potential Relay
Pressure Equalization
Selector Switch

REVIEW TEST

Name	Date	Grade

Circle the letter that indicates the correct answer.

1. **Single room air-conditioning units generally range in capacity from _____ Btu/h.**
   A. 1500 to 15,000
   B. 3000 to 20,000
   C. 4000 to 24,000
   D. 5000 to 30,000

2. **Most modern room air conditioners utilize _____ metering device.**
   A. an electronic expansion valve
   B. a thermostatic expansion valve
   C. an automatic expansion valve
   D. a capillary tube

3. **The evaporator typically operates _____ the dew point temperature of the room air.**
   A. below   B. above

4. **The compressor in a room air conditioner is _____ type.**
   A. an open
   B. a serviceable hermetic
   C. a fully sealed hermetic

5. **The condensate in a window unit is generally directed to:**
   A. a place where it will drain into the soil.
   B. a storm drain.
   C. the condenser area where it evaporates.
   D. a holding tank.

6. A heat-pump room air conditioner has the four major components in a cooling-only unit plus:
   A. a three-way valve.
   B. a four-way valve.
   C. an automatic expansion valve.
   D. an indoor coil.

7. _____ are used in heat-pump room air conditioners to ensure the refrigerant flow through the correct metering device at the correct time.
   A. Filter-driers   C. Globe valves
   B. Check valves   D. Electronic relays

8. During the heat pump heating cycle, the hot gas from the compressor discharge line is directed to the _____ coil.
   A. indoor   B. outdoor

9. In the heat pump heating cycle, the refrigerant condenses in the _____ coil.
   A. indoor   B. outdoor

10. All 230 V power cords have the same plug.
    A. true   B. false

11. The fin patterns on a coil:
    A. increase the heat exchange rate.
    B. are for decoration.
    C. make the air sweep the room.
    D. none of the above.

12. Two types of compressors commonly used for window and room units are:
    A. centrifugal.
    B. rotary.
    C. reciprocating.
    D. screw.

13. A regular window air conditioner can be used in:
    A. a double hung window.
    B. a casement window.

14. The thermostat bulb located close to the fins in the return air stream:
    A. keeps the unit running for long cycles.
    B. stops the unit when it is overheating.
    C. keeps the fins clean.
    D. helps prevent frost or ice buildup.

15. Window air conditioners typically have:
    A. two fan motors.
    B. two fan motors with different shaft sizes.
    C. one fan motor with two shafts.
    D. one fan.

Window Air-Conditioner Familiarization

Name	Date	Grade

Objectives: Upon completion of this exercise, you will be able to identify the various parts of a window air conditioner.

Introduction: Your instructor will assign you a window unit; you will remove it from its case, then identify and describe the major components.

Text References: Paragraphs 47.1 through 47.5.

Tools and Materials: Straight-blade and Phillips screwdrivers, $1/4$" and $5/16$" nut drivers, a flashlight, and a window or room air conditioner.

Safety Precautions: Make sure the unit is unplugged before starting the exercise. Be careful while working around the coil fins because they are sharp.

PROCEDURES

1. If the unit is installed in a window in a slide-out case, remove it from its case and set it on a level bench. If the unit is in a case fastened with screws, set the unit on a bench and remove enough screws to remove the case.

2. When you have the unit where all parts are visible, fill in the following information:

   Indoor Section:

   - Type of fan blade _____
   - What is the evaporator tubing made of? _____
   - What are the fins made of? _____
   - Where does the condensate drain to from the evaporator? _____

   Outdoor Section:

   - Type of fan blade _____
   - What is the condenser tubing made of? _____
   - What are the fins made of? _____
   - Does the condenser fan blade have a slinger ring for evaporating condensate? _____
   - Does the condenser coil appear to be clean for a good heat exchange? _____

   Fan Motor Information:

   - Size of fan shaft _____
   - Looking at the lead end of the motor, what is the rotation, clockwise or counterclockwise? _____
   - Fan motor voltage _____ V
   - Fan motor current _____ A
   - Type of fan motor, PSC or shaded pole _____

- If PSC, what is the MFD rating of the capacitor? _____ MFD
- If available on the nameplate of the motor, what are its speeds?_____

Unit Nameplate:

- Manufacturer _____
- Serial number _____
- Model number _____
- Unit voltage _____
- Unit full load current _____ A
- Test pressure _____ psi

General Unit Information:

- What ampere rating does the power cord have? _____ A
- Was this unit intended for a casement window or a double-hung window? _____
- What type of air filter is furnished with the unit, fiberglass or foam rubber? _____

3. Assemble the unit.

Maintenance of Workstation and Tools: Return all tools to their proper locations. Place the window or room unit in a storage area if instructed. Leave your work area clean.

Summary Statement: Describe how condensate is evaporated in the condenser portion of the window unit you worked with.

QUESTIONS

1. How many fan motors are normally included in a window cooling unit?

2. What are the two types of designs of window units relative to covers and access for service?

3. What is the most common refrigerant used in window units?

4. What is the most common metering device used in window units?

5. What is the purpose of the heat exchange between the metering device and the suction line?

6. Why must window cooling units operate above freezing?

7. At what temperature do window cooling unit evaporators typically boil the refrigerant?

8. What is the evaporator tubing in these units usually made of?

9. What are the evaporator fins made of?

10. How does the unit dehumidify the conditioned space?

Checking the Charge Without Gages

Name	Date	Grade

Objectives: Upon completion of this exercise, you will be able to evaluate the refrigerant charge for a window air conditioner without disturbing the refrigerant circuit with gage connections.

Introduction: You will operate a window unit and use temperature drop across the evaporator and temperature rise across the condenser to evaluate the refrigerant charge without gages.

Text Reference: Paragraph 47.8.

Tools and Materials: An electronic thermometer (or two glass thermometers), straight-blade and Phillips screwdrivers, $1/4$" and $5/16$" nut drivers, and a window cooling unit.

Safety Precautions: Make sure that when the unit panels are removed for the purpose of observing the suction line, the condenser and evaporator have full airflow over them. Ensure that all leads, clothing, and your hands are kept away from fan blades.

PROCEDURES

1. Remove the unit from its case and place it on a bench with a correct power supply nearby.

2. Make sure that the correct airflow passes across each coil before starting the compressor. Cardboard may be placed over any place where a panel has been removed to direct the airflow across the coil.

3. Place a thermometer at the inlet and outlet of the evaporator and start the compressor. DO NOT LET THE THERMOMETER LEAD GET IN THE FAN.

4. Cover about $2/3$ of the condenser surface, Figure 47-1. This will cause the head pressure to rise. DO NOT COVER THE ENTIRE CONDENSER. When the head pressure rises, the condenser then operates with the correct charge, pushing the correct refrigerant charge to the evaporator. The suction line leading to the compressor should then sweat.

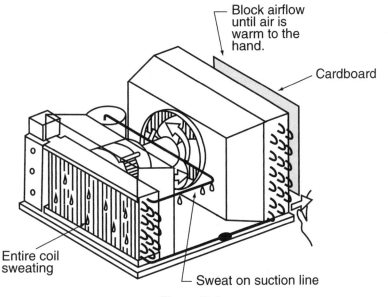

Block airflow until air is warm to the hand.

Cardboard

Entire coil sweating

Sweat on suction line

Figure 47-1

5. Let the unit run for about 15 minutes, then go to step 6.

6. How much of the evaporator is sweating, $1/4$, $1/2$, $3/4$, or all of it? How much of the suction line to the compressor is sweating? _____

7. Record the air temperature entering the evaporator: _____°F
   Record the air temperature leaving the evaporator: _____°F
   Record the temperature difference: _____°F

8. Now record the temperature difference across the condenser: _____°F. It will not be significant but is worth noting.

9. Shut the unit off and unplug it. Replace all panels and place the unit in its permanent location.

Maintenance of Workstation and Tools: Clean up the workstation and return all tools to their proper places.

Summary Statement: Evaluate the unit you worked on and tell how the above test indicates whether the charge is correct. (Be sure to read the text reference—Paragraph 47.8.)

QUESTIONS

1. What is the primary maintenance needed on room cooling units?

2. How do some manufacturers provide evaporator freeze protection?

3. Why should gages be installed only when absolutely necessary?

4. What two types of compressors may be used in room units?

5. What are the four major components in the room unit refrigeration cycle?

6. What are two purposes of the condenser?

7. How is the compressor heat used to evaporate condensate?

8. Describe how the condenser fan slinger provides for condensate evaporation.

9. How can the exhaust or fresh air options be controlled?

10. What are options the selector switch controls?

Charging a Window Unit by the Sweat-Line Method

Name	Date	Grade

Objectives: Upon completion of this exercise, you will be able to add a partial charge to a window unit using the sweat-line method.

Introduction: You will add refrigerant to a window unit through the suction-line service port until a correct pattern of sweat has formed on the suction line to the compressor.

Text Reference: Paragraph 47.8.

Tools and Materials: Straight-blade and Phillips screwdrivers, $1/4$" and $5/16$" nut drivers, a gage manifold, a cylinder of refrigerant (R-22 for most units), gloves, goggles, and a window unit with low- and high-side pressure service ports.

Safety Precautions: Gloves and goggles should be worn while making gage connections and transferring refrigerant. A high-side gage should be used while adding refrigerant.

PROCEDURES

1. Unplug the unit and remove it from its case. Set the unit on a bench with a power supply for the unit.

2. Put on your goggles and gloves. Fasten the gage lines to the high- and low-side connections. Purge the gage lines through the center line and connect it to the refrigerant cylinder.

3. Make sure the airflow across the evaporator and condenser is correct. Start the unit and observe the pressures. Record the following after 15 minutes: Suction pressure _____ psig. Discharge pressure _____ psig. Where is the last point that the suction line is sweating?

4. Restrict the airflow across the condenser as in Lab 47-2 until about $2/3$ of the condenser is blocked. Wait 15 minutes and record the following: Suction pressure _____ psig. Discharge pressure _____ psig. Where is the last point that the suction line is sweating?

5. If the unit has the correct charge, recover some refrigerant until the suction pressure is about 30 psig for R-22.

6. Record the following after 15 minutes of running time at about 30 psig of suction pressure. Suction pressure _____ psig. Discharge pressure _____ psig. Where is the last point that the suction line is sweating?

7. Start adding vapor refrigerant in short bursts through the suction line. Add a little and wait for about 5 minutes. Keep adding in short bursts until the sweat line moves to the compressor, waiting about 5 minutes after each addition of refrigerant. When the correct charge is reached, record the following: Suction pressure _____ psig. Discharge pressure _____ psig. See Figure 47-2.

8. Unplug the unit and remove the gages. Return the unit to its permanent location. Be sure to seal the unit gage ports.

Maintenance of Workstation and Tools: Return all tools to their proper locations and make sure the workstation is clean.

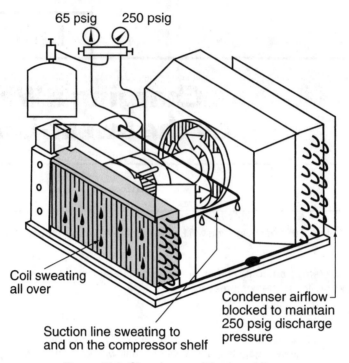

65 psig   250 psig

Coil sweating
all over

Suction line sweating to
and on the compressor shelf

Condenser airflow
blocked to maintain
250 psig discharge
pressure

Figure 47-2 Charging a unit by sweat line.

Summary Statement: Explain why the condenser airflow must be blocked in order to correctly add a partial charge to a unit.

QUESTIONS

1. Does the fan motor normally need periodic lubrication?

2. What will happen to the suction pressure if the indoor coil becomes dirty?

3. If the low-side pressure becomes too low, what will happen to the condensate on the evaporator coil?

4. Why should gages be installed only when absolutely needed?

5. What are line tap valves?

6. In this exercise, why is the airflow through the condenser blocked?

7. If only part of the evaporator gets cold in this test, what could the problem be?

8. If the charge is correct but the evaporator coil is only cool, what could the problem be?

Refrigeration & Air Conditioning Technology

Charging a Unit Using Scales or a Charging Cylinder

Name	Date	Grade

Objectives: Upon completion of this exercise, you will be able to add an accurate charge to a window unit using scales or a charging cylinder.

Introduction: You will evacuate a unit to a low vacuum and add a measured charge to the unit using scales or a charging cylinder.

Text References: Paragraphs 8.3, 8.4, 8.5, 8.6, 9.15, 9.16, 10.4, 10.5, and 47.8

Tools and Materials: Straight-blade and Phillips screwdrivers, $1/4$" and $5/16$" nut drivers, a gage manifold, a cylinder of correct refrigerant for the system, a vacuum pump, goggles, gloves, and a room cooling unit.

Safety Precautions: Goggles and gloves should be worn while making gage connections and transferring refrigerant.

PROCEDURES

1. Remove the unit to a bench with power. Put on your goggles and gloves. Fasten the gage lines to the gage ports and purge the gage lines. **NOTE:** IF THE UNIT HAS ONLY A SUCTION PORT, YOU CAN STILL DO THIS EXERCISE.

2. If the unit has a charge of refrigerant, your instructor may advise you to recover the refrigerant from the system using an approved recovery system.

3. When all of the refrigerant is removed from the system, unplug the unit and connect the vacuum pump. Check the oil in the vacuum pump. If it is not at the correct level, ask your instructor for directions. Start the pump to evacuate the system.

4. While the vacuum pump is running, get set up to charge the system either with scales or a charging cylinder. Determine the correct charge, make necessary calculations, and be ready to add the refrigerant as soon as the correct vacuum has been obtained.

5. **NOTE:** IF YOU ARE PULLING A VACUUM FROM BOTH THE LOW- AND HIGH-PRESSURE SIDES OF THE SYSTEM, YOU SHOULD SHUT THE HIGH-PRESSURE SIDE OF THE SYSTEM OFF AND REMOVE THE GAGE LINE WHEN A DEEP VACUUM HAS BEEN REACHED. IF THE GAGE PORT IS A SCHRADER VALVE, ALLOW THE PRESSURE IN THE SYSTEM TO RISE TO ABOUT 10 PSIG WHILE ADDING REFRIGERANT AND THEN REMOVE THE GAGE LINE. This procedure is to keep the measured refrigerant from moving into the high-pressure line and condensing, thereby affecting the charge.

6. When refrigerant is allowed to enter the system, the system pressure will soon equalize with the cylinder pressure and no more refrigerant will flow. You may start the unit and the rest of the charge may be charged into the system. **NOTE:** IF YOU ONLY HAVE A LOW-PRESSURE CONNECTION, BE CAREFUL THAT YOU ONLY CHARGE THE CORRECT AMOUNT INTO THE SYSTEM.

7. Let the unit run long enough to establish that it is running correctly. You may want to cover part of the condenser. The suction line should sweat and the suction pressure should be about 65 psig for R-22 and 114 psig for R-410a.

8. When the correct charge is in the system, shut the unit off and disconnect the low-side gage line from the unit. Place caps on all gage connections and replace the panels.

9. Return the unit to its permanent location.

Maintenance of Workstation and Tools: Return all tools to their proper locations.

Summary Statement: Describe what you found to be the most difficult step in the above charging sequence.

QUESTIONS

1. Why are larger ports better when evacuating a system?

2. What is the problem if, when charging a system, the head pressure rises and the suction pressure does not?

3. If there is a problem with the capillary tube, what other types of metering devices may be installed in a room cooling unit?

4. What must the compressor motor have for starting devices if one of the above metering devices is installed?

5. Why must the motor have the above starting devices?

6. Why is it important to check for a leak if the refrigerant is found to be low?

7. Why should an accurate dial or electronic scale be used when weighing in the refrigerant?

High-Pressure, Low-Pressure, and Absorption Chilled Water Systems

Unit Overview

Chilled water systems are used for larger applications for central air conditioning because of the ease with which chilled water can be moved in the system. A chiller refrigerates circulating water. As the water passes through the evaporator section of the machine, the temperature of the water is lowered. Typically, 45°F water is furnished to the building. It is then circulated throughout the building, where it picks up heat.

There are two basic categories of chillers, the compression cycle and the absorption chiller. The compression type of chiller uses a compressor to provide the pressure differences inside the chiller in order to boil and condense refrigerant. The absorption chiller uses heat, salt, and water to accomplish the same results.

The compression cycle chiller has the same four basic components as a typical residential air conditioner; a compressor, an evaporator, a condenser, and a metering device. There are high-pressure and low-pressure compression cycle chillers. The compressors used in high-pressure systems are the reciprocating, scroll, screw, and centrifugal. The centrifugal is the only one used in low-pressure systems. Chiller reciprocating compressors operate in a similar manner to those found in residential air-conditioning systems. However, they will usually have additional features for capacity control and other requirements of this type of system. Scroll compressors are usually in the 10- to 15-ton range but operate the same as the smaller scroll compressors. The rotary screw compressors may be used for larger capacity chillers, using high-pressure refrigerants. These are capable of handling large volumes of refrigerant. The centrifugal compressor is like a large fan that creates a pressure difference from one side of the compressor to the other using centrifugal force. To produce the pressure needed, these compressors are often operated at very high speed and are turned by means of a gear box generating speeds up to 30,000 rpm. They may also utilize more than one stage of compression.

Building air-conditioning requirements vary by the season, so unit capacity control is used to match the air-conditioning load.

Chiller evaporators may be of the direct expansion (dry) type or they may be flooded evaporators.

Condensers for high-pressure chillers may either be water cooled or air cooled.

Four types of metering devices may be used; thermostatic expansion valve, orifice, high- and low-side float, and the electronic expansion valve.

All low-pressure chillers use centrifugal compressors. These are similar to those used in the high-pressure systems. Evaporators for these systems are similar to those found in high-pressure systems, condensers are water cooled. Two types of metering devices are typically used, the orifice and the high- or low-side float.

When a centrifugal uses a low-pressure refrigerant, the low-pressure side of the system is always in a vacuum when in operation. If there is a leak in the system, air will enter. This can cause many problems. These chillers have purge systems to remove this air.

Absorption air-conditioning chillers are considerably different from compression cycle chillers. They utilize heat as the driving force instead of a compressor. This type of equipment utilizes water for the refrigerant and certain salt solutions that have enough attraction for water to create a pressure difference. The salt solution used as the absorbent or attractant is lithium-bromide(Li-Br). This liquid solution is diluted with distilled water. This solution is very corrosive and the system must be kept very clean.

These systems operate in a vacuum, so it is possible that air and other noncondensables from the atmosphere may enter the system if there is a leak. Both motorized and nonmotorized purge systems are used to remove these noncondensables from the system.

Key Terms

Absorbent (attractant)
Absorber
Absorption Chiller
Anti-Recycle Control
Autotransformer Start
Blocked Suction
Centrifugal Compressor
Chilled Water Systems

Concentrator
Condenser Subcooling
Compression Cycle
Corrosive
Crystallization
Cylinder Unloading
Direct Expansion Evaporator
Direct Fired Systems

Electronic Expansion Valve
Electronic Starter
Float-Type Metering Device
Flooded Evaporator
Guide Vanes
High Pressure Chiller
Lithium-Bromide
Low Pressure

Marine Water Box
Operating-Performance Log
Orifice
Part-Winding Start
Phase Failure Protection
Phase Reversal

Purge Unit
Reciprocating Compressor
Rotary Screw Compressor
Salt Solution
Scroll Compressor
Suction Valve Lift Unloading

Surge
Thermostatic Expansion Valve
Water Box
Wye-Delta

REVIEW TEST

Name	Date	Grade

Circle the letter that indicates the correct answer.

1. A chiller refrigerates:
   A. circulating air.
   B. circulating water.
   C. an oil-based refrigerant.
   D. nitrogen.

2. Two basic categories of chillers are the:
   A. compression cycle and absorption.
   B. purge and absorption.
   C. absorption and lithium-bromide.
   D. purge and lithium-bromide.

3. Cylinder unloading for a high-pressure chiller:
   A. provides for a better heat exchange.
   B. produces capacity control.
   C. produces a more concentrated salt solution.
   D. all of the above.

4. Suction valve lift unloading:
   A. provides for capacity control on a rotary screw compressor.
   B. is normally used only on centrifugal compressors.
   C. produces hot gas in the evaporator.
   D. provides for capacity control on a reciprocating compressor.

5. Centrifugal compressors are used:
   A. only in high-pressure chillers.
   B. only in low-pressure chillers.
   C. only in absorption chillers.
   D. in both high- and low-pressure chillers.

6. Speeds of up to _____ are used in centrifugal compressors.
   A. 15,000
   B. 20,000
   C. 25,000
   D. 30,000

7. Capacity control is accomplished by means of _____ in centrifugal compressors.
   A. guide vanes at the impeller eye
   B. suction valve lift at the impeller eye
   C. motorized purge units
   D. all of the above

8. Direct expansion evaporators have:
   A. expansion joints between sections of the evaporator.
   B. an established superheat at the outlet of the evaporator.
   C. subcooling at the outlet of the evaporator.
   D. an established superheat at the inlet of the evaporator.

9. Condensers for high-pressure chillers are:
   A. always water cooled.
   B. always air cooled.
   C. either water cooled or air cooled.

10. Thermostatic expansion valves are the only metering devices used on high pressure chillers.
   A. true
   B. false

11. All low-pressure chillers use _____ compressors.
   A. reciprocating   C. rotary screw
   B. scroll   D. centrifugal

12. When water is introduced into the end of a chiller it is contained in:
   A. water boxes.
   B. a combination filter receiver.
   C. a receiver.
   D. none of the above.

13. **The orifice is a metering device that:**
    A. has several moving parts.
    B. provides a restriction in the liquid line.
    C. provides superheat at the outlet of the evaporator.
    D. provides subcooling at the compressor end of the suction line.

14. **When a centrifugal compressor uses a low-pressure refrigerant, the low-pressure side of the system:**
    A. always shows a positive pressure.
    B. is always in a vacuum.
    C. contains a mixture of oil and air.
    D. rejects oxygen and moisture.

15. **A purge system:**
    A. helps to provide subcooling in the condenser.
    B. helps to provide superheat in the suction line.
    C. relieves noncondensable gases from the system to the atmosphere.
    D. provides capacity control for a rotary screw compressor.

16. **An absorption air-conditioning chiller:**
    A. utilizes absorption and high compression to provide cooling.
    B. uses a scroll compressor to pump refrigerant.
    C. utilizes heat as the driving force rather than a compressor.
    D. utilizes a salt solution in the compression cycle.

17. **The absorption system uses _____ for the refrigerant.**
    A. water
    B. R-22
    C. R-11
    D. a salt solution

18. **Lithium-bromide is used in absorption systems as:**
    A. a salt solution.
    B. an absorbent (attractant).
    C. a solution that creates a pressure difference.
    D. all of the above.

19. **The capacity can be controlled for a typical absorption system by controlling the supply of _____ to the concentrator.**
    A. Li-Br
    B. R-11
    C. water
    D. heat

20. **Absorption systems operate:**
    A. under a high pressure.
    B. under a low pressure.
    C. in a vacuum.
    D. under no significant pressure.

Cooling Towers and Pumps

Unit Overview

The heat that is absorbed by the chiller from the conditioned space must be rejected. The cooling tower is the unit in a water-cooled system that rejects this heat to the atmosphere. It must reject more heat than is absorbed from the structure because the heat of the compression produced by the compressor adds about 25% additional heat. The cooling towers reduce the temperature of the water by means of evaporation. The water is pumped from the chiller to the tower. In the tower, the surface area of the water is increased to enhance the evaporation.

Two types of cooling towers are the natural draft and forced or induced draft. The natural draft may be anything from a spray pond in front of a building to a tower on top of a building. These rely on the prevailing winds to increase the evaporation of the water. Forced and induced draft towers use a fan to move the air through the tower. Centrifugal fans may be used; larger towers generally use propeller-type fans. Two methods used to spread the water are the splash method and the fill or wetted surface method.

All cooling towers will need service on a regular basis. The tower must have access doors to the fill material for cleaning and possible removal. It must have a sump for the water to collect in. Sumps, depending on the climate, will normally need some method of heating the water to prevent freezing unless the tower is drained in the winter. There must be a makeup water system to replace the water that is evaporated. Sediment and other materials will become concentrated in the sump. A system known as blowdown is used to bleed off part of the water and replace it with supply water. This reduces the concentration of minerals in the sump.

Key Terms

Blowdown
Cavitation
Fill or Wetted Surface Method
Forced or Induced Draft
Heat of Compression
Natural Draft

Splash Method
Spray Pond
Sump
Vortexing
Wet-Bulb Temperature

REVIEW TEST

Name	Date	Grade

Circle the letter that indicates the correct answer.

1. **Heat that the cooling tower must reject from the water from the compression cycle chiller is the heat absorbed from the structure and the:**
   A. heat of absorption.
   B. heat of compression.
   C. heat absorbed as a result of vortexing.

2. **Cooling towers are generally designed to reduce the entering water temperature to within _____ of the air wet-bulb temperature.**
   A. 35°F
   B. 14°F
   C. 7°F
   D. 21°F

3. **Spray ponds are considered _____ draft applications.**
   A. forced
   B. induced
   C. natural

4. **Larger cooling towers generally use _____ type fans.**
   A. centrifugal
   B. propeller

5. **Two methods used to aid in the evaporation of the water in the forced and induced draft tower are:**
   A. vortexing and cavitation.
   B. spray pond and prevailing wind.
   C. float-type valve and tower bypass valve.
   D. the splash and the fill or wetted surface methods.

6. **A blowdown system for a cooling tower:**
   A. relieves the concentration of minerals in the sump water.
   B. provides extra draft for greater evaporation.
   C. prevents vortexing.
   D. provides a means for balancing the water flow.

7. **The condenser water pump is generally a _____ pump.**
   A. reciprocating
   B. rotary screw
   C. centrifugal
   D. compressor-assisted

8. **Vortexing in the cooling tower sump is:**
   A. a whirlpool-type action.
   B. caused by a high head pressure.
   C. a system whereby water bypasses the pump.
   D. an absorption process.

9. **The purpose of the cooling tower bypass valve is to:**
   A. route the debris away from the tower sump.
   B. prevent cavitation at the pump inlet.
   C. provide makeup water to the tower sump.
   D. help maintain the correct tower water temperature.

10. **Two types of shaft seals used for centrifugal pumps are:**
    A. stuffing box and mechanical types.
    B. stuffing box and splash types.
    C. ceramic and splash types.
    D. ceramic and bellows types.

11. **Cooling tower fill material must be:**
    A. fireproof.   C. small.
    B. large.   D. porous.

12. **Cooling tower systems typically use _____ pumps.**
    A. rotary   C. reciprocating
    B. screw   D. centrifugal

13. **Algae in cooling towers is controlled by:**
    A. chemicals.
    B. filters.
    C. changing the water.
    D. hydraulic cylinders.

14. **One psi of pressure will support a column of water:**
    A. 1 foot high.
    B. 3.3 feet high.
    C. 10 feet high.
    D. 2.31 feet high.

Operation, Maintenance, and Troubleshooting of Chilled Water Air-Conditioning Systems

Unit Overview

The operation of chilled water systems involves starting, running, and stopping chillers in an orderly and prescribed manner. However, the brief discussion in the text should not be used as the only guide to the starting, operating, and stopping of chiller systems. This type of equipment is very expensive, and the specific manufacturer's manuals should be followed in all aspects of the operation of this equipment. The text provides some guidelines, but these guidelines should be used only in conjunction with manufacturer's literature.

Key Terms

Algae
Eddy Current Test
Field Control Circuit
Flow Switch
Light Emitting Diode (LED)
Load Limiting Control

Megger (megohmmeter)
Mercury Manometer
Net Oil Pressure
Pressure Test Point
Refrigerant Recovery

REVIEW TEST

Name	Date	Grade

Circle the letter that indicates the correct answer.

1. **The first step in starting a chilled water system is to:**
   A. determine that compressor unloading capabilities exist.
   B. establish that there is chilled water flow.
   C. determine that the blocked suction option is operating.

2. **A flow switch is:**
   A. typically a paddle placed in a water stream that moves with the water flow, operating a switch.
   B. a switch placed in an airstream that is activated by the airflow.
   C. a switch in the suction line of a chiller that is activated by the vapor refrigerant flow.

3. **A chiller field control circuit:**
   A. is often called an interlock circuit.
   B. is a circuit that starts systems one at a time, all of which must be started before the compressor starts.
   C. ensures that appropriate fans and pumps are started in the correct sequence.
   D. all of the above.

4. **Which of the following compressors are considered positive displacement compressors?**
   A. scroll and rotary-screw
   B. centrifugal
   C. all of the above
   D. none of the above

5. **The reciprocating, scroll, and rotary-screw compressors:**
   A. have separate external oil pumps.
   B. are lubricated from within or have internal oil pumps.

6. **The rotary-screw chiller is:**
   A. air cooled.
   B. water cooled.
   C. either air or water cooled.

7. **Reciprocating chillers have crankcase heat to:**
   A. preheat the chiller water for better efficiency.
   B. prevent refrigerant migration to the crankcase.
   C. preheat the compressor bearings for longer wear.
   D. minimize condenser subcooling.

8. **Water-cooled chillers must have water treatment to:**
   A. increase the pH content of the water.
   B. decrease the pH content of the water.
   C. prevent minerals and algae from forming.
   D. help preheat the water in the condenser circuit.

9. **Centrifugal chillers range in size from _____ tons.**
   A. 10–50   C. 50–500
   B. 10–80   D. 100–10,000

10. **A megohmmeter (megger) is used to check:**
    A. amperage to 100 amperes.
    B. amperage to 1000 amperes.
    C. voltage to 10,000 volts.
    D. ohms to 1,000,000+ ohms.

11. **When a reciprocating compressor is sweating at the crank case:**
    A. normal, good lubrication is taking place.
    B. it is flooding liquid back to the crankcase.
    C. there is too much superheat.
    D. the heat exchange is working correctly.

12. **The instrument used to check for electrical grounds in a large motor is:**
    A. an ohmmeter.
    B. an ammeter.
    C. a thermistor.
    D. a megger (megohmmeter).

13. **Tube failure can often be detected before the failure occurs using:**
    A. an ammeter.
    B. a megger.
    C. an eddy current test instrument.
    D. pressure drop.

14. **Noncondensable gases are removed from an operating absorption chiller by means of:**
    A. a vacuum pump.
    B. a reciprocating compressor purge unit.
    C. a megger.
    D. an ohmmeter.